Molecular Mechanisms in Bioorganic Processes

Molecular Mechanisms in Bioorganic Processes

Edited by

Christine Bleasdale and Bernard T. Golding
University of Newcastle-upon-Tyne

British Library Cataloguing in Publication Data
Royal Society of Chemistry. *Joint Perkin Division – Bioorganic Group Meeting*, (1989; University of Newcastle-upon-Tyne)
Molecular mechanisms in bioorganic processes.
1. Organisms. Molecules. Structure & properties
I. Title II. Bleasdale, Christine III. Golding, Bernard T.
574.88

ISBN 0-85186-946-7

Proceedings of the Royal Society of Chemistry Joint Perkin Division – Bioorganic Group Meeting, July 17–20 1989, The University of Newcastle-upon-Tyne

Published by The Royal Society of Chemistry, Thomas Graham House, Science Park, Cambridge CB4 4WF

Printed and bound in Great Britain by Bookcraft (Bath) Ltd.

PREFACE

This book contains all the lectures and selected posters from an International Symposium ('Molecular Mechanisms in Bioorganic Processes') held at the University of Newcastle upon Tyne, July 17–20, 1989. The stated aim of this Symposium was to bring together chemists, biochemists and medical scientists to address a variety of current topics relevant to the title of the Symposium. We believe the book to be a timely statement of contemporary bioorganic chemistry and associated topics, and hope that it will be a useful teaching aid and guide to future research.

We thank all the speakers for their lectures and manuscripts, and also the presenters of posters. The 2nd Clemo Memorial Lecture, sponsored by the Society of Chemical Industry, was presented at the Symposium by Professor Grant Buchanan (pp 225–243). The chapter by Professor Steven Benner (pp 166–187) does not correspond to the lecture he presented. Professor Benner felt that the present contribution was better suited to the book than the subject matter of his lecture. The chapter by Professor Adrian Harris (and Dr Ian D Hickson, pp 83–99) is an extension of a previous publication by Professor Harris (in TIG April 1988, vol 4, no 4, Elsevier Publications, Cambridge) and is published here with the permission of Elsevier. The three chapters based on posters (see pp 343, 350, and 359) were chosen by us for the excellence of the posters. Dr R M Horak's poster won the Symposium's poster prize. All contributions have been re–typed to a common format, although the original diagrams of the author(s) have mainly been used. We are grateful for assistance from Miss Emma Bennett and Ms Kate Palmer in the re–typing of manuscripts.

We wish to thank the many people who assisted us with the organisation of the Symposium, especially Ms Gina Howlett and Dr John F Gibson (Royal Society of Chemistry), and Mrs Maureen Craig and Mrs Joan Trett (Department of Chemistry, University of Newcastle upon Tyne).

Finally, we very much regret to have to record that Fritz Gerhart, one of the Symposium's speakers, died on 26 January 1990.

April 1990 Christine Bleasdale and Bernard Golding

CONTENTS*

*The author who presented the paper at the Symposium is italicised

†Deceased 26 1 90

§These contributions were presented as posters

MECHANISMS OF CARBON-SULPHUR BOND FORMATION IN ENZYMIC PROCESSES

Robert M Adlington

The Dyson Perrins Laboratory, University of Oxford, South Parks Road, Oxford OX1 3QY, UK

1 INTRODUCTION

Carbon-sulphur bond formation occurs in many biosynthetic processes. Of particular interest to the research group at Oxford are such processes which occur during the biosynthetic pathway to penicillins and cephalosporins.[1] This chapter concentrates on the mode of action of the enzyme isopenicillin N synthase (IPNS), which is responsible for the direct conversion of the so-called Arnstein tripeptide, LLD-ACV-SH* (1), into isopenicillin N (2) (see Scheme 1). This remarkable enzyme catalyses a desaturative cyclisation process which involves loss of four hydrogen atoms, two of which come from strong carbon-hydrogen bonds, and other than molecular oxygen, does not require an additional co-substrate such as α-ketoglutarate, typical of many other desaturase/oxygenase enzymes (*eg* as required during the conversion of deacetoxycephalosporin C into deacetylcephalosporin C).

IPNS was initially isolated and purified to homogeneity from a fungal source *Cephalosporium acremonium* CO728.[2] Its *N*-terminal sequence was established and this subsequently led by genetic engineering to the construction of a new strain of *Escherichia coli* which overproduces a single recombinant IPNS in high yield (*ca* 20% of soluble protein).[3] Today hundreds of milligrams of > 90% pure recombinant IPNS are routinely available by a simple three-step purification process [(i) lysis of the cell wall; (ii) ion-exchange chromatography; (iii) FPLC].

* unless otherwise stated *all* tripeptides referred to in this chapter have LLD configuration.

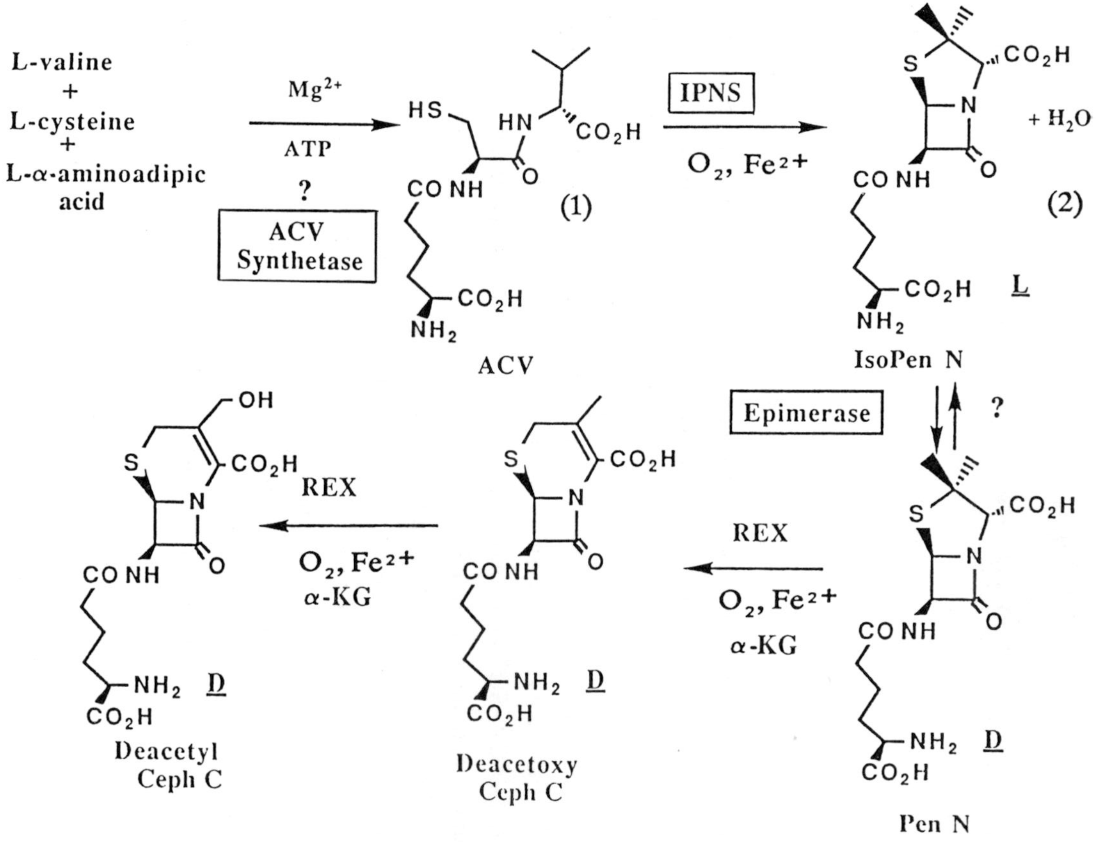

L-valine + L-cysteine + L-α-aminoadipic acid

Mg^{2+}, ATP, ?, ACV Synthetase

ACV (1)

IPNS, O_2, Fe^{2+}

IsoPen N (2) L + H_2O

Epimerase ?

Pen N D

REX, O_2, Fe^{2+}, α-KG

Deacetoxy Ceph C D

REX, O_2, Fe^{2+}, α-KG

Deacetyl Ceph C D

Scheme 1: Biosynthetic pathway to penicillins and cephalosporins.

IPNS is a single chain polypeptide (MW 38,000) consisting of 336 amino acids and does not contain any heme unit. It has a natural substrate K_m value of 0.20 mM, a pI value of 4.8 and an operational optimal pH of *ca* 7.6. The peptide chain contains 2 cysteines at position 106 and 255 from the *N*–terminus, which if replaced by serines (*via* application of genetic engineering techniques) leads to reduced enzymic activity.[4] Other IPNS genes have now been isolated and sequenced, such as those from the eukaryotes *Penicillium chrysogenum*[5] and *Aspergillus nidulans*,[6] and from the prokaryotes *Streptomyces lipmanii*[6] and *Streptomyces clavuligerus*.[7] A comparison of their deduced amino acid sequences shows the expected strong sequence homology.

For the conversion of ACV (1) into isopenicillin N (2) molecular oxygen is required as a co–substrate whilst ferrous ions, ascorbic acid and dithiothreitol (DTT) are required as essential co–factors. The latter two co–factor requirements probably reflect the *in vitro* nature of the observed enzymic activity: DTT is needed to maintain the substrate in its active thiol form and ascorbate perhaps enables efficient restoration of over–oxidized iron to its active ferrous state. An *in vitro* turnover number of *ca* 200–400 has been estimated although these figures are believed to be much larger *in vivo*.[8]

The stoichiometry of oxygen consumption,[9] ACV + $1O_2$ → Isopen N + $2H_2O$, merits comment. It is unprecedented amongst enzymes of the desaturase/monooxygenase class and implies that the total oxidative potential of molecular oxygen is used to form the two strained rings of isopenicillin N *ie* the β–lactam and thiazolidine rings. It should be noted that each of these ring closures requires cleavage of a strong carbon–hydrogen bond. Thiazolidine ring formation leads directly to an isolable new carbon–sulphur σ–bonded species, whereas the β–lactam closure operates, we believe, by formation and cleavage of a weak carbon–sulphur π–bond along the reaction pathway.

2 EVIDENCE FOR INITIAL β–LACTAM FORMATION

To understand the mechanism of IPNS action a key question is the order of the two ring closures. Two proposals were initially forwarded. The first of these suggested closure to a four–membered

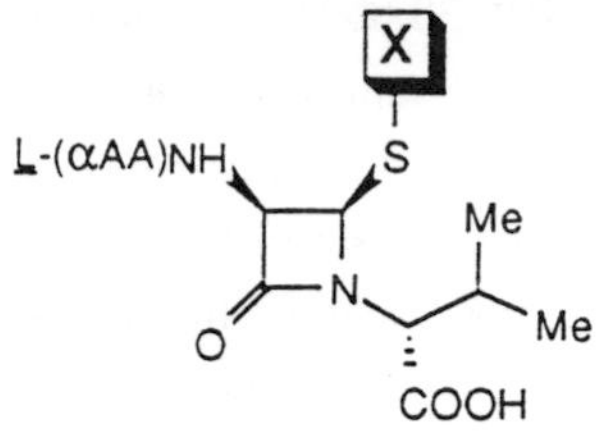

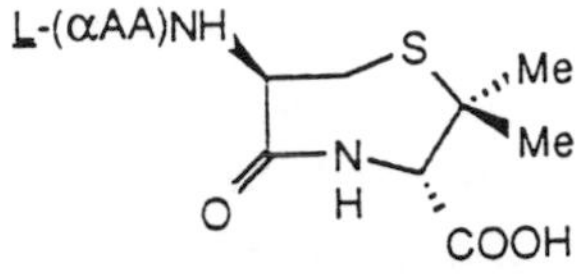

(3) X = H or ? (4)

ring thiol (**3**) as an enzyme-free intermediate, which could then lead *via* closure of the second ring to isopenicillin N.[10] A second proposal suggested initial formation of a 7-membered monocyclic peptide (**4**) as intermediate followed by synchronous dual ring formation.[11] Both of these potential intermediates were synthesised and incubated with IPNS; neither gave isopenicillin N (**2**). The β-lactam thiol was found to be unstable at the optimal pH for IPNS activity, being rapidly ring opened to a thioaldehyde and subsequently to other non β-lactam degradation products.

As a direct 'intermediate' incubation approach failed to reveal the order of the two ring closures we next turned to kinetic isotope effects. As mentioned earlier each of the two ring closures requires cleavage of a strong C–H bond, specifically a cysteinyl C3–H and a valinyl C3–H. Thus, we synthesised substrates where either or both of these bonds were replaced by C–D bonds. In a simple system, *eg* a chemical process, substitution of a C–D for a C–H would only lead to an observable isotope effect, and hence overall rate lowering, if that C–H bond cleavage were the rate limiting step. In an enzymic process, *ie* a catalytic process, the effect of the change from C–H to C–D becomes very complicated because more than a single step on the reaction profile can be rate limiting.

Initially, we evaluated *competitive mixed isotope effects* V_{max}/K_m (V_{max} maximum velocity, K_m Michaelis constant) which reflect events up to and including the first irreversible event.[12a] In these experiments a 1:1 mixture of labelled (S_D) and unlabelled (S_H) substrate were converted by IPNS with various levels of substrate consumption. Isotopic discrimination was then looked for by mass spectral analysis of the unreacted substrate and where possible, by analysis of the so-formed isopenicillin N. The results of such experiments revealed that a V_{max}/K_m effect was only observed by substitution of deuterium at the cysteinyl 3–position *and not* by substitution at the valinyl 3–position[12b]

Secondly, we measured individual V_{max} effects at either (and both) positions. In order to do so a mixed enzyme assay was developed. Thus, the production of isopenicillin N was coupled to cleavage of this penam by a β–lactamase enzyme from *Bacillus cereus* (*Note*: rate of β–lactam cleavage was much greater than rate of β–lactam formation). Titration of the so–formed penicilloate with standard hydroxide solution enabled a *continuous assay* for isopenicillin N production, and thus enabled initial velocity (V_{max}) measurements to be obtained from both the natural and specifically deuteriated substrates. The results of such experiments revealed isotope effects:

$$D_V = V_{max}(H)/V_{max}(D) \quad \text{of } 5.6 \pm 0.3,\ 13 \pm 2 \text{ and } 18 \pm 2$$

to be calculated for the cysteinyl C3, valinyl C3, and a combination of both positions, respectively.

What, if any, conclusion can be drawn from these isotope effect studies? The V_{max}/K_m mixed substrate results imply firstly that initial binding is reversible as discrimination against the cysteinyl 3–D substrate was observed. Secondly, the first irreversible event occurs at the cysteinyl C3 position with the valinyl event, which did not

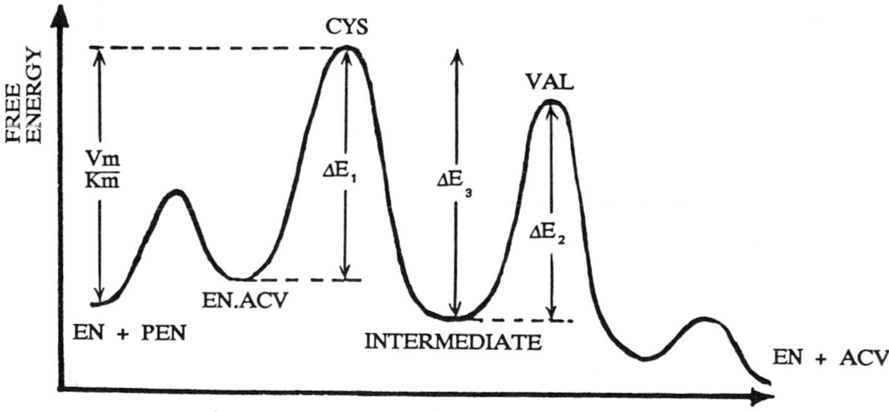

Figure 1: Free energy profile for the conversion of ACV (1) into Isopen N (2).

show V_{max}/K_m discrimination, downstream of this irreversible event (*Note*: the observation of individual D_V effects at both the cysteinyl C3 and valinyl C3 positions demonstrated that a V_{max}/K_m mixed substrate effect could have been observed at either position). The V_{max} data is consistent with a stepwise process for the two C–H cleavages. If the cleavages were synchronous a much larger V_{max} value for the substrate labelled at both the Cys C3 and Val C3 positions would have been expected. A stepwise process thus implies the presence of an intermediate derived from cysteinyl C3–H cleavage which, due to the observed V_{max} values of the three labelled substrates, must remain enzyme–bound.[13] If the intermediate were enzyme free then a V_{max}/K_m effect may be expected at the Val C3 position. From this data we propose that such an intermediate is a substituted β–lactam and also propose a simplistic free energy profile (Figure 1) for the conversion ACV (1) → Isopen N (2). The free energy profile is consistent with the observed reversible binding, the observation of two V_{max} effects at both Cys and Val C3–H positions ($\Delta E_1 \sim \Delta E_2$), and the production of an intermediate after an irreversible event ($\Delta E_3 > \Delta E_1$) of a stepwise process.

As an intermediate β–lactam was proposed to result from a first irreversible event at the Cys C3–position we decided to re–investigate the stereospecificity of this process. Intact cell experiments performed by Young and Aberhart[14] using tritiated cysteines indicated that *overall* the cysteinyl 3–pro–*S* hydrogen was selectively removed during penicillin formation, *eg* 21% ^{3}H retention from (2*RS*, 3*S*)–[U–^{14}C, 3–^{3}H]Cys, and 76% ^{3}H retention from (2*RS*, 3*R*)–[U–^{14}C, 3–^{3}H]Cys (values of 0% and 100% ^{3}H retention, respectively, are predicted from a stereospecific process). In order to refine the intact cell experiments we prepared two specifically deuteriated tripeptides (**5**) and (**6**) and incubated them with IPNS. We found that the (3*S*)–monodeuterio cysteinyl tripeptide (**5**) gave *fully* unlabelled isopenicillin N, whereas the (3*R*)–monodeuterio cysteinyl tripeptide (**6**) cyclised with *complete* retention of the deuterium label.[15] These results indicate that in the cyclisation to the proposed β–lactam intermediate there is stereospecific removal of the Cys 3–pro–*S* hydrogen of the natural substrate [*Note*: the substrate (**5**) shows that this preference overrides the operation of a bond strengthening isotope effect].

IPNS

(5)

IPNS

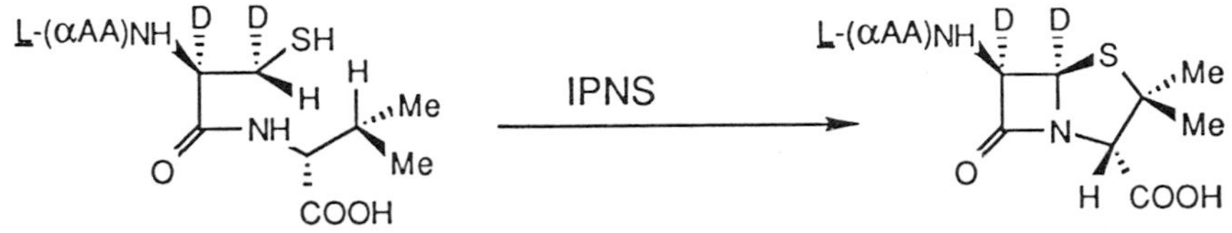

(6)

Next we evaluated other cysteinyl analogues of ACV to see if they would be consistent with the cysteinyl C3–stereospecificity of the natural substrate. Firstly, substitution of L–cysteine by L–alanine, or L–serine, or D–cysteine, did not give an acceptable substrate. Secondly, functionalisation of the L–cysteinyl residue was performed. We found that a (3*S*)–3–methylcysteinyl–tripeptide was not a substrate, but (3*R*)–3–methylcysteinyl–,[16] (2*R*)–2–methylcysteinyl–[16] and (2*S*)–2–methoxycysteinyl–tripeptides[17] were penicillin–forming substrates. From these results we concluded that:

(i) a cysteinyl 3–pro–*S* hydrogen is essential for penam formation and,

(ii) functionalisation at either C2 or C3 of the cysteinyl residue can be tolerated [provided condition (i) is met].

A mechanistic proposal for the overall conversion of ACV (1) into Isopen N (2) by IPNS, consistent with the experimental data presented so far, is indicated in Scheme 2. Initially a reversibly bound high energy ACV–Ferryl (Fe^{IV})–Oxygen–Enzyme complex (7) is formed (ferrous being oxidised to a ferryl state by dioxygen). The sulphur atom of the cysteinyl moiety provides a conducting bridge for electron flow from the peptide *via* the ferryl entity into the dioxygen molecule. In this Scheme removal of the 3–pro–*S* cysteinyl hydrogen as the first irreversible event is seen as an acid–base reaction

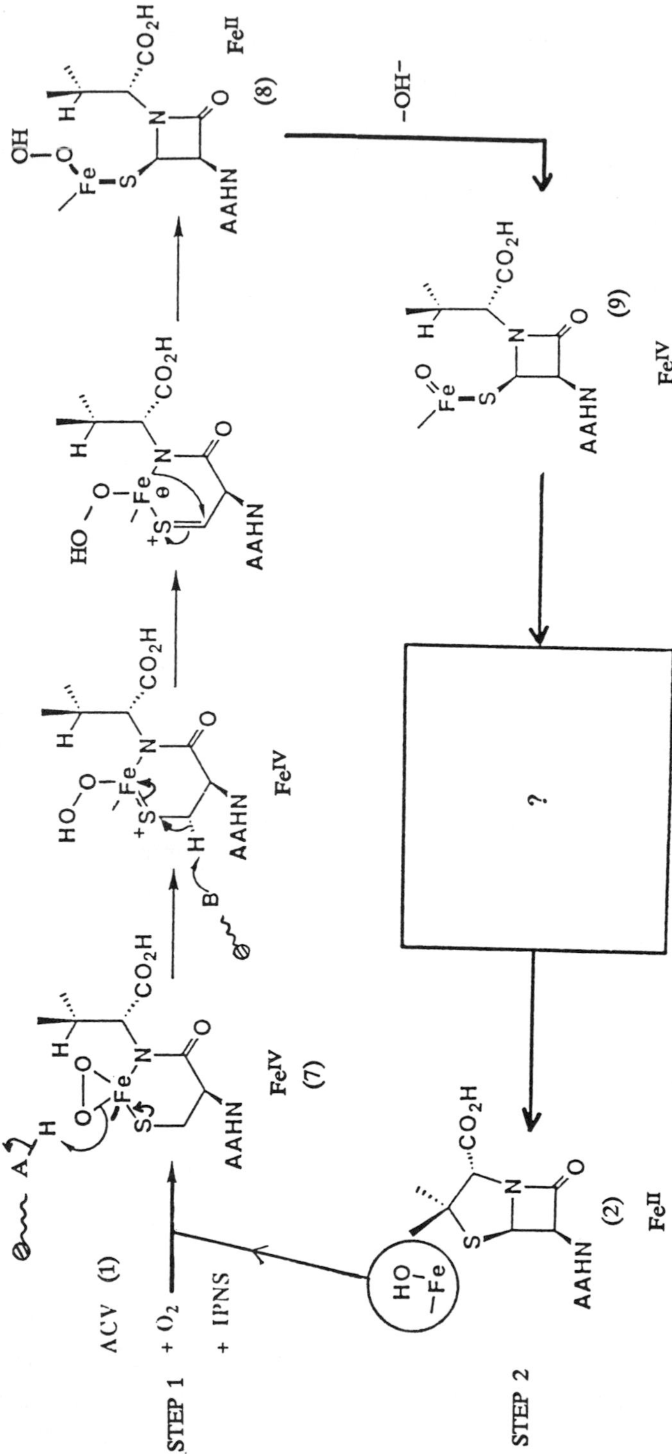

Scheme 2: Mechanism for the conversion of ACV (1) into Isopen N (2).

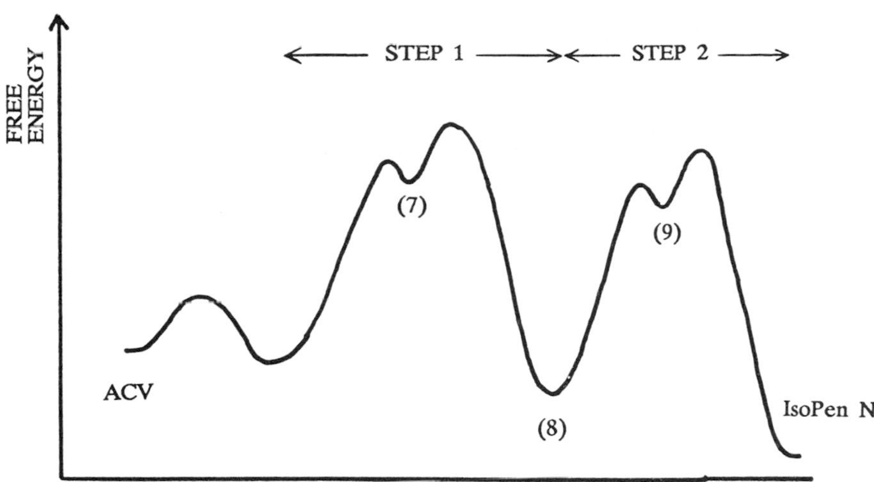

Figure 2: 'Superior' free energy profile for the conversion of ACV (1) into Isopen N (2).

mediated by a suitably placed basic site on the enzyme and leads to a weak carbon–sulphur π–bond. β–Lactam formation then follows *via* 1,3–sigmatropic shift.[18] To this point β–lactam formation has been coupled to the mildly exothermic reduction of dioxygen to a ferrous (Fe^{II}) hydroperoxide species (**8**), which we postulate to be a low energy intermediate. Cleavage of the peroxide bond of (**8**) then generates a high energy β–lactam–*S*–ferryl oxene intermediate (**9**) as required for C–H (*ie* Val C3–H) cleavage during the second step of the process. A superior free energy profile is represented in Figure 2. It is noteworthy that ferryl (Fe^{IV}) oxene generation *via* reaction of ferrous (Fe^{II}) with hydrogen peroxide is documented in cytochrome P450 model systems.[19]

If subsequent to the first irreversible event a β–lactam–*S*–ferryl oxene intermediate (**9**) is formed, then maybe a shunt metabolite could be obtained by bifurcation from the natural pathway. Such a shunt could confirm our views of the ordering of ring closures. As our V_{max} studies had indicated appreciable slowing down of the overall turnover of the 3–deuteriovalinyl tripeptide we used this tripeptide as a probe for a shunt pathway. Thus we compared the total products obtained from IPNS action on A[3–^{13}C]Cys–[3–^{2}H]Val (**10**) and A[3–^{13}C]Cys–[3–^{1}H]Val (**11**) by ^{13}C NMR. In addition to the expected production of large amounts of 5–^{13}C isopenicillin N

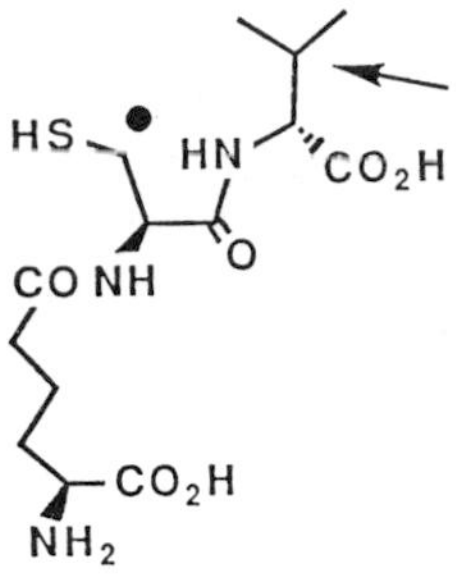

(10) ● = ^{13}C, ^{2}H at arrowed position

(11) ● = ^{13}C, ^{1}H at arrowed position

(δ_C 67 ppm) we observed an enhanced level of an unknown species with δ_C 89 ppm [*ca* 1–5% from (10), *ca* < 1% from (11)] from the 3–deuteriovalinyl tripeptide. At high resolution the 'δ 89' species was shown to consist of two resonances: either a doublet or two singlets. Secondly, we prepared and incubated A[3–^{13}C]Cys–[^{15}N, 3–^{2}H]Val and again observed the 'δ 89' signal as only two resonances, *ie* no $J^{13}C–^{15}N$ splittings (*est* 0.04 ppm) were observed. This result demonstrated that a β–lactam species could not account for 'δ 89'. In order to establish the structure of 'δ 89', further NMR and degradative studies were performed. We showed by DEPT experiments in D_2O that there was no hydrogen at the original cysteinyl C2 position, whilst a ^{13}C CW experiment indicated that the carbon giving 'δ 89' bore one hydrogen atom. Treatment of 'δ 89' with $NaBH_4$ in D_2O gave a δ 62 ppm signal as two poorly resolved singlets [identically given by (LLD)– and (LDD)–A.serinyl.V].

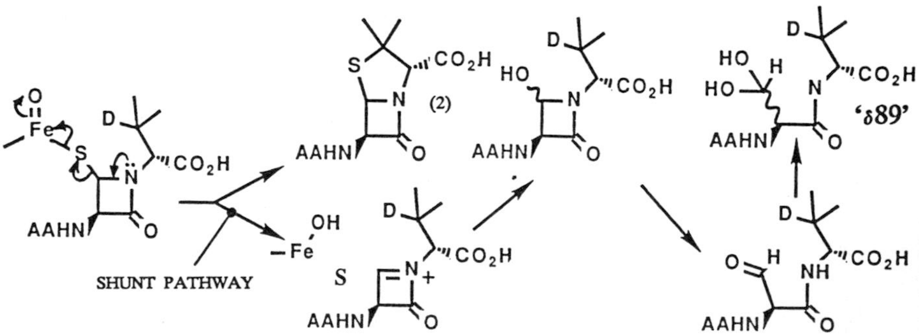

SHUNT PATHWAY

Scheme 3: Mechanism of formation of 'δ 89' *via* a shunt pathway.

At that point M Bradley suggested that 'δ 89' would be consistent with a hydrated aldehydic carbon; thus 'δ 89' should be A.[3–hydroxyserinyl].V. The two signals for 'δ 89' were explained as due to epimerisation at the C2 position [hydrate $\rightleftharpoons$ aldehyde $\rightleftharpoons$ enol forms], which also accounted for the lack of C2–hydrogens for 'δ 89' in deuteriated media. Reduction of the aldehyde form would give (LLD)– and (LDD)–A.serinyl.V. Final proof of the identity of 'δ 89' followed from synthesis of an authentic sample (from ozonolysis of A.vinylglycine.V) and subsequent HPLC, MS, ^{1}H and ^{13}C NMR comparisons.[20]

Is 'δ 89' a shunt downstream of a β–lactam–S–ferryl intermediate? In our opinion it can be considered so as the ferryl entity, which is slow to react with a valinyl C3–^{2}H bond, can act as an electron sink with respect to the β–lactam nitrogen lone pair electrons. Such an electron flow would generate a 4–membered acyliminium ion which could be subsequently intercepted by water to give a hydroxy–β–lactam capable of ring opening to an acyclic aldehydic form (see Scheme 3). This mechanism also returns the iron species from the Fe^{IV} to Fe^{II} oxidation level and, in this respect, is equivalent to the natural penicillin–forming pathway.[21]

3 MECHANISMS FOR SECOND RING CLOSURE

The second step of isopenicillin N production, the formation of the thiazolidine ring, is dependent upon the reactivity of the proposed β–lactam–S–ferryl intermediate (**9**). In this section it will be demonstrated that this species, itself formed subsequent to the first irreversible event of the pathway, is highly reactive and capable of forming bicyclic β–lactam products *via* four distinct mechanistic pathways.

Initially, the stereochemistry of the C–S bond formation of the natural substrate was reinvestigated. Thus AC[3S, 4–^{13}C]V was synthesised and found to give *exclusively* [2α–^{13}C]–isopenicillin N with IPNS *via* a stereospecific retention pathway.[22] Parallel retention pathways to penam products were also found with the isoleucinyl,[22] *allo*–isoleucinyl[22] and O–methyl–*allo*–threonyl tripeptides.[23] From these results our earlier proposal of a homolytic method for thiazolidine formation in penams seemed questionable.[24] The next substrate investigated was AC.aminobutyrate tripeptide; this represents a conservative change with respect to valine to give a sterically less demanding substrate. Incubation of AC.aminobutyrate (12) with IPNS gave three products: a major 2β-methylpenam, a minor 2α-methylpenam and a novel cepham product.[25] This cepham product was the first non-penam bicyclic product isolated, and gave the first indication of the highly reactive, and thus promiscuous, nature of the β-lactam-*S*-ferryl species (9) [It is interesting to note that an epimer of this cepham was subsequently isolated from a *Streptomyces sp* by an ICI group[26]]. The cepham formation from the aminobutyrate tripeptide, but not from ACV, can be rationalised by consideration of the C–H bond dissociation energies for methyl (98 kcal mol^{-1}), methylene (95 kcal mol^{-1}) and methine (92 kcal mol^{-1}). Thus the difference in the valinyl case, $\Delta E = +6$ kcal mol^{-1}, cannot be overcome, whereas for the aminobutyrate peptide this difference is smaller ($\Delta E = +3$ kcal mol^{-1}) and may be compensated, in part, from a more stable cepham-forming transition state.

The bond dissociation energy for an allylic C–H bond of 87 kcal mol^{-1} led to the prediction that allylic tripeptides could be effective substrates for IPNS. Thus, three allylic substrates, AC.$\beta\gamma$-dehydrovaline (13), AC.allylglycine (14) and AC.(5,6-dehydronorleucine) (15), were prepared. The first (13) gave two products: a cepham derived from allylic abstraction in a desaturase (–4H) pathway, and an α-hydroxymethyl penam,[27] whose hydroxyl group was shown to be derived from molecular oxygen, *via*

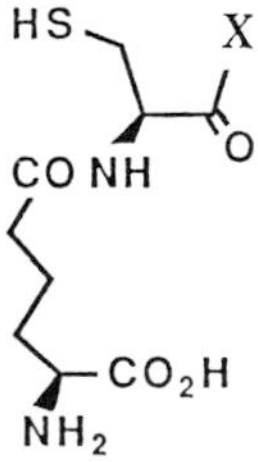

(12) X = $-NHCHEtCO_2H$

(13) X = $-NHC(=CMe_2)CO_2H$

(14) X = $-NHCH(CH_2CH=CH_2)CO_2H$

(15) X = $-NHCH[=C(CH_3)CH_2CH_3]CO_2H$

a monooxygenase (–2H + 1O) pathway.[28] The second (**14**) gave six enzymic products: three derived by the desaturase (–4H) pathway, and three *via* the monooxygenase (–2H + 1O) pathway[29] (again molecular oxygen was shown to be the source of the novel oxygen incorporated into the monooxygenase products[28]). With the third a limit was reached: (**15**) gave two cephams *via* operation of the desaturase pathway only.[27]

At this stage we wanted to confirm that we were truly observing the effect of multiple pathways from a single enzyme. Thus, the aminobutyrate (**12**) and allylglycinyl (**13**) tripeptides were each incubated with the same quantity (IU's) of both fungal (the only protein source pre–1987) and recombinant (the major source since 1987) IPNS. Identical product ratios were obtained for both tripeptides from either IPNS source.[3b] As the recombinant source can only be a single protein we are confident that we are observing the results of multiple pathways from a single enzyme.

Our next approach was to use specifically deuteriated unnatural substrates to reveal the mechanistic pathways possible for the second ring closure. Four distinct pathways from a common β–lactam intermediate were identified, which we have termed:

A	Hydrogen atom abstraction/Recombination	Desaturase (–4H)
B	Oxy–ene type process	
C	(2π + 2π) Cycloaddition followed by reductive insertion	Monooxygenase (–2H + 1O)
D	Epoxide formation	

The hydrogen atom abstraction/recombination pathway (*eg* Scheme 4) operates, we believe, during the conversion of the natural substrate (and related C3 methinyl analogues). Here the ferryl entity enables valinyl C3–H homolysis to give a carbon free radical. Provided this radical is restrained from free rotation, by virtue of its binding to the enzyme, closure directly onto sulphur or *via* recapture to a six–membered metallocycle, followed by reductive–insertion with retention, would give isopenicillin N with the observed retention of stereochemistry from the valinyl moiety. The aminobutyrate tripeptide (**12**) gave the first example of firm evidence for a free radical pathway to a penam product. When stereospecifically labelled with deuterium at the aminobutyrate C–3 position, the (3*R*)–substrate (**16**)

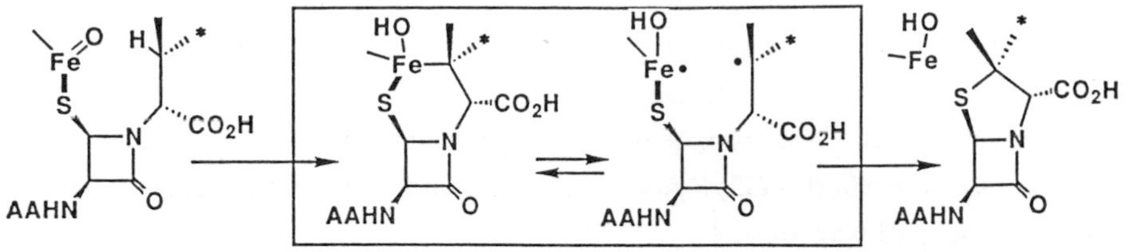

Scheme 4: Hydrogen atom abstraction/recombination pathway.

3 R (16)

3 S (17)

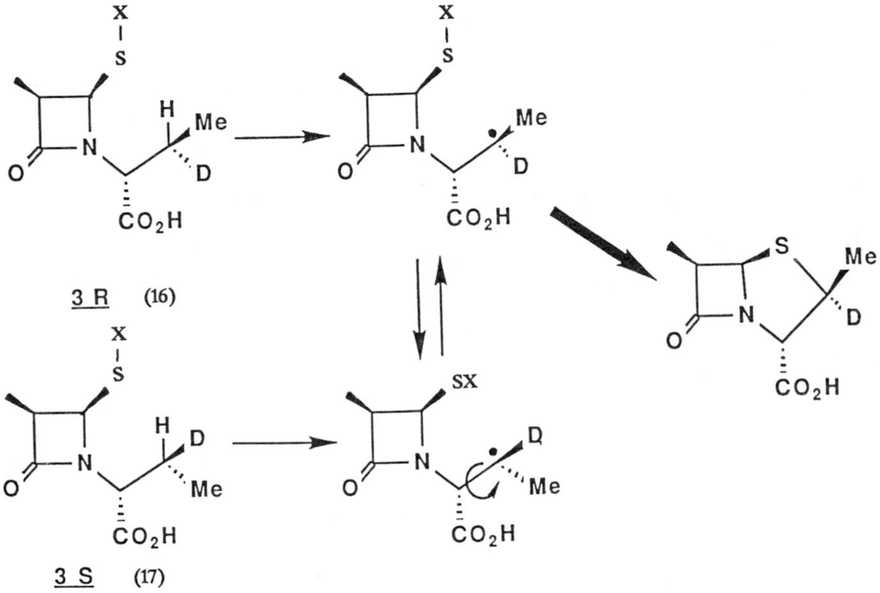

gave a β–methyl–α–deuteriopenam *via* a retention pathway, whereas the (3*S*)–substrate (17) gave the *same penam via* an inversion pathway.[30] These results, we believe, are best explained by operation of the *hydrogen atom abstraction/recombination pathway* which cleaves the weaker C3–H bond of either epimeric substrate to provide a common pair of rapidly equilibrating free radicals. Closure of the radical form (or its metallocycle equivalent) onto sulphur, controlled by the enzyme's active site, then provides the observed penam product. Probably the cepham product observed from this substrate is another example of a homolytic free radical pathway.

The *oxy–ene* and $2\pi + 2\pi$ pathways were revealed from a study of specifically deuteriated olefinic substrates. Firstly, the

(18)

O_2

Oxy–Ene

2+2

S

(19)

R

(20)

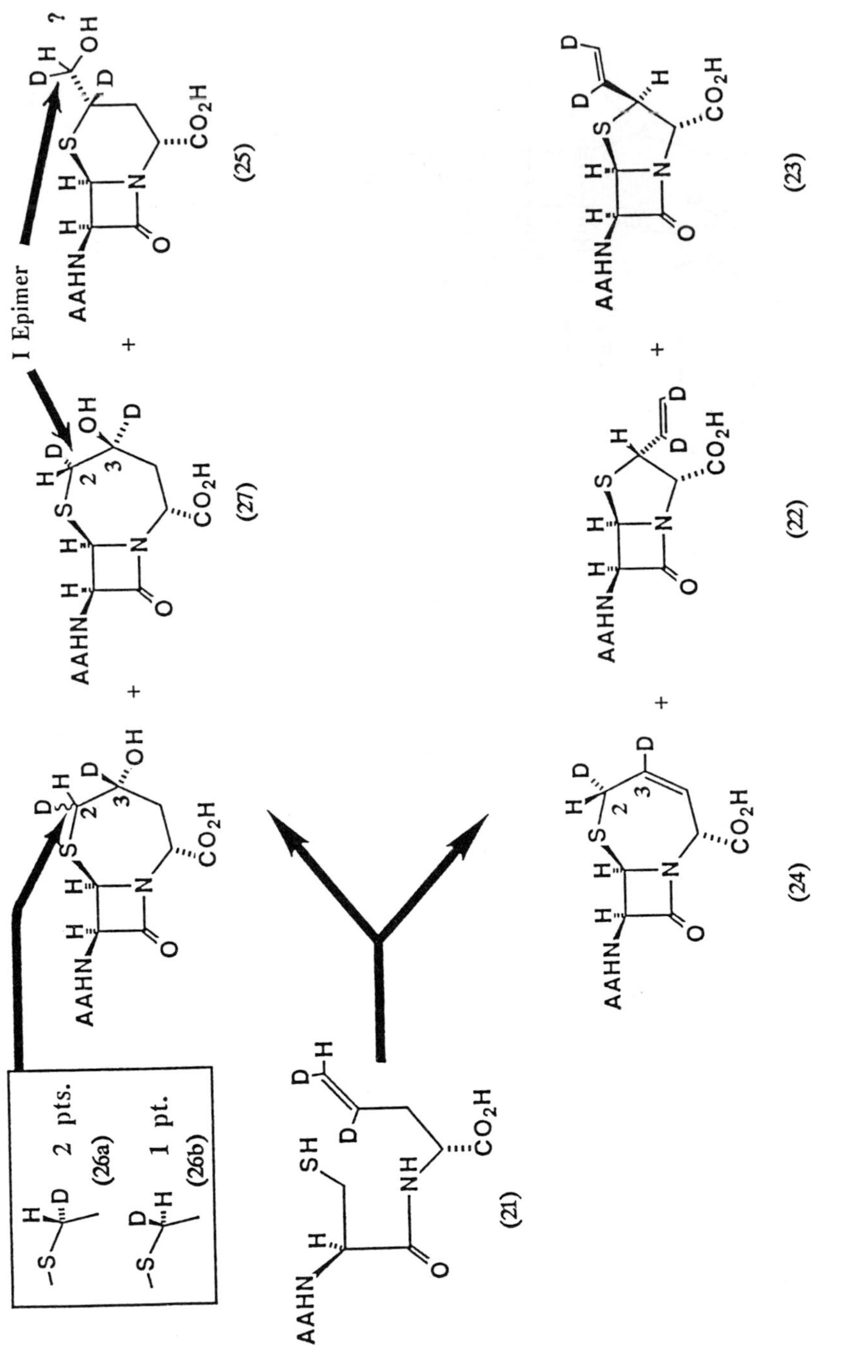

(26a) 2 pts.

(26b) 1 pt.

I Epimer

(21)

(24) + (22) + (23)

(25) + (27) + (26)

2+2

EPOXIDE

2+2

(25)

(26b)

(26a)

Z-4-deutero-$\beta\gamma$-dehydrovalinyl tripeptide (**18**) gave stereospecifically a (2*S*)-2-deuterioexomethylene cepham (**19**) and an α-(*R*-deuterio-methylenehydroxy) penam (**20**).[31] These assignments followed from nOe measurements made directly on the cepham and on the lactone derived from the α-hydroxymethyl penam. We believe that these two stereospecifically derived products result from two distinct conformations of the olefinic moiety at the active site. One conformation allows a 4 + 2 process between the ferryl (**9**) and olefin functions, *ie oxy–ene*, which enables *complete* allylic transfer with stereospecific iron–carbon bond formation. Subsequent stereospecific reductive insertion *via* a retention process then gives the observed (2*S*)-cepham (**19**). The alternative conformation leads to a *2π + 2π cycloaddition* of the ferryl (**9**) and olefin functions with stereospecific incorporation of oxygen (initially from molecular oxygen). Subsequent stereospecific reductive insertion *via* a retention process then gives the observed α-(*R*-deuteriomethylenehydroxy) penam (**20**).

Secondly, the (*E*)-4,5-dideuterio-allylglycinyl peptide (**21**) was incubated with IPNS and the stereochemistry of the six products established where possible[29b,32] (*via* nOe measurements). The desaturase (–4H) products again resulted from either the *hydrogen atom abstraction pathway* to give the α- and β- vinyl-penams (**22**), (**23**) [*Note*: the olefinic stereochemistry is retained in these products, which may reflect the geometric stability of the resonance hybrid of the initially formed C3-allylic radical] or from the *oxy–ene* process to give the (2*R*)-2,3-dideuteriohomocepham (**24**). Two of the products of the monooxygenase (–2H + 1O) pathways can be explained by the *2π + 2π cycloaddition* pathway, which can give either the 2α-(deuteriomethylenehydroxy)cepham (**25**) [formed *via* stereospecific C–O bond formation but of unknown absolute configuration] or the major (2*R*)-2,3-dideuterio-3α-hydroxyhomocepham (**26a**).

An *epoxide formation pathway* whereby the ferryl species delivers a bridging oxygen atom to the olefin *via* a *syn*-addition pathway to form an epoxide, which is subsequently opened by sulphur attack in a S_N2 mode from the least hindered end, accounts for the minor (2*S*)-2,3-dideuterio-3α-hydroxyhomocepham (**26b**) as well as the third product of the monooxygenase pathway, the (2*R*)-2,3-dideuterio-3β-hydroxyhomocepham (**27**).

AAHN H H S–Fe O D D N O CO_2H → AAHN H H S Fe O D D N O CO_2H

AAHN H H S Fe D D N O CO_2H → AAHN H H S D D OH N O CO_2H (27)

4 SUMMARY

All products derived by second ring closure from a β–lactam–*S*–ferryl species (**9**) can be rationalised by four mechanistic pathways (Scheme 5) which proceed with variable stereochemical outcomes. Of the desaturase pathways the *hydrogen atom abstraction/recombination* pathway can give either stereospecific or stereorandom C–S bond formations, whereas the *oxy–ene process* gives stereospecific C–S bond formation. For the monooxygenase pathways both the *2π + 2π cycloaddition* pathway and the *epoxide* pathway lead to stereospecific C–O and C–S bond formation. In all cases the reaction mode is controlled in part by the conformation of the valinyl analogue which allows for more than one regiochemical outcome. It should be noted that all the reaction modes operate from a *single* steric location of the analogous ferryl species (**9**) with respect to the various valinyl analogues.

Scheme 6 shows the total pathway of the natural substrate. The ferryl entity (**9**) allows valinyl C3–H homolysis and the subsequent reductive insertion process reduces the iron back to the ferrous (Fe^{II}) state ready to re–enter the catalytic cycle. It should be noted that the iron species is formally at the ferryl (Fe^{IV}) oxidation state during both C–H bond cleavage steps and an iron species, ferryl or ferrous in its oxidation level, remains *directly bound to the substrate* subsequent to cleavage of the cysteinyl 3–pro–*S* hydrogen as the first

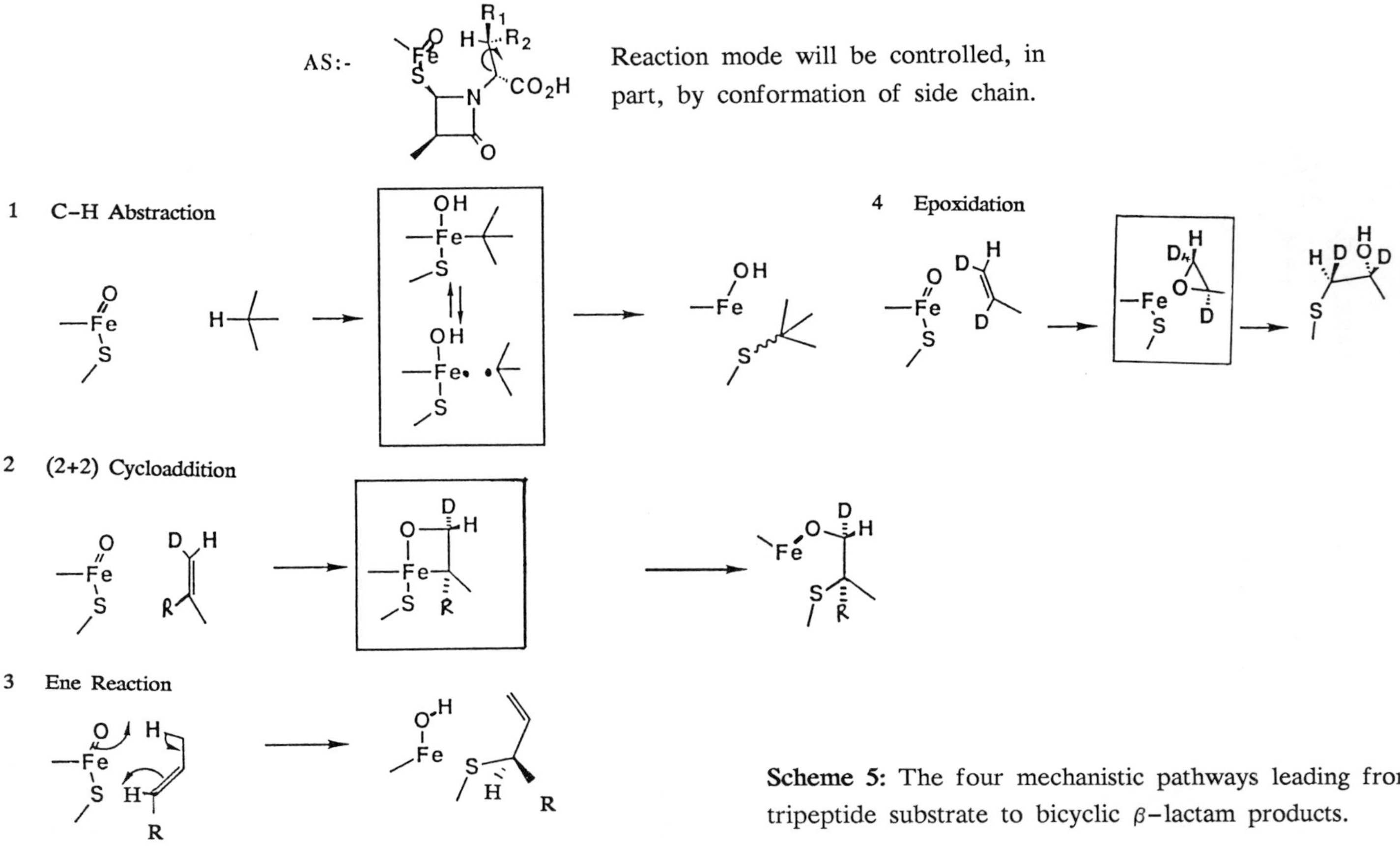

AS:- Reaction mode will be controlled, in part, by conformation of side chain.

1 C–H Abstraction

2 (2+2) Cycloaddition

3 Ene Reaction

4 Epoxidation

Scheme 5: The four mechanistic pathways leading from tripeptide substrate to bicyclic β–lactam products.

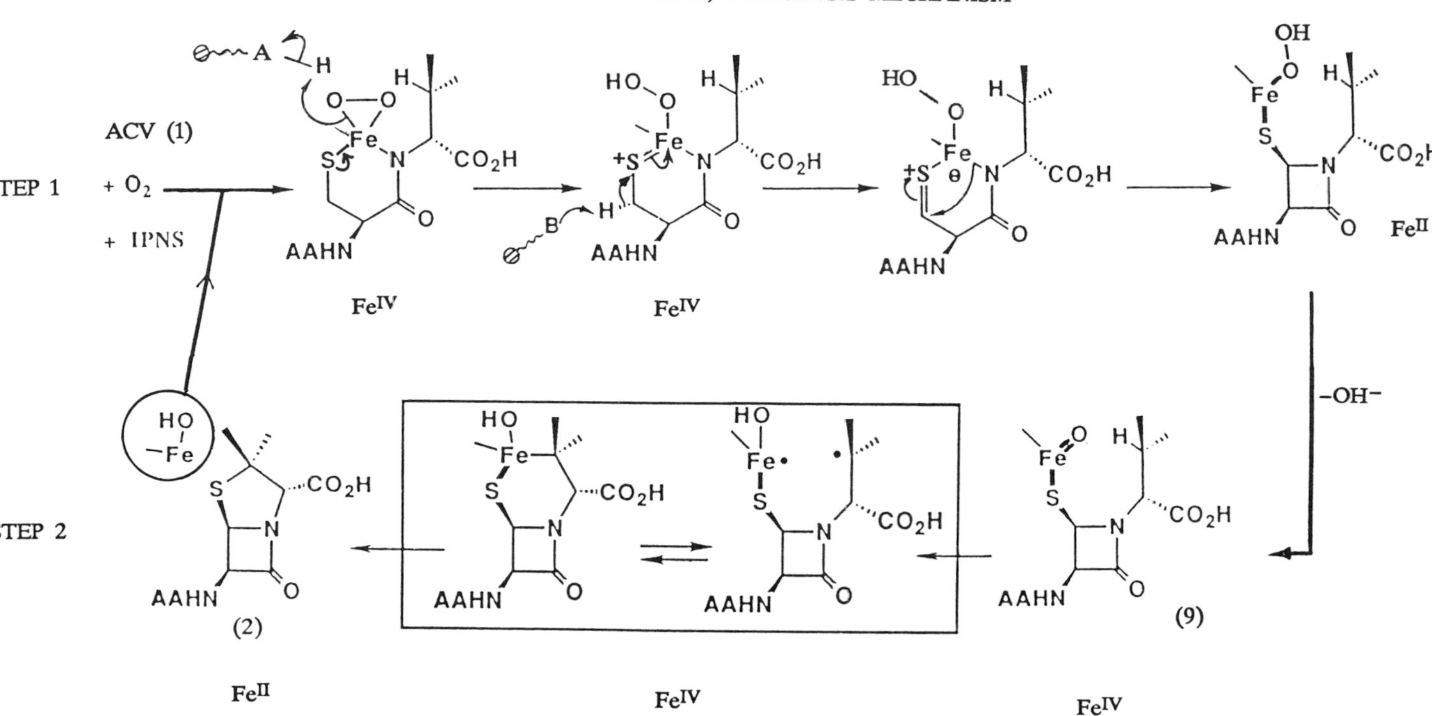

Scheme 6: Complete pathway from ACV (1) to Isopen N (2).

irreversible event up to the point where isopenicillin N is released. This close proximity and the ferryl oxidation states must, at least in part, facilitate the electron transfer necessary during the ring forming (and C–H cleavage) steps. The overall exothermic reduction of dioxygen to two molecules of water (ΔG −77 kcal mol^{-1}) provides the thermodynamic driving force for the desaturase process leading to isopenicillin N. The weaker cysteinyl C3–H bond is cleaved during the less exothermic (ΔG −14.1 kcal mol^{-1}) reduction of dioxygen to a hydroperoxide equivalent and the unactivated, and therefore strong valinyl C3–H bond, is cleaved during the more exothermic (ΔG −62.5 kcal mol^{-1}) process associated with peroxide cleavage to two molecules of water. Whether the iron is directly bound to the enzyme during the catalytic cycle is not clear at this stage, although it seems highly likely.

ACKNOWLEDGEMENTS

I thank Professor J E Baldwin for both the opportunity to work for him over the past nine years and to give the lecture related to this chapter. Also, I wish to express my appreciation to the numerous researchers who obtained the results described herein: they are cited in the reference section. We thank the IRC, SERC, British Technology Group and Lilly Research Centre, Indianapolis for support.

REFERENCES

1 For recent reviews see a) J E Baldwin, and E P Abraham, *Natural Product Report*, 1988, **5**, 129; b) J E Baldwin, 'Recent Advances in the Biosynthesis of Penicillins and Cephalosporins', Proc of the 4th Int Symp on 'Recent Advances in the Chemistry of β–Lactam Antibiotics', ed P H Bentley and R Southgate, The Royal Society of Chemistry, 1989, p1.

2 a) J E Baldwin, E P Abraham, R M Adlington, B Chakravarti, G S Jayatilake, C–P Pang, H–H Ting, and R L White,

Biochem J, 1984, **222**, 789; b) J E Baldwin, J Gagnon, and H–H Ting, *FEBS Letters*, 1985, **188**, 253.

3 a) S M Samson, R Belagaje, D T Blankenship, J L Chapman, D Perry, P L Skatrud, R M VanFrank, E P Abraham, J E Baldwin, S W Queener, and T D Ingolia, *Nature*, 1985, **318**, 191; b) J E Baldwin, S J Killin, A J Pratt, J D Sutherland, N J Turner, M J C Crabbe, E P Abraham, and A C Willis, *J Antibiotics*, 1987, **40**, 652.

4 S M Samson, J L Chapman, R Belagaje, S W Queener, and T D Ingolia, *Proc Natl Acad Sci USA*, 1987, **84**, 5705.

5 L G Carr, P L Skatrud, M E Scheetz, S W Queener, and T D Ingolia, *Gene*, 1986, **48**, 257.

6 B J Wiegel, S G Burgett, V J Chen, P L Skatrud, C A Frolik, S W Queener, and T D Ingolia, *J Bacteriol*, 1988, **170**, 3817.

7 B K Leskiw, Y Aharonowitz, M Mevarech, S Wolfe, L C Vining, D W S Westlake, and S E Jensen, *Gene*, 1988, **62**, 187.

8 J D Sutherland, private communication, Oxford 1989.

9 J E Baldwin, R L White, E–M M John, and E P Abraham, *Biochem J*, 1982, **203**, 791.

10 J E Baldwin, Sir Edward P Abraham, R M Adlington, M J Crimmin, L D Field, G S Jayatilake, R L White, and J J Usher, *Tetrahedron*, 1984, **40**, 1907.

11 J E Baldwin, G Bahadur, T Wan, M Jung, Sir Edward P Abraham, J A Huddleston, and R L White, *J Chem Soc Chem Commun*, 1981, 1146.

12 a) H Simon and D Palm, *Angew Chem Int Ed Engl*, 1966, **5**, 920 b) J E Baldwin, R M Adlington, S E Moroney, L D Field, and H–H Ting, *J Chem Soc Chem Commun*, 1984, 984.

13 For a more detailed account consult ref 1a.

14 a) D J Aberhart, L J Lin, and J Y–R Chu, *J Chem Soc Perkin Trans 1*, 1975, 2517; b) D W Young, D J Morecombe, and P K Sen, *Eur J Biochem*, 1977, **75**, 133; c) J A Huddleston, E P Abraham, D W Young, D J Morecombe, and P K Sen, *Biochem J*, 1978, **169**, 705.

15 J E Baldwin, R M Adlington, N G Robinson, and H–H Ting, *J Chem Soc Chem Commun*, 1986, 409.

16 J E Baldwin, R M Adlington, N Moss, and N G Robinson, *J Chem Soc Chem Commun*, 1987, 1664.

17 J E Baldwin, R M Adlington, and N Moss, *Tetrahedron*, 1989, **45**, 2841.

18 For a recent chemical precedent see Y Kita, O Tamura, T Miki, H Tono, N Shibata, and Y Tamura, *Tetrahedron Lett*, 1989, 729.

19 For example, see J T Groves, and M Van Der Puy, *J Am Chem Soc*, 1974, **96**, 5274.

20 J E Baldwin, R M Adlington, M Bradley, W J Norris, N J Turner, and A Yoshida, *J Chem Soc Chem Commun*, 1988, 1125.

21 For related examples of analogous shunt pathways from A.Homocysteinyl.Valine see a) J E Baldwin, W J Norris, R T Freeman, M Bradley, R M Adlington, S Long-Fox, and C J Schofield, *J Chem Soc Chem Commun*, 1988, 1128; b) J E Baldwin, J M Blackburn, M Sako, and C J Schofield, *ibid*, 1989, 970.

22 J E Baldwin, R M Adlington, B P Domayne-Hayman, H-H Ting, and N J Turner, *J Chem Soc Chem Commun*, 1986, 110.

23 J E Baldwin, R M Adlington, A Basak, S L Flitsch, S Petursson, N J Turner, and H-H Ting, *J Chem Soc Chem Commun*, 1986, 975.

24 J E Baldwin and T S Wan, *J Chem Soc Chem Commun*, 1979, 249.

25 a) J E Baldwin, E P Abraham, R M Adlington, B Chakravarti, A E Derome, J A Murphy, L D Field, N B Green, H-H Ting, and J J Usher, *J Chem Soc Chem Commun*, 1983, 1317; b) S L Flitsch, *D Phil Thesis*, Oxford University, 1985.

26 D C Aldridge, D M Carr, D H Davies, A J Hudson, R D Nolan, J P Poyser, and C J Strawson, *J Chem Soc Chem Commun*, 1985, 1513.

27 J E Baldwin, R M Adlington, A Basak, S L Flitsch, A K Forrest, and H-H Ting, *J Chem Soc Chem Commun*, 1986, 273.

28 J E Baldwin, R M Adlington, S L Flitsch, H-H Ting, and N J Turner, *J Chem Soc Chem Commun*, 1986, 1305.

29 a) J E Baldwin, R M Adlington, A E Derome, H-H Ting, and N J Turner, *J Chem Soc Chem Commun*, 1984, 1211; b) J E Baldwin, R M Adlington, M Bradley, N J Turner, and A R Pitt, *J Chem Soc Chem Commun*, 1989, 978.

30 J E Baldwin, E P Abraham, R M Adlington, J A Murphy, N B Green, H–H Ting, and J J Usher, *J Chem Soc Chem Commun*, 1983, 1319.

31 J E Baldwin, R M Adlington, L G King, M F Parisi, W J Sobey, J D Sutherland, and H–H Ting, *J Chem Soc Chem Commun*, 1988, 1635.

32 J E Baldwin, M Bradley, A R Pitt, N J Turner, R M Adlington, and A E Derome, *Tetrahedron*, in press.

STEREOCHEMISTRY OF ONE-CARBON METABOLISM IN AEROBES AND ANAEROBES

Heinz G Floss, Thomas Frenzel, David R Houck, Lai-Duien Yuen, Pei Zhou, Lynne D Zydowsky, and John M Beale

Department of Chemistry, University of Washington, WA 98195, and Department of Chemistry, Ohio State University, Columbus, OH 43210, USA

1 INTRODUCTION

In 1969 the groups of Cornforth and Arigoni broke new ground in the study of enzyme reaction stereospecificity by reporting methodology for the synthesis and configurational analysis of enantiomeric samples of acetic acid which were chiral by virtue of the presence of the three isotopes of hydrogen, ^{1}H, deuterium, and tritium, in the methyl group.[1,2] This methodology has allowed the analysis of the steric course of many enzymatic reactions in which methyl groups are transferred between biochemical molecules, are generated from methylene groups or converted into methylene groups by addition or removal of a hydrogen.[3] While the synthesis of the enantiomers of a stereogenic (so-called chiral) methyl group can be achieved by logical extensions of methods used for stereospecific labelling of prochiral systems, the conceptually most novel contribution of the work of Cornforth and Arigoni consisted of the development of a methodology allowing the distinction of an *R* from an *S* version of such a methyl group. Their method is based on abstraction of one hydrogen from the methyl group, in the form of acetyl-coenzyme A, and its replacement by a carbon substituent. The resulting methylene group contains tritium in both hydrogens, but in one position it is flanked by deuterium whereas in the other position it has normal hydrogen as its neighbour. In addition, if the reaction abstracting the hydrogen exhibits a sizeable primary kinetic isotope effect, the remaining tritium will also be distributed unsymmetrically between the two methylene hydrogens. The isotope distribution in this methylene group can be analysed in a number of ways, for example by a stereospecific enzyme reaction or by tritium NMR spectroscopy. In practical terms, the reaction chosen for the abstraction of hydrogen from acetyl-CoA was catalysed by malate synthase and involved the condensation of acetyl-CoA with glyoxylic acid with an isotope effect $k_H/k_D = 3.8$; the tritium distribution in the methylene group of the resulting malate was then determined by incubation with fumarase, which effects a stereospecific dehydration of malate to fumarate. The tritium retention in the fumarase reaction (called the F value) is a measure of the configuration and optical purity of the chiral methyl group (F = 21 = 100% ee *S*; F = 79 = 100% ee *R*; F = 50 = racemic).[3]

Table 1: Enzymatic methyl transfer reactions
NB: all proceed with Inversion.

Enzyme or Product	Site of Methylation
Arigoni group	
Vitamin B_{12}	Carbon
Loganin	Oxygen
Homocysteine *S*-methyltransferase	Sulphur
Phenylethanolamine *N*-methyltransferase	Nitrogen
Floss group	
Indolmycin	Carbon
Indolmycin	Nitrogen
Catechol *O*-methyltransferase	Oxygen
Pectin	Oxygen
Histamine *N*-methyltransferase	Nitrogen
Aplasmomycin	Carbon
4'-*O*-Methylnorlaudanosoline 6-*O*-methyltransferase	Oxygen
Norreticuline *N*-methyltransferase	Nitrogen
*Eco*RI DNA methyltransferase	Nitrogen
*Hha*I DNA methyltransferase	Carbon

2 METHYLTRANSFERASE REACTIONS

S_N2 Methyl Transfers

Using the enzymatic analysis methodology developed by Cornforth, Arigoni and their coworkers, both our laboratory[4] and that of Arigoni[5] studied a number of methyl transfers from *S*-adenosylmethionine (AdoMet) to various nucleophilic substrates. The nucleophiles serving as acceptors of the methyl group include carbon, nitrogen, oxygen, and sulphur. For these studies a synthesis of methionine carrying a chiral methyl group was developed, and following the enzymatic transfer, the methyl group was carved out of the various products and converted by a sequence of reactions of known stereochemistry into acetic acid for chirality analysis. Table 1 summarizes reactions which have been studied in this way: *all proceed with inversion of methyl group configuration.* This suggests a direct transfer of the methyl group from the sulphur of AdoMet to the acceptor nucleophile and excludes ping-pong mechanisms in which the methyl group is transiently transferred to a site on the enzyme.

Methylation with Retention of Configuration: Thienamycin

While the previous work suggests an empirical rule that methyltransferases operate by a single direct transfer of the methyl group and that these occur with inversion of configuration, a different stereochemistry was observed in a methylation occurring in the biosynthesis of the antibiotic thienamycin.[6] This remarkably potent, non-classical β-lactam antibiotic (Figure 1) is formed from one

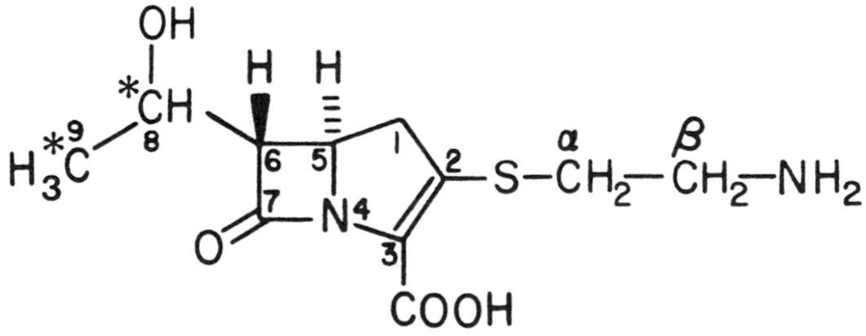

Figure 1: Structure of thienamycin (the asterisks denote the two carbons derived from methionine).

molecule each of cysteine, glutamic acid and acetic acid; the hydroxyethyl group arises by successive transfer of two methyl groups from methionine. A feeding experiment with methionine carrying a $^{13}CD_3$ methyl group showed that all four carbon-bound hydrogens of the hydroxyethyl group arise from the methyl group of methionine. Subsequent feeding experiments with methionine carrying an *R* or *S* methyl group gave thienamycin in which the methyl group had the same configuration as in the starting methionine. This intriguing result suggests that in this particular case the methyl group of methionine must undergo two sequential transfers to become the C-9 methyl group of thienamycin. A possible analogy may be seen in the transfer of the methyl group of *N*-methyltetrahydrofolate to the sulphur of homocysteine to generate methionine, catalysed by the B_{12}-dependent methionine synthase from *Escherichia coli.*[7] Unfortunately, for a variety of practical reasons, the thienamycin system does not lend itself to a more detailed examination of this unusual methyl transfer reaction. In fact, in the absence of a cell-free system, it is not even clear whether the substrate for the methyltransferase is AdoMet or whether the methionine methyl group is transferred in a different way.

Tryptophan 2-Methyltransferase

A second methyl transfer with net retention of configuration was recently discovered in the biosynthesis of the quinaldic acid moiety of the modified thiopeptide antibiotic, thiostrepton (Figure 2).[8] The quinaldic acid moiety arises by a skeletal rearrangement of tryptophan, cleaving the C-2/N bond in the indole ring and reconnecting the indole nitrogen to C-2' of the side chain. Tryptophan accounts for all of the carbon atoms of the quinaldic acid moiety except the methyl group, C-12, which arises intact from the methyl group of methionine. Again, feeding of methionine carrying a chiral methyl group to *Streptomyces laurentii* followed by Kuhn-Roth oxidation of the resulting thiostrepton samples and chirality analysis of the acetic acid showed that the C-12 methyl group had the same configuration as the methyl group in the precursor. Further experiments revealed that the methyl group is introduced early in the biosynthetic reaction sequence by a methylation of tryptophan in the 2-position of the indole ring. An AdoMet : tryptophan 2-methyltransferase activity was detected in crude cell-free extracts of *S laurentii.*[9] The transfer of the methyl group of AdoMet to give 2-methyltryptophan was indeed shown to occur with retention of configuration. The enzyme is particulate but is partly solubilized during cell rupture. All attempts to purify the soluble protein have so far been unsuccessful,

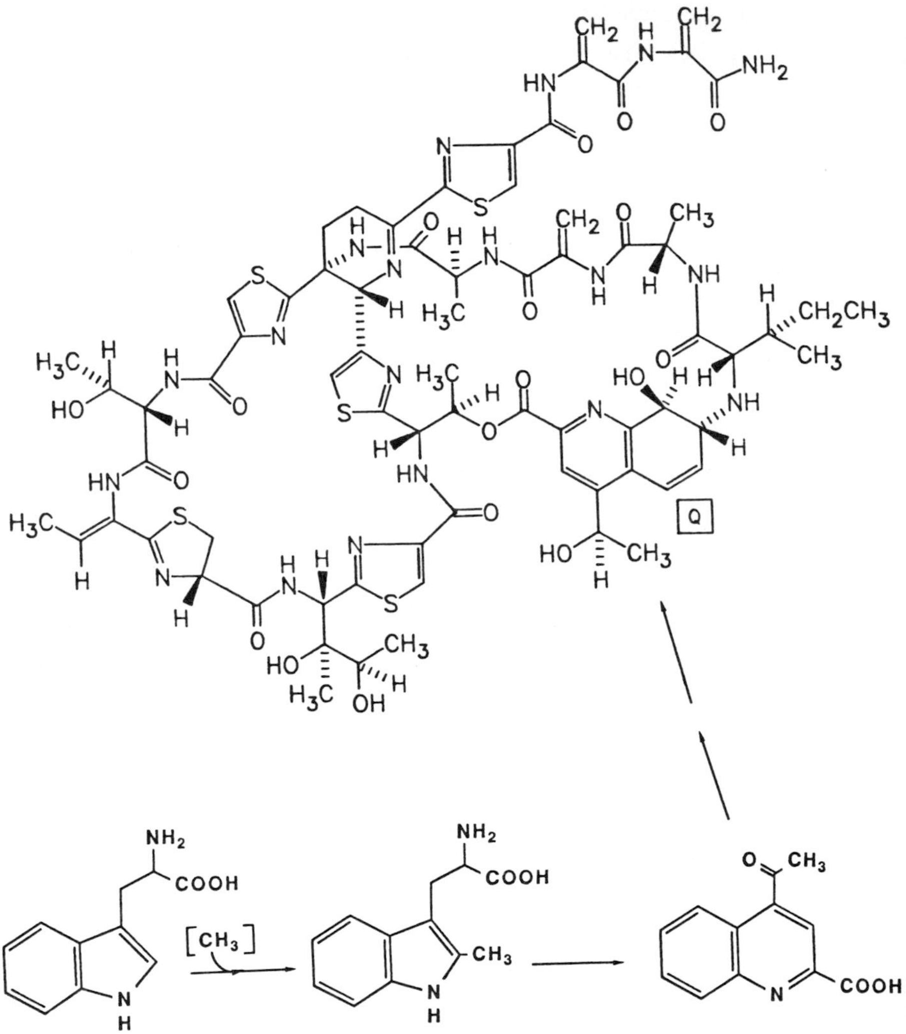

Figure 2: Structure of thiostrepton and biosynthesis of the quinaldic acid (Q) moiety.

Substrates:	Adenosylmethionine (K_m 120μM) (not L-methionine, methyl-tetrahydrofolate or methylcobalamin)
	D- or L-tryptophan, indolepyruvate (not indole or 1-*N*-methyltryptophan)
Inhibitors:	Adenosylhomocysteine (K_i 480μM)
	1-*N*-methyltryptophan (not non-intrinsic factor)

NB1: The enzyme is apparently at least partly membrane-bound.

NB2: The dialysed soluble enzyme is not reactivated by B_{12}, PLP, NH_4^+, K^+, Na^+, Mg^{2+}, Ca^{2+}, Fe^{2+}, Fe^{3+}, Mn^{2+}, Co^{2+}, Ni^{2+}, Cu^{2+}, Zn^{2+}

Figure 3: Some properties of crude AdoMet : tryptophan 2-methyltransferase from *S laurentii.*

but the enzyme was partially characterized in the crude cell-free system (Figure 3). The enzyme does show some features which are unusual compared to other methyltransferases, particularly the rather weak inhibition by adenosylhomocysteine. No evidence for involvement of cobalamin in the reaction could be obtained, but the intermediacy of N_1-methyltryptophan has been excluded. Further work is obviously needed to unravel the mechanism of action of this unusual methyltransferase.

3 PROTOBERBERINE ALKALOID BIOSYNTHESIS

In collaboration with the research group of Professor M H Zenk in Munich we have studied two sequential transformations of AdoMet-derived methyl groups in the course of the biosynthesis of protoberberine-type isoquinoline alkaloids by enzymes from plant cell

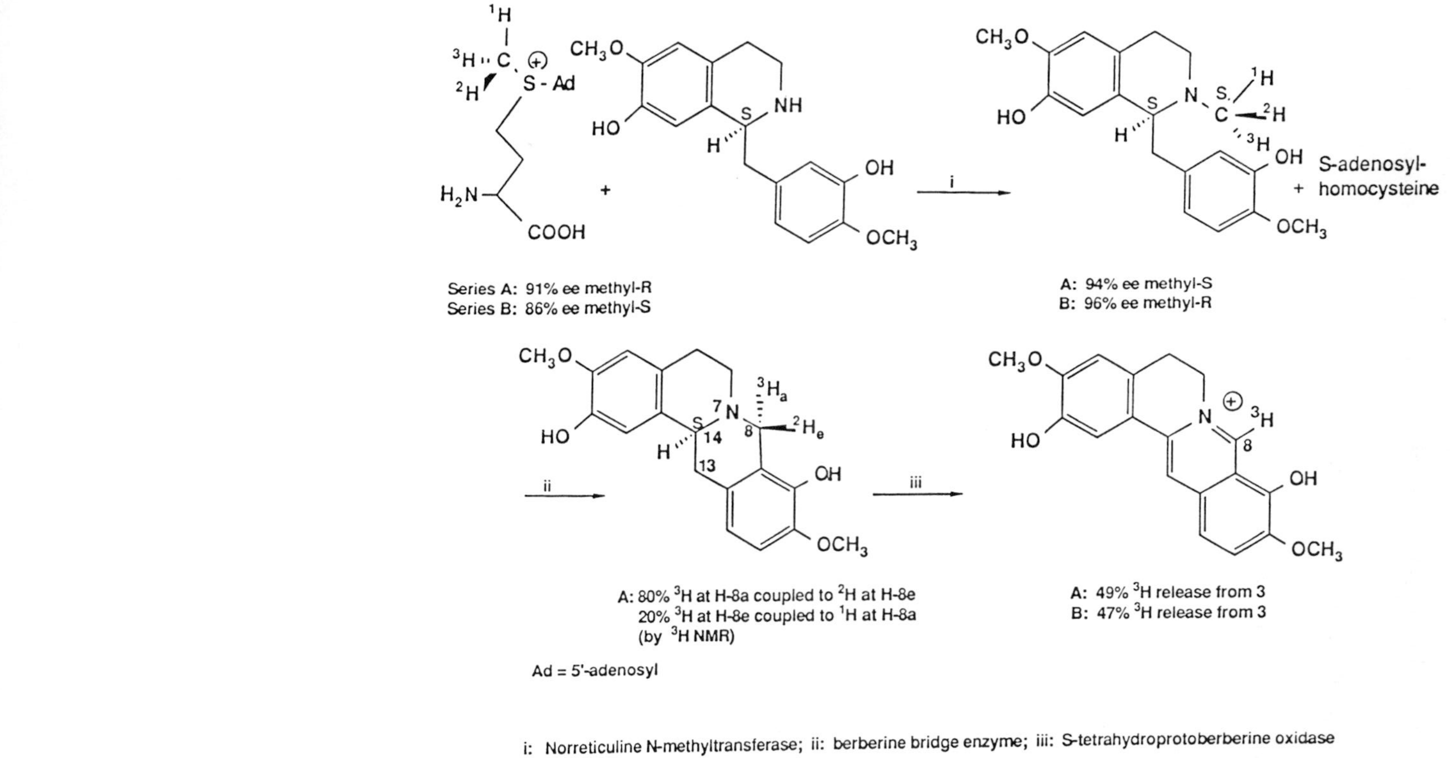

Series A: 91% ee methyl-R
Series B: 86% ee methyl-S

S-adenosyl- + homocysteine

A: 94% ee methyl-S
B: 96% ee methyl-R

A: 80% ^{3}H at H-8a coupled to ^{2}H at H-8e
20% ^{3}H at H-8e coupled to ^{1}H at H-8a
(by ^{3}H NMR)

Ad = 5'-adenosyl

A: 49% ^{3}H release from 3
B: 47% ^{3}H release from 3

i: Norreticuline N-methyltransferase; ii: berberine bridge enzyme; iii: S-tetrahydroprotoberberine oxidase

Figure 4: Reaction sequence involved in the generation of the berberine bridge and its further oxidation.

cultures. One of these involves the sequential transformation of an *N*-methyl group into a methylene bridge between nitrogen and carbon and the subsequent oxidation to an sp^2 methine group, whilst the other is the transformation of a methoxy into a methylenedioxy group, followed by reductive opening to a methoxy function.

Formation of the Berberine Bridge

The reaction sequence involved in the formation and further oxidation of the berberine bridge is shown in Figure 4. As demonstrated by degradation of the intermediate, reticuline, transfer of the methyl group from AdoMet to the nitrogen proceeds cleanly with inversion of configuration.[10] A sample of reticuline carrying an *S*-methyl group was then converted with the purified berberine

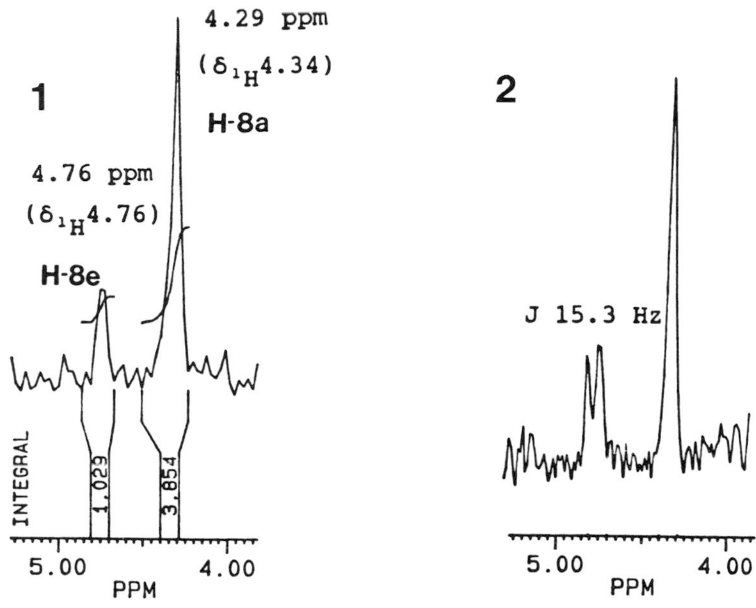

Figure 5: Tritium NMR spectra of scoulerine generated from (methyl-*S*)-[methyl-$^2H_1$,^3H]AdoMet with berberine bridge enzyme. The sample contained 153 μCi ^3H, solvent CD_3OD, repetition time 1.0 s: (1) composite pulse broadband ^1H decoupled, 76472 acquisitions; (2) ^1H gated decoupled, WALTZ-16 ^2H broadband decoupled, 47347 acquisitions.

bridge enzyme into scoulerine which was analysed by tritium NMR spectroscopy. The spectra (Figure 5) reveal that 80% of the tritium is present in the axial position at C–8, flanked by deuterium, whereas 20% is present in the equatorial position coupled to a proton. From this analysis it follows that the replacement of a hydrogen of the *N*–methyl group by the aromatic ring carbon has occurred in an inversion mode, and that the hydrogen abstraction has an isotope effect k_H/k_D of 4. Finally, the subsequent oxidation of the berberine bridge carbon by *S*–tetrahydroprotoberberine oxidase (STOX) was examined by following the tritium release from each enantiomer of chiral methyl reticuline upon first addition of berberine bridge enzyme and subsequently STOX. The results[10] showed that berberine bridge enzyme released about 8% of the tritium from each isomer, consistent with the observed isotope effect of 4. Subsequent addition of STOX led again to equal release of nearly half of the remaining tritium from the two diastereomeric substrates. This indicates that the reaction catalysed by STOX must be non–stereospecific and involves little or no isotope effect. This is consistent with the view[11] that STOX only catalyses the introduction of a Δ–7,14 double bond into the substrate and that the subsequent aromatization occurs spontaneously.

Formation of the Methylenedioxy Bridge

The second reaction sequence, shown in Figure 6, involves the transfer of a methyl group from AdoMet to oxygen to give *N*–norreticuline with complete inversion of configuration. The isomers of norreticuline were then fed to callus cultures of *Berberis koetineana*, which converted them *via* the methylenedioxy compound (berberine) into jatrorrhizine. The latter, which carries the methyl group now on the adjacent oxygen, was degraded and the methyl groups were found to have the same configuration as in norreticuline, but a configurational purity of only about one–fourth that of the precursor.[12] From these results it follows that one of the two reactions, probably the oxidative closure of the methylenedioxy bridge, must occur with retention of configuration and the other, probably the reductive opening of the methylenedioxy bridge, with inversion of configuration. The migration of the methyl carbon from

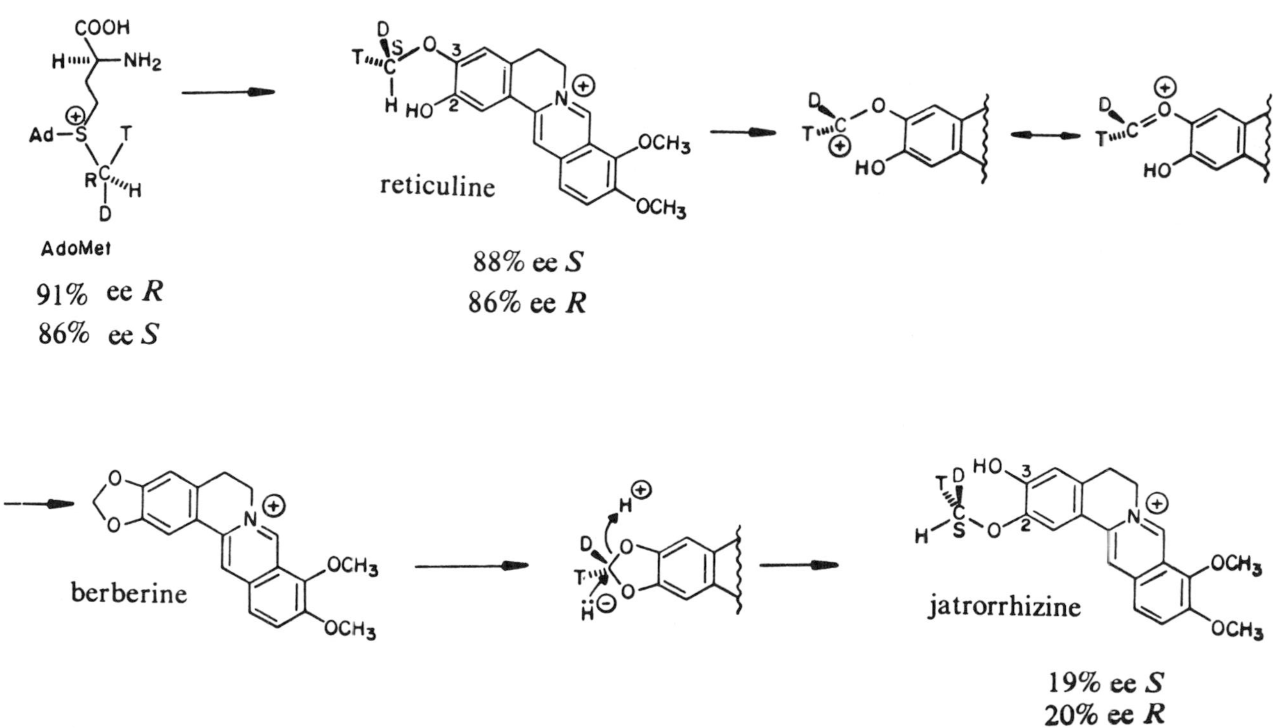

AdoMet

91% ee *R*
86% ee *S*

reticuline

88% ee *S*
86% ee *R*

berberine

jatrorrhizine

19% ee *S*
20% ee *R*

Figure 6: Reaction sequence leading to the formation of the methylenedioxy bridge of berberine, and its subsequent reductive opening to give the hydroxy/methoxy functions of jatrorrhizine.

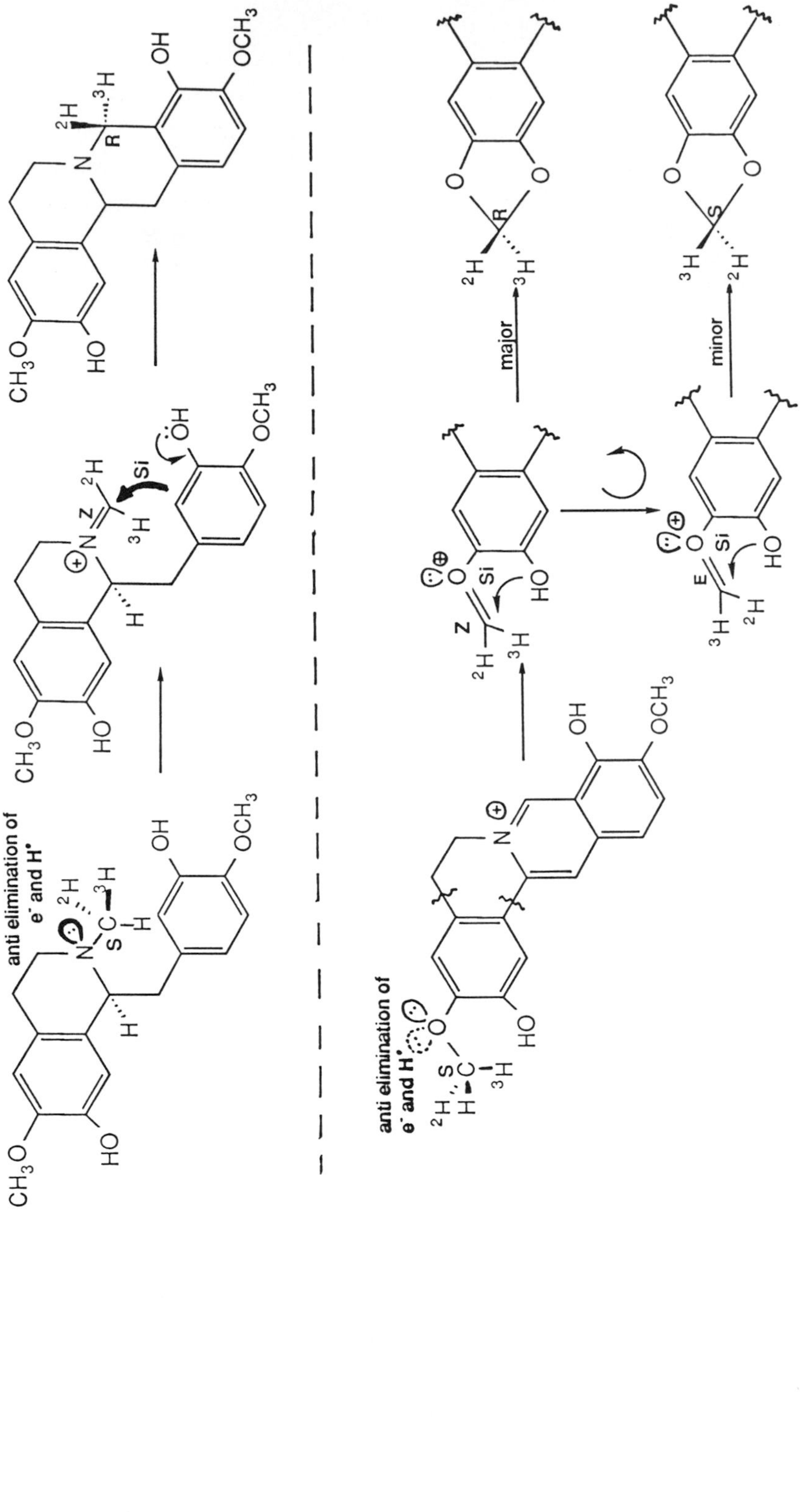

Figure 7: Proposed stereochemical mechanism for the formation of the

the original to the adjacent oxygen atom is formally equivalent to another inversion. A complete interpretation of these results obviously requires the stereochemical analysis of the intermediate methylenedioxy bridge generated from a chiral methyl group. This is currently in progress.

As a general mechanism for the two bridge-forming reactions, which is consistent with the stereochemical results, we suggest the sequence shown in Figure 7. It is proposed that in both cases *anti*-elimination of an electron from the heteroatom and a hydrogen atom from the methyl group leads to formation of a methylene iminium or methylene oxonium ion. While the former is configurationally stable and reacts stereospecifically by attack of carbon on the face opposite the one from which hydrogen has departed, the methylene oxonium ion has a substantially lower barrier to rotation and can to a significant extent undergo configurational isomerization prior to attack of oxygen on the methylene carbon, accounting for the partial racemization. In the unisomerized intermediate oxygen attacks on the same face from which the hydrogen has departed. Obviously, in order to put the latter interpretations on a firm basis, the stereochemical analysis of the intermediate methylenedioxy bridge must be completed.

4 ACETOGENESIS

Acetogenic bacteria, like *Clostridium thermoaceticum*, have the remarkable ability to convert one mole of a hexose virtually quantitatively into three moles of acetic acid. Detailed studies of this unusual fermentation have revealed that two moles of acetate are formed by fermentation of the sugar to pyruvate, which is then oxidatively decarboxylated to acetate and CO_2. The third mole of acetate, remarkably, is generated reductively from two moles of CO_2.

One mole is reduced sequentially *via* formate, 5,10-methenyltetrahydrofolate and 5,10-methylenetetrahydrofolate to 5-methyltetrahydrofolate. Another mole of CO_2 is reduced to CO. Condensation of the two one-carbon units in the presence of coenzyme A then produces acetyl-CoA, which gives acetate with formation of one mole of ATP. The formation of acetyl–CoA from methyltetrahydrofolate requires several enzymes, including CO–dehydrogenase, a B_{12}–enzyme, and a methyltetrahydrofolate:B_{12} methyltransferase. It was originally thought that the condensation of the two one–carbon units occurs as a carbonyl insertion into the methyl–cobalt bond of a methyl cobalamin. Recent work by Wood and coworkers[13] has, however, shown that CO dehydrogenase has a more extensive role in this reaction sequence. In addition to the reduction of CO_2 it also seems to catalyse the assembly of the methyl group, CO and coenzyme A into acetyl–CoA. These two possibilities can be distinguished by a stereochemical probe, determining the steric fate of the methyl group of methyltetrahydrofolate during the conversion into the methyl group of acetate. When this experiment was carried out with a cell–free extract of *C thermoaceticum*, in collaboration with the groups of Professors H Simon in Munich and S J Benkovic at Penn State,[14] it was found that the methyl group of the acetic acid formed had the same configuration as the methyl group of the starting methyltetrahydrofolate. This result supports Wood's mechanism which would involve two methyl transfers, one from methyl–tetrahydrofolate to B_{12} and another from methyl–B_{12} to CO–dehydrogenase followed by the carbonyl insertion reaction which, based on chemical precedent, must occur with retention of configuration. The earlier assumption, carbonylation of methyl–B_{12}, would predict one methyl transfer with inversion followed by carbonylation with retention to give overall inversion of configuration, contrary to the experimental observation. It was also shown in collaboration with investigators at the MIT[15] that the assembly and disassembly of acetyl–CoA by CO–dehydrogenase proceeds with complete preservation of the stereochemical integrity of the methyl group. This follows from the observation that chiral methyl acetyl–coenzyme A undergoes up to 70% exchange of ^{14}C with $^{14}CO_2$ without any loss in stereochemical purity of the methyl group.

5 METHANOGENESIS

One of the largest 'commercial' fermentations is the production of methane in sewage plants by various anaerobic methanogenic bacteria.

The majority of methanogens produce methane from CO_2 and H_2 by a sequence of reductions in which CO_2 is ultimately converted into methyltetrahydromethanopterin, which transfers its methyl group to the sulphur of the unique coenzyme M, mercaptoethanesulphonic acid. Methyl–CoM is then the substrate for a complex enzyme, methylreductase, which converts the methyl group into methane.

Formation of Methyl–Coenzyme M

While the majority of methanogens use only CO_2 and H_2 as substrates, some species, such as *Methanosarcina barkeri*, can utilize other C1 donors, such as methylamine, methanol, or the methyl group of acetic acid. Presumably, the methyl group from these compounds enters the sequence at the same point as the methyl group from methyltetrahydromethanopterin. The exact sequence of reactions is not certain, although some evidence points to the involvement of two methyltransferases in the conversion of methanol to methyl–CoM. This would presumably mean that the methyl group undergoes two transfers, which should result in net retention of configuration. This question was probed by preparing the two enantiomers of chiral methanol and incubating them with a cell–free extract from *M barkeri*. The reaction sequence was blocked at the stage of methyl–CoM by adding bromoethanesulphonate, an inhibitor of methylreductase, to the incubation mixture. The resulting methyl–CoM was degraded by a stereospecific sequence of reactions to give acetic acid for chirality analysis. The results showed that the methyl group in methyl–CoM has the same configuration as the starting material. Hence, the data do indeed indicate two transfers of the methyl group.[16] Similarly, chiral acetic acid was incubated with a cell–free extract of *M barkeri* adapted for growth on acetate, and the methyl–CoM degraded in the same fashion. Again, both the substrate and the product had the same configuration. The disassembly of acetate into a methyl group, therefore, presumably occurs with retention of configuration, and the methyl group then undergoes two sequential transfers to give methyl–CoM.[17]

Formation of Methane

The final task remaining in this project is the stereochemical analysis of the reduction of methyl–CoM to methane by methylreductase. This reaction is thought to involve the transfer of the methyl group to the nickel of a unique cofactor, coenzyme F_{430}, followed perhaps by protolytic cleavage of the methyl–nickel bond. Obviously, the stereochemical analysis of this reaction presents a problem, because four isotopes of hydrogen are required to generate a chiral version of methane, but nature only provides three. Thus, we have to work out methodology to place a different group in the position of the fourth hydrogen. Fortunately, it is known that methylreductase will utilize ethyl–CoM at a substantial rate. The strategy we are developing, therefore, involves the synthesis of ethyl–CoM labelled stereospecifically with deuterium and tritium in the methylene group. Reduction of this substrate will then produce ethane carrying one chiral methyl group. The ethane can be degraded by light–catalysed chlorination followed by permanganate oxidation to give acetic acid retaining half the radioactivity in the

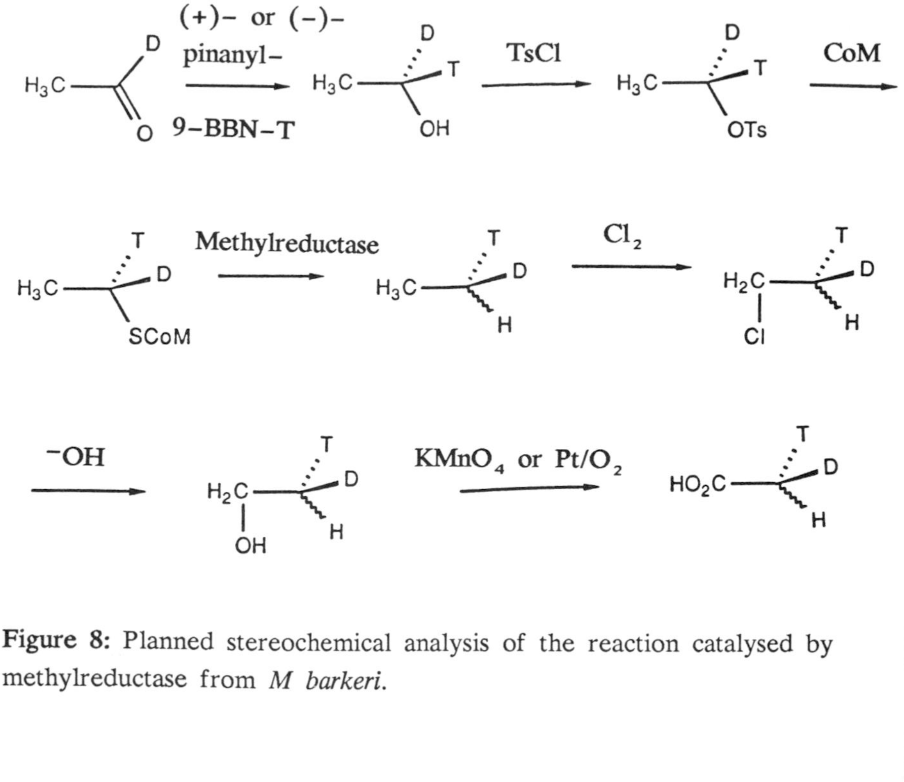

Figure 8: Planned stereochemical analysis of the reaction catalysed by methylreductase from *M barkeri.*

form of an intact chiral methyl group. The chemistry necessary for both the synthesis and degradation (Figure 8) has been worked out and we are now in the process of implementing this reaction sequence with labelled material.

6 DIRECT NMR ANALYSIS OF CHIRAL METHYL GROUPS

In collaboration with Professor F A L Anet we have recently demonstrated the feasibility of determining the configuration and optical purity of a chiral methyl group by direct NMR analysis of the intact methyl group.[18] This method is based on work by Anet and Kopelevich,[19] who demonstrated that the two diastereotopic protons in a rapidly rotating CH_2D group in a suitable chiral molecule can have an observable chemical shift difference. A suitable system is 1,2-dimethylpiperidine carrying one deuterium in the *N*-methyl group, in which the two remaining protons of the *N*-methyl group showed a chemical shift difference of 0.014 ppm. To demonstrate the utility of this system for the configurational analysis of chiral methyl groups, we[20] reacted (*R*)-[methyl-$^{2}H_{1}$,^{3}H]*N*,*N*-ditosylmethylamine (approximately 80 to 90% ee, prepared from chiral acetic acid[21]), with (*R*)- and (*S*)-2-methylpiperidine (100% and 98% ee, 20-fold excess). The crude reaction mixtures (without purification) were diluted with CD_2Cl_2 and analysed directly by tritium NMR spectroscopy. The spectra, shown in Figure 9, indicate a 4.4 Hz chemical shift difference (at 320 MHz) for the two diastereotopic tritons. As predicted,[18] the signal for the (2*R*,7*S*)/(2*S*,7*R*) pair comes at higher field (lower frequency) than that for the (2*S*,7*S*)/(2*R*/7*R*) pair. Quantitative evaluation reveals that the chiral methyl group in 1,2-dimethylpiperidine has an enantiomeric purity of 86% ee. The decrease in optical purity compared to the starting acetic acid is probably encountered as a result of exchange and racemization in the Schmidt reaction. Consistent with this notion the spectra reveal the presence of approximately 3% CH_2T species. Although the present method is obviously less sensitive than the enzymatic procedure, it has the advantage over the latter that, given adequate amounts of tritium, the configurational purity of the methyl group can be determined with much greater accuracy.

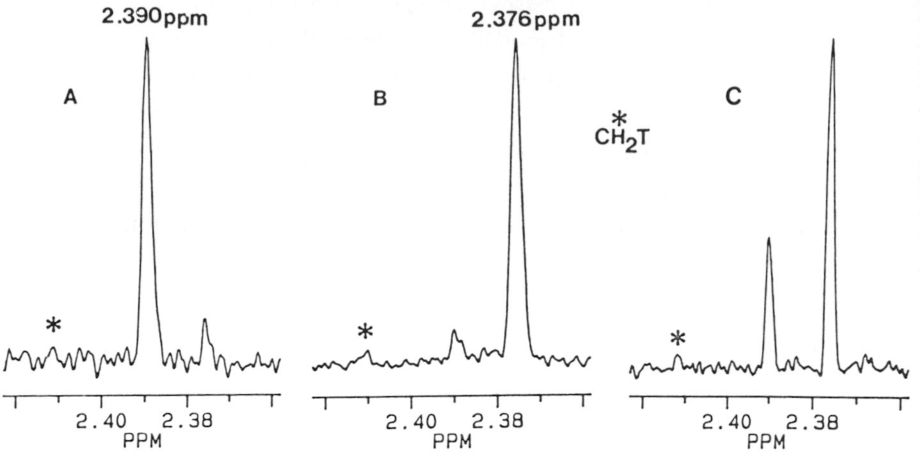

Figure 9: 320 MHz ^{3}H–NMR spectra of:
(**A**) (2*S*,7*S*)-[7–$^{2}H_{1}$,^{3}H]1,2–dimethylpiperidine (1.1 mCi),
(**B**) (2*R*,7*S*)–[7–$^{2}H_{1}$,^{3}H]1,2–dimethylpiperidine (2.1 mCi), and (**C**) a 0.7:1 mixture of (a) + (b).
Spectral accumulations at 2.8 s repetition time were: (a) 4000, (b) 2190, (c) 1800 pulses.

ACKNOWLEDGEMENTS

We acknowledge with pleasure the fruitful collaborations with the laboratories of Professors F A L Anet, J N Reeve, H Simon, and M H Zenk, and the financial support of this work by the National Institutes of Health and by a US Senior Scientist Award to HGF from the Alexander von Humboldt Foundation.

REFERENCES

1 J W Cornforth, J W Redmond, H Eggerer, W Buckel, and C Gutschow, *Nature*, 1969, **221**, 1212; *idem*, *Eur J Biochem* 1970, **14**, 1.
2 J Lüthy, J Rétey, and D Arigoni, *Nature*, 1969, **221**, 1213.
3 H G Floss and M D Tsai, *Adv Enzymol*, 1979, **50**, 243.

4 H G Floss in 'Mechanisms of Enzymatic Reactions: Stereochemistry', ed P A Frey, Elsevier, 1986, p 71.

5 D Arigoni, *Ciba Foundation Symposium*, 1978, **60**, 243.

6 D R Houck, K Kobayashi, J M Williamson, and H G Floss, *J Am Chem Soc*, 1986, **108**, 5365.

7 T M Zydowsky, L F Courtney, V Frasca, K Kobayashi, H Shimizu, L-D Yuen, R G Matthews, S J Benkovic, and H G Floss, *J Am Chem Soc*, 1986, **108**, 3152.

8 P Zhou, D O'Hagan, U Mocek, Z Zeng, L-D Yuen, T Frenzel, C J Unkefer, J M Beale, and H G Floss, *J Am Chem Soc*, 1989, **111**, 7274.

9 T Frenzel, P Zhou, and H G Floss, manuscript submitted for publication.

10 T Frenzel, J M Beale, M Kobayashi, M H Zenk, and H G Floss, *J Am Chem Soc*, 1988, **110**, 7878.

11 M Amann, N Nagakura, and M H Zenk, *Tetrahedron Lett*, 1985, 953.

12 M Kobayashi, T Frenzel, J P Lee, M H Zenk, and H G Floss, *J Am Chem Soc*, 1987, **109**, 6184.

13 H G Wood, S W Ragsdale, and E Pezacka, *Biochem Int*, 1986, **12**, 421.

14 H Lebertz, H Simon, L F Courtney, S J Benkovic, L D Zydowsky, K Lee, and H G Floss, *J Am Chem Soc*, 1987, **109**, 3173.

15 S A Raybuck, N R Bastian, L D Zydowsky, K Kobayashi, H G Floss, W H Orme-Johnson, and C Walsh, *J Am Chem Soc*, 1987, **109**, 3171.

16 L D Zydowsky, T M Zydowsky, E S Haas, J W Brown, J N Reeve, and H G Floss, *J Am Chem Soc*, 1987, **109**, 7922.

17 L D Zydowsky, S Lee, E S Haas, J N Reeve, and H G Floss, manuscript in preparation.

18 F A L Anet, D O'Leary, J M Beale, and H G Floss, *J Am Chem Soc*, 1989, **111**, 8935.

19 F A L Anet and M Kopelevich, *J Am Chem Soc*, 1989, **111**, 3429.

20 K Kobayashi, P K Jadhav, T M Zydowsky, and H G Floss, *J Org Chem*, 1983, **48**, 3510.

21 R W Woodard, L Mascaro, R Horhammer, S Eisenstein, and H G Floss, *J Am Chem Soc*, 1980, **102**, 6314.

MECHANISM-BASED β-LACTAM INHIBITORS OF HUMAN LEUKOCYTE ELASTASE

Raymond A Firestone,[a] Peter L Barker,[b] and Judith M Pisano[c]

a Bristol-Myers Co, PRDD, PO Box 5100, Wallingford, CT 06492-7660, USA
b Genentech, 460 Point San Bruno Blvd, San Francisco, CA 94080, USA
c Merck Sharp & Dohme Research Laboratories, PO Box 2000, Rahway, NJ 07065, USA

1 INTRODUCTION

Human leukocyte elastase (HLE, EC 3.4.21.37), a serine protease, is implicated in the etiology of emphysema and other disorders.[1,2] Therefore, inhibitors of this enzyme might be valuable anti-inflammatory agents. Similarities between HLE and β-lactamases, serine proteases that can be inactivated by certain β-lactams, prompted Dr M Zimmerman at Merck to propose that β-lactams might also inhibit HLE.

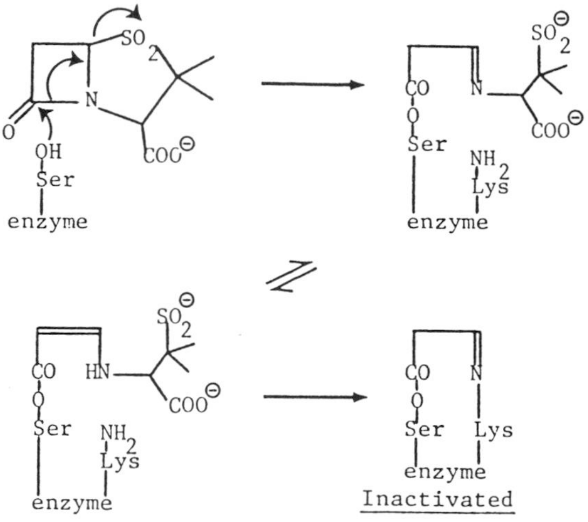

Scheme 1: Inhibition of RTEM β-lactamase.

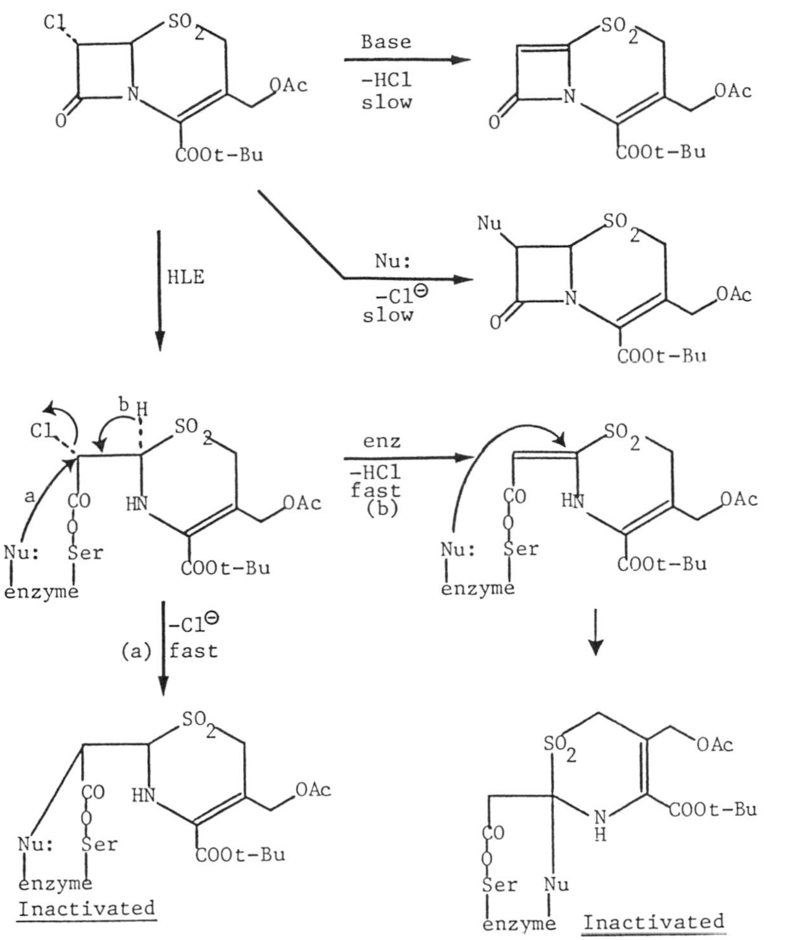

Scheme 2: Proposed inhibition of HLE by 7α-chlorocephalosporin sulphone.

The mechanism of inhibition of β-lactamase by a typical β-lactam is shown in Scheme 1.[3] Acylation of the active-site serine by the β-lactam (1st hit, reversible) triggers elimination of a sulphinic acid and unmasks a new electrophile, a Schiff base, which captures active-site lysine (2nd hit, irreversible). The second step is an example of mechanism-based or 'suicide' inactivation.[4]

2 BICYCLIC β-LACTAM INHIBITORS OF HUMAN LEUKOCYTE ELASTASE

Initial efforts at Merck,[5] following the penicillin sulphone model, focused on cephalosporin sulphones as potential inhibitors of HLE. Unlike β-lactamase, HLE responds only to cephalosporin esters. An early lead compound (Table 1, X = OMe) gave fairly good inhibition, but it was not irreversible, and the mechanism was thus in doubt. It occurred to us that the chloro analog (Table 1, X = Cl) might provide irreversible inhibition *via* the mechanism in Scheme 2. The α-haloamide of the parent cephalosporin is a potential electrophile, either in itself or as Michael acceptor after elimination of HCl. However, the reactivity of the chloro group in both substitutions and eliminations will be low owing to increased I-strain within the 4-membered ring in the transition states for these reactions. Once the ring has been opened by acylation of the active-site serine, the I-strain prohibition is lifted, releasing the full reactivity of the Cl toward either substitution (S_N2) or elimination. The electrophile thus created captures a second active-site nucleophile and the overall mechanism, then, has two hits. The first is active-site-directed, and presumably reversible, whilst the second is mechanism-based, and hopefully irreversible.

The compound (Table 1, X = Cl) proved to be a potent, irreversible inactivator of HLE. Information about its mechanism of action was obtained by Dr M Navia *et al*[6] using the closely related enzyme porcine pancreatic elastase (PPE). A solution of the inhibitor was soaked into crystalline PPE, and the structure of the enzyme-inhibitor complex was established by *X*-ray crystallography.

Table 1: Inhibitors of HLE.

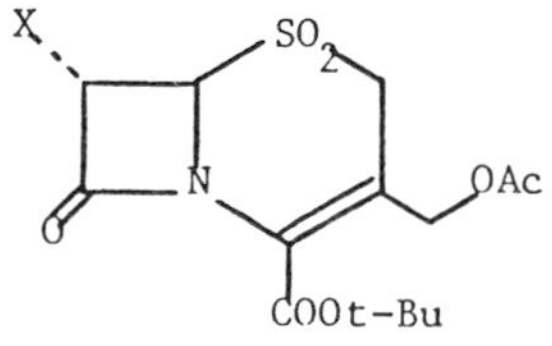

X, SO_2, N, O, OAc, COOt-Bu

X	IC_{50} μg/ml
OMe	0.5
Cl	0.02

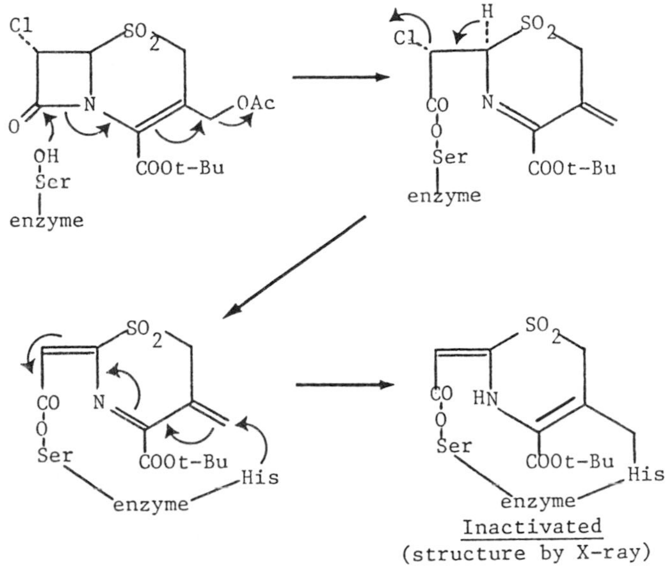

Scheme 3: Inhibition of HLE by 7α–chlorocephalosporin sulphone.

This indicated that acylation of serine had triggered elimination of HCl as desired, but it had also caused loss of the 3'–acetoxy, a common event in the acylation of cephalosporins (see Scheme 3, extrapolated for HLE). The resulting Michael acceptor therefore had more than one good acceptor site, and the enzyme's active–site histidine chose to attack at the 3'–position rather than at the C–6 (Scheme 3).

3 MONOCYCLIC β–LACTAM INHIBITORS OF HLE

We then considered whether a two–stage mechanism such as that in Scheme 1 required the complete bicycle structure of a penicillin or cephalosporin. Two classes of monocyclic β–lactams were designed bearing leaving groups X whose departure from the intact inhibitor is impeded by I–strain. However, their expulsion should be rapid once

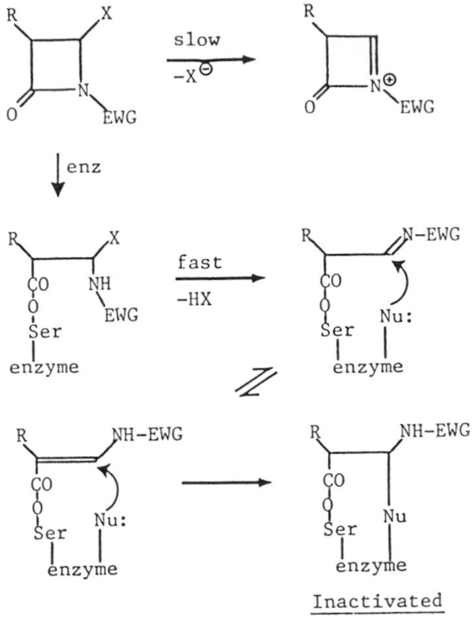

Scheme 4: Proposed inhibition of HLE by monocyclic β-lactams, class I.

the 4-membered ring has been opened and unmasks the second electrophile. The two classes differed in the placement of the leaving group. Both contained electron-withdrawing groups (EWG's) on the lactam nitrogen to compensate for the lower acylating power of monocyclic β-lactams *vis-a-vis* their more strained bicyclic counterparts. Schemes 4 and 5 show the proposed mechanisms of action of the two classes of inhibitors. In both classes, the group R at the 3-position lies in the P_1 binding pocket of the enzyme. Powers and his coworkers[7-9] have studied this pocket (Table 2), finding that isopropyl is good, but propyl is much better. Therefore, we made both classes of inhibitor with various alkyl and aryl groups in this position. The results are given in Tables 3 and 4 and show that both classes provide very potent irreversible inhibitors. In accord with data of refs 7–9, small n-alkyl groups R are best. If R is too small, or branched, or phenyl, the activity is much lower. The presence of an EWG on nitrogen to enhance the lactam's reactivity is beneficial.

Our monocyclic β-lactams are racemates with one exception (see Table 4). It is likely that the chiral enzyme recognises only one enantiomer. If that is so, then the proper enantiomers of the best inhibitors have IC_{50} = 0.005 μg/ml, which is only ten molecules of inhibitor per molecule of enzyme.

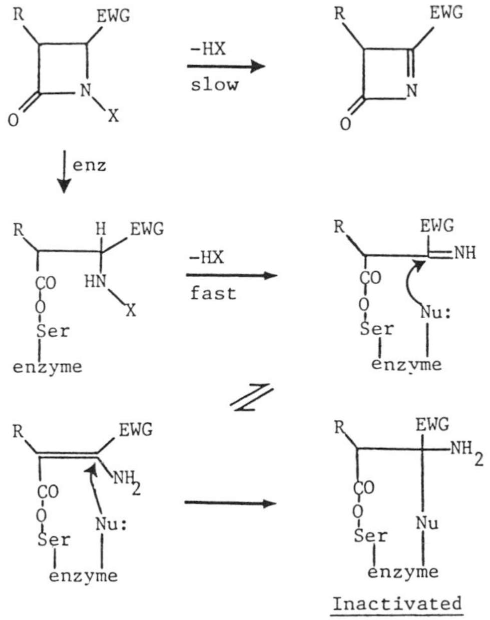

Scheme 5: Proposed inhibition of HLE by monocyclic β-lactams, class II.

If hydroxylamine is added to HLE one minute after inactivation with a Class I inhibitor, the enzyme recovers full activity, but if it is added after ten minutes' inactivation, no reactivation occurs. This fits the mechanism in Scheme 4, wherein the first step, acylation of active-site serine, is reversed by hydroxylamine, but the second hit is not reversible. This mechanism has recently been supported by *X*-ray crystallography.[10] What cannot be discerned from the crystal structure analysis is whether the active form of the electrophile responsible for the second hit is the imine or enamine tautomer. There is, however, one compound, the 3,3-dimethyl derivative, which can function only as the imine tautomer, and this has good activity.

No *X*-ray work was done with Class II inhibitors, but their behaviour with hydroxylamine (same as Class I) and their structure activity relationships are all in accord with the mechanism of Scheme 5.

Table 2: HLE has a binding pocket that likes a small alkyl group in the P_1 position.

Best substrates	k_{cat}/K_m	Ref
Boc-Ala-Pro-Nva-SBzl (4-Cl)	130	7
MeOSucc-Ala-Ala-Pro-Val-SBzl	1.2	8

Best inhibitors	Ref
2-(SO_2F)-C$_6$H$_4$-$NHCOC_2F_5$	9
2-C_3F_7-4H-3,1-benzoxazin-4-one	9

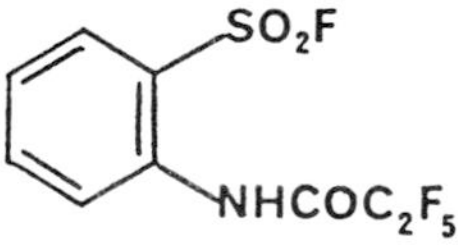

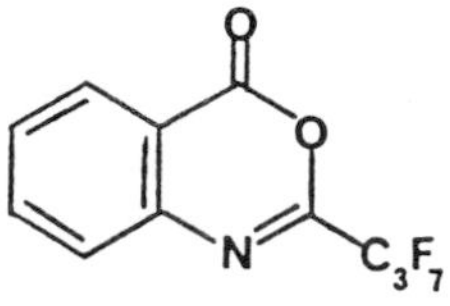

Table 3: Inhibitors of HLE, Class I

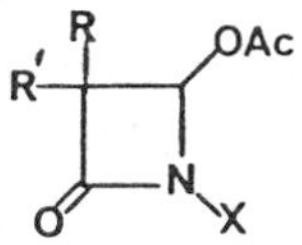

R	R'	*cis* or *trans*	IC_{50} (μg/ml) X = H	X = Ac
H	H	-	-	3
Me	H	*c/t* = 1	20	3
Me	Me	-	20	0.5
Et	H	*c/t* = 1	15	0.1
Pr	H	*t*	15	0.1[a]
Bu	H	*t*	0.4	0.1

a 0.01 after 5 min

Table 4: Inhibitors of HLE, Class II

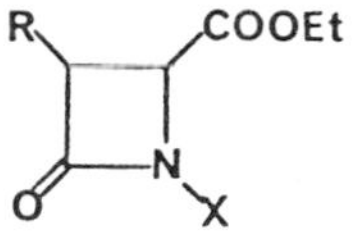

R	*cis* or *trans*[a]	IC_{50} (μg/ml) X = H	X = PNBS
H[b]	-	-	3
Et	t	3	0.01
	t[c]	>20	0.2, 0.6
	c	1	0.05
vinyl	c	10	0.02
Pr	c	2	0.06
allyl	c	20	0.05
2-propenyl	c	5	0.5
Ph	c	>20	3

a *cis* always more active than *trans*, except for R = Et

b benzyl instead of ethyl ester

c optically active

REFERENCES

1 A Janoff, 'Pulmonary Emphysema and Proteolysis', ed C Mittman, Academic Press, New York, 1972, p 1; *idem*, *J Lab Clin Med*, 1977, **115**, 461; *idem*, *Chest*, 1983, **83**, 54.

2 H Menninger, R Putzier, W Mohr, B Hering, and H D Mierau, 'Biological Functions of Proteases', ed H Holtzer and H Tschesche, Springer, Berlin, 1979, p 196; G D Virca, G Matz, and H P Schnebli, *Eur J Biochem*, 1984, **144**, 1.

3 G D Brennan and J R Knowles, *Biochemistry*, 1981, **20**, 3680; *idem*, *ibid*, 1984, **23**, 5833.

4 K Bloch, *The Enzymes (3rd Edn)*, 1972, **5**, 441; R Rando, *Science*, 1974, **185**, 320.

5 J B Doherty, B M Ashe, L W Argenbright, P L Barker, R J Bonney, G O Chandler, M E Dahlgren, C P Dorn, P E Finke, R A Firestone, D Fletcher, W K Hagmann, R Mumford, L O'Grady, A L Maycock, J M Pisano, S K Shah, K R Thompson, and M Zimmerman, *Nature*, 1986, **322**, 192.

6 M A Navia, J P Springer, T-Y Lin, H R Williams, R A Firestone, J M Pisano, J B Doherty, P E Finke, and K Hoogsteen, *Nature*, 1987, **327**, 79.

7 J W Harper, R R Cook, C J Roberts, B M Laughlin, and J C Powers, *Biochemistry*, 1984, **23**, 2995.

8 M J Castillo, K Nakajima, M Zimmerman, and J C Powers, *Anal Biochem*, 1979, **99**, 53.

9 T Yoshimura, L N Barker, and J C Powers, *J Biol Chem*, 1982, **257**, 5077; T Teshima, J C Griffin, and J C Powers, *ibid*, p 5085.

10 Unpublished experiments by M A Navia *et al.*

AZIRIDINO-DAP : A POTENT IRREVERSIBLE INHIBITOR OF DIAMINOPIMELATE EPIMERASE

Fritz Gerhart† and William Higgins

Merrell Dow Research Institute, BP 447 R/9, 16 Rue d'Ankara, 67009 Strasbourg Cedex, France

1 INTRODUCTION

Over many years, our Institute has developed a number of fluoromethyl and difluoromethyl analogues of amino acids.[1-3] These have proved useful as enzyme-activated irreversible inhibitors of pyridoxal phosphate-dependent enzymes and, in the case of difluoromethylornithine (DFMO), to have therapeutic value against trypanasomiasis or sleeping sickness.[4] As part of our programme on the design of enzyme-activated irreversible inhibitors aimed at specific bacterial pathways, we prepared an analogue of *meso*-2,6-diaminopimelic acid (1, DAP), α-fluoromethyldiaminopimelic acid **(2a)** as a potential inhibitor of diaminopimelate decarboxylase (E.C. 4.1.1.20). This enzyme converts DAP into lysine and is therefore important in the DAP-lysine biosynthetic pathway (Scheme 1).

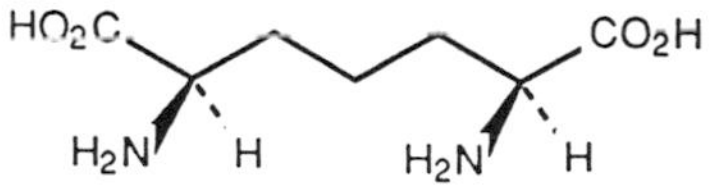

HO_2C ... CO_2H, H_2N H, H_2N H

1

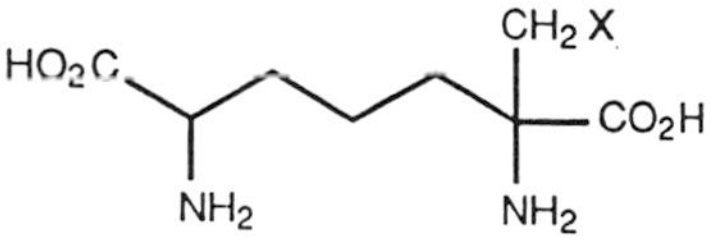

HO_2C ... CH_2X, CO_2H, NH_2, NH_2

2

2a : X = F
2b : X = Cl
2c : X = Br

HO_2C ... CO_2H, NH_2, NH

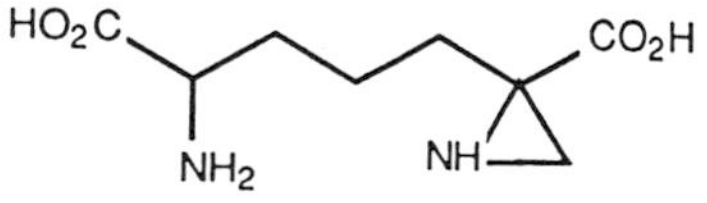

3

† deceased 26 January, 1990.

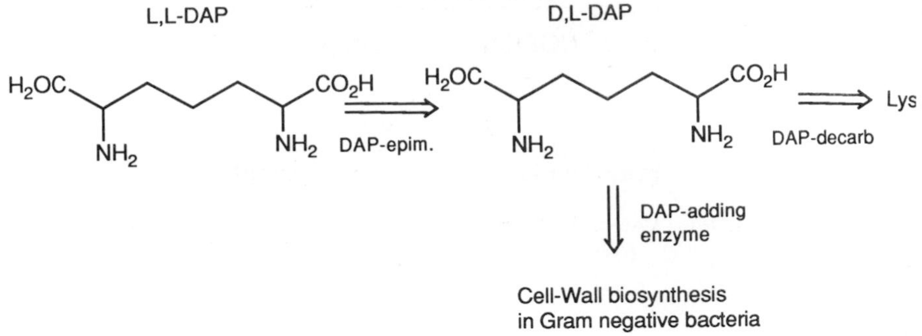

Scheme 1: Biosynthesis of lysine (Lys) from (L,L)-2,6-diaminopimelic acid *via* *meso*-DAP.

Although **(2a)** was a relatively poor inhibitor of the target enzyme *in vitro* ($t_{1/2}$ = 20 min at 100 μM concentration), it was nevertheless found to inhibit the growth of log-phase cultures of *Escherichia coli.* This result was atypical of other DAP-decarboxylase inhibitors (unpublished results of the authors) and interference with another enzyme in the DAP-lysine pathway, DAP-epimerase (E.C. 5.1.1.7) (Scheme 1) was suspected.

When **(2a)** was incubated with partially purified DAP-epimerase from *E coli,* time-dependent and irreversible inhibition (as judged by dilution and dialysis experiments) was observed. Previously, the facile formation of aziridinecarboxylic acids from α-monofluoromethylamino acids (but not difluoromethyl derivatives) had been observed.[5] Therefore, spontaneously formed 'aziridino-DAP' **(3)** was suspected to be the actual inhibitor.

2 KINETICS OF FORMATION OF AZIRIDINO-DAP

Following our preliminary results with the α-fluoromethyl derivative **(2a)**, we prepared both the α-chloro- **(2b)** and α-bromomethyl **(2c)** derivative.[6] Even faster inhibition was found with these [see Figure 1 for a comparison of **(2a)** with **(2b)**]. The inactivation kinetics were found to be second order, but plots of $-\ln A_t/A_0$ *versus* t were linear.[6] This supported our hypothesis of a spontaneous (or

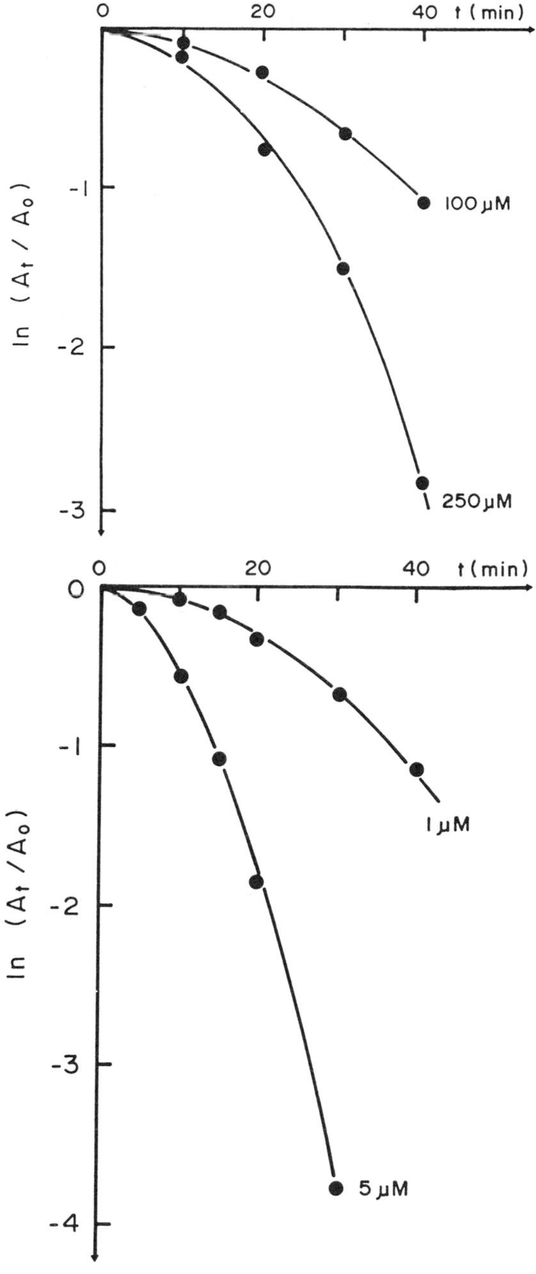

Figure 1: Inactivation of DAP-epimerase with α-halomethydiaminopimelic acids; Top: α-fluoro-derivative **(2a)**; Bottom: α-chloro-derivative **(2b)**.

enzymatic) conversion of α-halomethyl-DAP's into an intermediate aziridino-DAP **(3)**, which is the actual irreversible inhibitor.

Enzyme catalysis for this conversion was ruled out because no difference in the rate of formation of **(3)** was observed in the presence or absence of the enzyme. For the case of spontaneous conversion of α-halomethyl-DAP's into aziridino-DAP **(3)**, the kinetic situation has been derived[6] and simplifies to:

$$-\ln (A_t/A_0) = \frac{k_i k_0 [S]_0 t^2}{2K_i}$$

NB A_0 and A_t are enzyme activities at times t and 0, respectively. Furthermore:

$$S \xrightarrow{k_0} I \quad \text{and} \quad E + I \underset{}{\overset{K_i}{\rightleftharpoons}} EI \xrightarrow{k_i} \text{inhibited enzyme}$$

with $[E] + [EI] = [E]_0$ where E is enzyme and [E] its concentration; $[I]_t = [S]_0 [1-\exp(-k_0 t)]$.

Therefore, plots of $-\sqrt{\ln (A_t/A_0)}$ *versus* t should be linear (and this indeed was found[6]), with a slope y given by:

$$y = -\sqrt{\frac{k_0 k_i [S]_0}{2K_i}}$$

A plot y^2 *versus* $[S]_0$ (Figure 2) gave values for $k_i k_0/K_i$ of 1.45 x 10^{-5} μM^{-1} min^{-2} and 2 x 10^{-3} μM^{-1} min^{-2} for **(2a)** and **(2b)**, respectively. Therefore, spontaneous conversion into the inhibitor **(3)** is about 140 times faster for **(2b)** than for **(2a)**. These calculations encouraged us to study in more detail the kinetics of conversion of halomethyldiaminopimelic acids into the putative aziridino-DAP and to attempt the preparation of the aziridine.

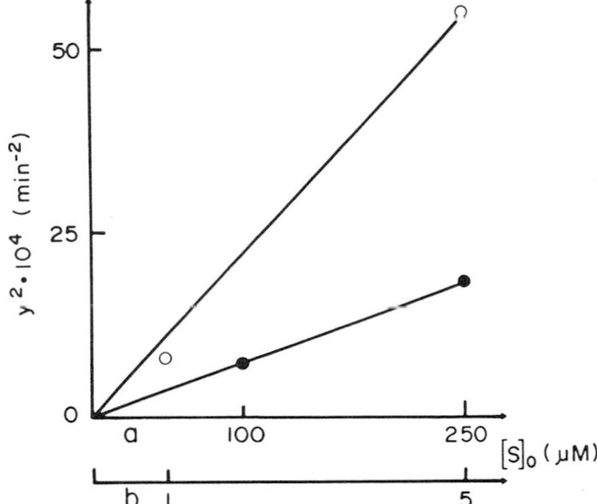

Figure 2: Tertiary plots of inhibition by **(2a)** [O, abscissa a] and by **(2b)** [●, abscissa b].

3 RATE OF HALIDE LIBERATION FROM α-HALOMETHYLDIAMINOPIMELIC ACIDS

Under conditions of complete deprotonation of the amino functions (aqueous alkali), halide liberation was measured at 70.0 °C and 80.0 °C for **(2a)** and at 20.5 °C and 28.0 °C for **(2b)**. ^{19}F-NMR was used to measure loss of F^- from **(2a)**, whilst liberation of Cl^- from **(2b)** was monitored with a chloride-specific electrode.[6] Halide liberation was found to be first order: free enthalpies and entropies of activation are given in the Summary. From these parameters, the following rate constants were calculated for halide release at 37 °C : 2.6 x 10^{-4} min^{-1} **(2a)** and 6.3 x 10^{-2} min^{-1} **(2b)**. Corresponding $t_{1/2}$ values are 45 h and 11 min, respectively.

Employing the values for $k_i k_0/K_i$ obtained previously from enzyme kinetics, the K_i/k_i values for the anticipated common intermediate are calculated to be 18 μM min and 31 μM min for **(2a)** and **(2b)**, respectively. Obviously, these values are imprecise because k_0 and $k_0 k_i/K_i$ were determined at a different pH and therefore at different degrees of protonation. Nevertheless, the nearness of these values (ideally they should be identical) supports both the postulate of a common inhibitory intermediate and the validity of the kinetic scheme.

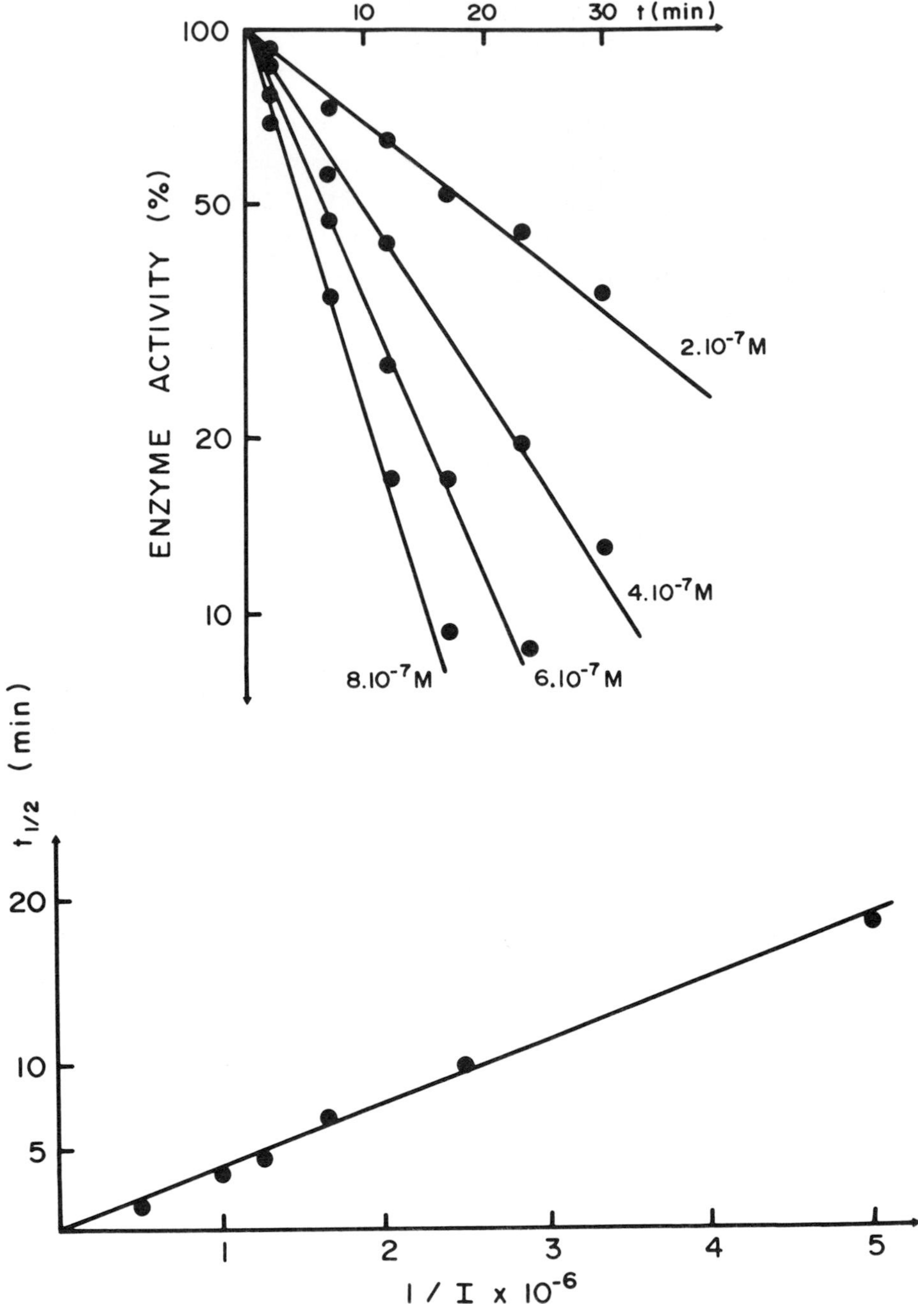

Figure 3:
Above: Inhibition of DAP-epimerase by **(3)** at concentrations shown;
Below: Plots of $t_{1/2}$ *versus* reciprocal inhibitor concentration.

4 PREPARATION OF AZIRIDINO-DAP AND RECOMBINANT DAP-EPIMERASE

α-Fluoromethyldiaminopimelic acid **(2a)** was prepared by standard methods.[6] When **(2a)** was kept at room temperature for a sufficient time in aqueous alkaline solution, aziridine **(3)** was formed in quantitative yield.[6] Although remarkably stable, its isolation in salt-free form proved difficult. Howcvcr, it was characterized by both ^{13}C NMR and FAB-MS.[6] For enzyme studies, solutions of **(3)** obtained as above were neutralized with HCl to pH 7 and were used as such.[7]

Treatment of DAP-epimerase, either the enzyme from a variety of bacterial sources, or the recombinant enzyme described next, led to first order inactivation kinetics with a short $t_{1/2}$ even at concentrations < 10^{-6}M (Figure 3, above). A Kitz and Wilson plot (Figure 3, below) gave K_i/k_i = 5.5 μM min. The values of K_i and k_i could not be determined separately because the curve intercepts the ordinate very close to the origin.[6]

Recombinant DAP-epimerase has recently been described,[8] and is obtained as 1% of soluble protein when expressed from plasmid pdF3. More recently, further sub-cloning has increased this overproduction of DAP-epimerase to 5% and now enables a simple two-step purification procedure to be employed.[9]

When such purified recombinant DAP-epimerase was treated with 50 μM aziridino-DAP **(3)** for 30 min at 37 ^{o}C, greater than 99% inactivation ensued. If an aliquot was removed, diluted 100-fold and tested for recovery of activity, none occurred. Furthermore, denaturation prior to carboxymethylation (necessary for peptide mapping) did not remove the aziridino-DAP. We therefore concluded that inactivation was irreversible.

5 PEPTIDE MAPPING

A comparison of tryptic peptide maps of carboxymethylated enzyme (Figure 4, top) before and after inactivation by aziridino-DAP (Figure 4, bottom), showed that they were virtually identical. Only one peptide (now known to be T-6, numbering tryptic peptides from

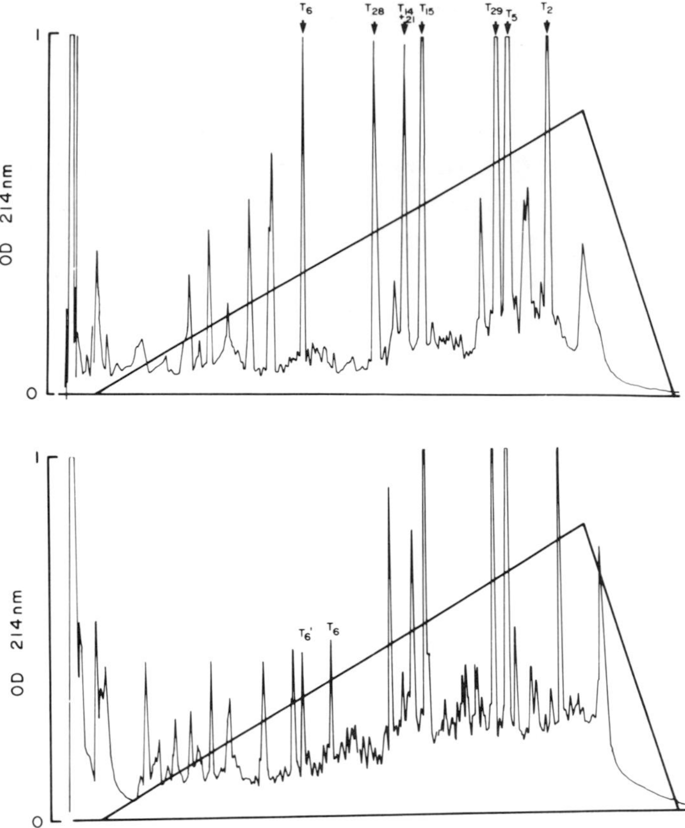

Figure 4: Tryptic peptide maps of DAP-epimerase obtained as described;[8]
Above: carboxymethylated DAP-epimerase;
Below: aziridino-DAP-labelled enzyme treated as above.

the amino-terminal[10]) was displaced to yield a unique peptide (T-6': see Figure 4, bottom). Both amino acid and sequence analysis confirmed that these two peptides were identical except at one position.[8] In the case of peptide T-6 this position was assigned as PTH-(*S*-carboxymethyl)-cysteine, whereas for T-6' we obtained a blank at the corresponding position (NB the Cys adduct of aziridino-DAP has not yet been synthesised). The protein sequencing results were confirmed by subsequent DNA sequence data on the dapF gene.[11] This also confirmed the cysteine residue by positioning it at residue-73 in the overall sequence of DAP-epimerase.

6 MECHANISTIC IMPLICATIONS

Based on kinetic evidence, Wiseman and Nichols had previously postulated a 'two-base mechanism' for DAP-epimerase and also a mechanistic relationship between this enzyme and proline racemase.[12] They predicted that one of these bases be a monoprotic base, probably the active-site thiol residue which had previously been identified by group-specific thiol reagents.[17] We suggest that one of these bases is indeed Cys-73, and that the other is the Cys-73 residue from an identical subunit forming a composite active-site.[8] Studies on proline racemase[13] had suggested such a composite active

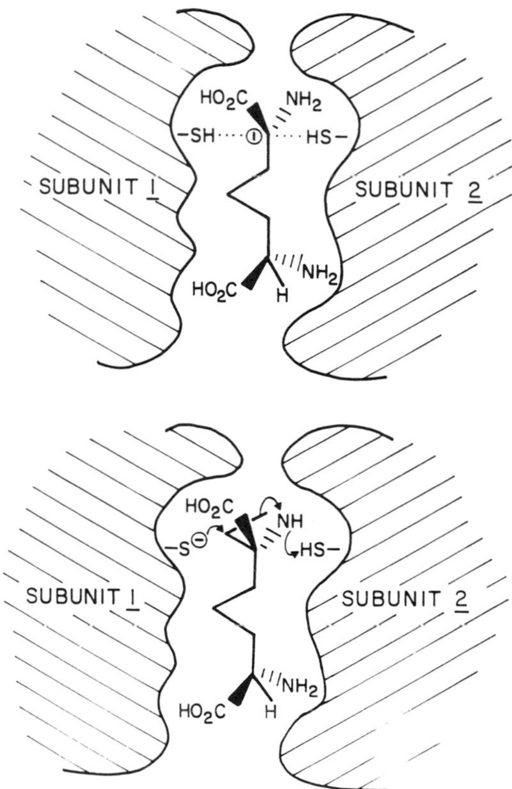

Figure 5: Schematic representation of composite active site of DAP-epimerase:

Above: 'transition-state' based on recent mechanistic study of proline racemase;[14]

Below: our proposed concerted mechanism for aziridino-DAP inhibition.

site and we feel that our isolation of only one major cysteine-containing peptide labelled by aziridino-DAP is further support for this thesis. In addition, a recent series of elegant kinetic studies on proline racemase[14] has suggested the 'transition-state' shown schematically in Figure 5, in which we have also indicated our conception for aziridino-DAP inhibition. We also compared the very limited sequence information available for proline racemase[13] and 4-hydroxyproline epimerase[15] around their active site Cys residues, with our sequence data on DAP-epimerase. We deduced the DNA sequence for these two enzymes, using the most frequently used codons, and found that changes in 5 of 21 bases for 4-hydroxyproline epimerase and 3 of 15 bases for proline racemase gave identical DNA sequences.[8] We proposed that all three of these non-pyridoxal phosphate racemases/epimerases may be descended from a common ancestral gene.[8]

SUMMARY

Aziridino-DAP **(3)** is remarkably stable, does not exhibit random alkylating properties and is generally non-toxic.[6] It is however an extremely potent irreversible inhibitor of DAP-epimerase; even with sophisticated stopped-flow equipment we cannot determine the kinetic parameters precisely. We may conclude that $K_i > 25$ μM and that $I/k_i < 10$ s, *ie* conversion of the enzyme-inhibitor complex into the inactivated enzyme is extremely fast. The formation of **(3)** from α-halomethyldiaminopimelic acids under physiological conditions is spontaneous with values for $K_i/k_i = 5.5$ μM min for **(3)**, and $K_i k_0/K_i$ values of 1.45×10^{-5} μM^{-1} min^{-2} and 2×10^{-3} μM^{-1} min^{-2} for **(2a)** and **(2b)**, respectively. Calculated rate constants k_0 for aziridine formation at 37 ^{o}C, pH 7.2, are 7.9×10^{-5} min^{-1} ($t_{1/2}$ 146 h) and 1.1×10^{-2} min^{-1} ($t_{1/2}$ 63 min) for **(2a)** and **(2b)**, respectively. Steric requirements are higher for the chloro-compound **(2b)** than for the fluoro-derivative **(2a)** ($S^{\#}$ -22.8 eu and -8.2 eu, respectively). The converse is true for $\Delta H^{\#}$ (23.5 kcal mol^{-1} and 15.5 kcal mol^{-1}, for **(2a)** and **(2b)**, respectively).

ACKNOWLEDGEMENTS

We wish to thank our colleagues Alan Cardin, Jean-Bernard Ducep, Robert Ireland, Peter Lewis, Pierre Mamont, Catherine Richaud, Chantal Tardif, and Joseph Wagner for their help and encouragement during this work.

REFERENCES AND NOTES

1 B W Metcalf, P Bey, C Danzin, M J Jung, P Casara, and J P Vévert, *J Am Chem Soc,* 1978, **100**, 2551.

2 N Seiler, M J Jung, and J Koch-Weser, eds in 'Enzyme-Activated Irreversible Inhibitors', Elsevier/North Holland Inc, 1978.

3 M J Jung and J Koch-Weser, in 'Molecular Basis of Drug Action', ed Singer and Ondarza, Elsevier/North Holland Inc, 1981, 135.

4 P J Schechter, J L R Barlow, and A Sjoerdsma, 'Clinical Aspects of Inhibition of Ornithine Decarboxylase with Emphasis on Therapeutic Trials of Eflornithine (DFMO) in Cancer and Protozoan Diseases' in 'Inhibition of Polyamine Metabolism', ed P P McCann, A E Pegg, and A Sjoerdsma, Academic Press, 1987, 345.

5 α-Fluoromethyltyrosine and α-fluoromethylhistidine give the corresponding aziridinecarboxylic acids in quantitative yield when heated in basic medium at 100 ^{o}C for about 30 min: F Gerhart, unpublished.

6 F Gerhart, W Higgins, C Tardif, and J-B Ducep, *J Med Chem,* 1990, in press.

7 The purity and stability of these solutions were checked by HPLC; no significant decomposition occurred in 10 mM stock solutions, stored at 4 ^{o}C for one year.

8 W Higgins, C Tardif, C Richaud, M A Krévanek, and A Cardin, *Eur J Biochem,* 1989, **86**, 137.

9 W Higgins, C Tardif, and C Richaud, unpublished.

10 R L Heinrikson, E T Kreuger, and P S Keim, *J Biol Chem,* 1977, **252**, 4913.

11 C Richaud and C Printz, *Nucleic Acid Res*, 1988, **16**, 10367.
12 J S Wiseman and J S Nichols, *J Biol Chem*, 1986, **259**, 8907.
13 G Rudnick and R H Abeles, *Biochemistry*, 1975, **14**, 4515.
14 W J Albery and J R Knowles, *Biochemistry*, 1986, **25**, 2572.
15 S G Ramaswamy, *J Biol Chem,* 1984, **259**, 249.

STEREO- AND REGIOSELECTIVE AGGREGATION OF *N*-OCTYL-ALDONAMIDES TO AMPHIPHILIC FIBRES

Jürgen-Hinrich Fuhrhop, Sönke Svenson, Christoph Boettcher, Reinhard Bach, and Peter Schnieder

Institut für Organische Chemie der Freien Universität Berlin, Takustraße 3, D–1000 Berlin 33, West Germany

1 INTRODUCTION

The lipid bilayers of cell membranes and the collagen-type protein fibres play the major part in the build-up of biological tissues. The lipid bilayer is a product of the hydrophobic effect, which forces water-insoluble amphiphiles into more or less curved bilayered planes. The protein fibre structures depend on a covalent polymer backbone, on regular substitution patterns and on linear chains of amide hydrogen bonds. In attempts to synthesise plasma-like complex organisates we use simple synthetic amphiphiles, which form well-defined aggregate structures in aqueous media. Carbohydrates are chosen as chiral head groups, since they are commercially available in a wide variety of diastereomers and enantiomers. Although carbohydrates are only minor constituents in most living cells, they are responsible for stereoselective recognition processes on cell surfaces as well as for the longevity of biological gels.

Aldonamides **1–16** provide both water-insoluble amphiphiles which tend to arrange in bilayers *and* amide bonds which favour fibre formation.[1–3] They also allow detailed investigations of such carbohydrate interactions, since the stereochemistry of the open-chain head groups is variable, and may mimic some of the selectivity patterns, which are found in receptor molecules. Furthermore the flexibility of the aldonamide chain allows the formation of highly entangled fibre structures with high surface energy. The tendency to crystallize is low for most of these compounds. In the following we present an overview of the relationships between growth mechanisms and molecular structures of the observed superstructures.

2 MICELLES AND WHISKER-TYPE FIBRE GROWTH

From the molecular structures of amphiphiles **1–16** with a 'narrow' oligomethylene chain and a 'broad' hydrated head group one may deduce at once micelle formation in water.[4] Only aggregates of high curvature should be formed. One would also expect that the hydrophobic aggregates cannot be observed by electron microscopy.

D-Gly 1 | L-Gal 9

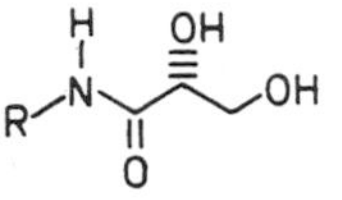

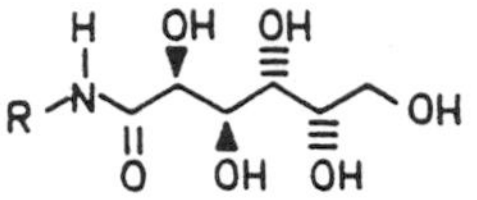

D-Ara 2 | L-Man 10

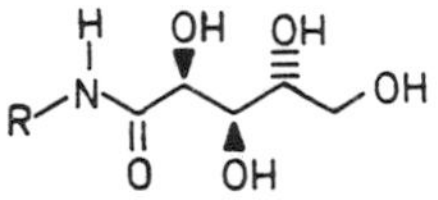

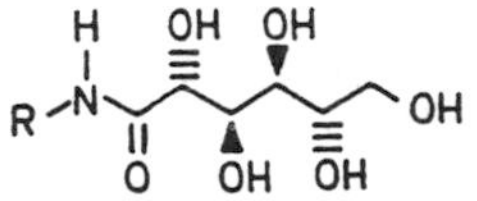

D-Lyx 3 | L-Glu 11

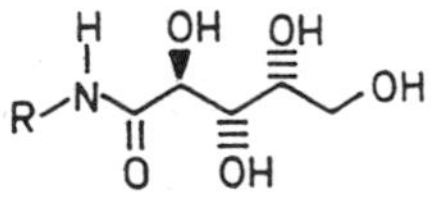

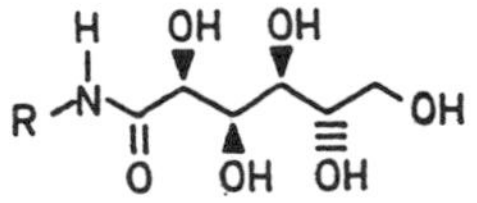

D-Rib 4 | D-Tal 12

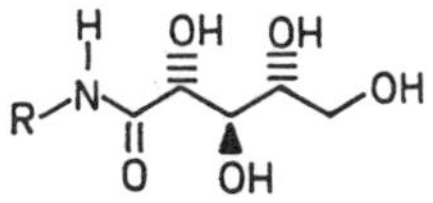

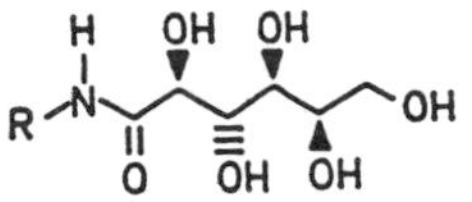

D-Xyl 5 | D-Gul 13

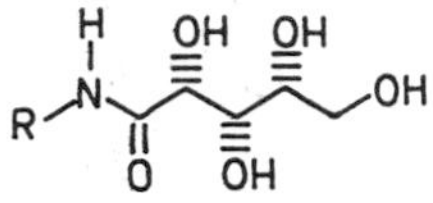

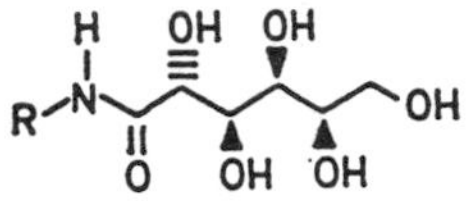

D-Gal 6 | D-Alt 14

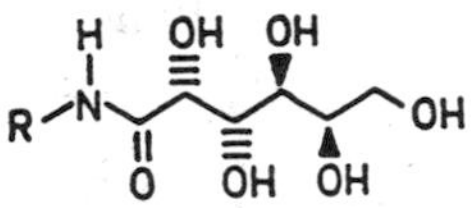

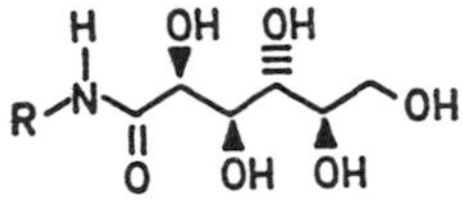

D-Man 7 | D-All 15

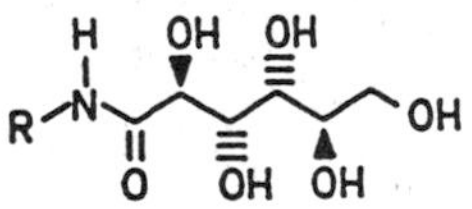

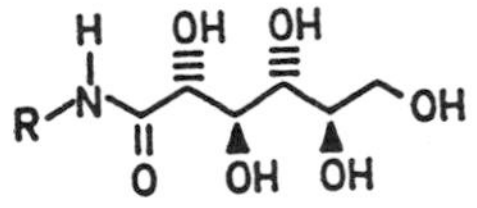

D-Glu 8 | D-Ido 16

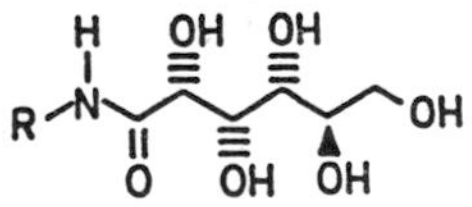

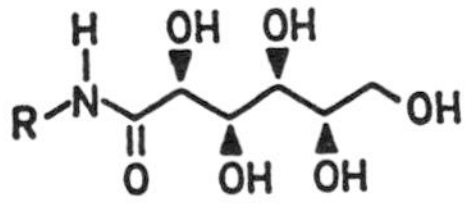

R–= $CH_3(CH_2)_7$–

8a R–= $CH_3(CH_2)_7$–
8b R–= $CH_3(CH_2)_{11}$–

11a R–= $CH_3(CH_2)_7$–
11b R–= $CH_3(CH_2)_{17}$–
11c R–= $CH_3(CH_2)_{11}$–

Neither micelles nor liquid crystals with high curvatures can be fixated with the usual techniques. The reason for the short lifetime of micellar aggregates is the relatively high solubility of the monomers which favours rapid fractionation processes of the aggregates.[5]

In aqueous aldonamide solutions one finds micelles of type (A, Figure 1) at temperatures above 70–95°C, depending on the stereochemistry of the head group and the chain length of the hydrophobic part.[3] On rapid cooling a sharp solidification point in differential scanning calorimetry is observed and fibres of uniform configurations are seen in electron micrographs of the solid gels (see next Section). Longevity and electron optical detectability are presumably caused by the formation of chain–like amide hydrogen bonds. Micellar rods (B) and tubules (C) are observed. The tubules are often accompanied by vesicles (D) and/or planar sheets (E). The analogous aldonesters do form liquid crystals, but no fibres have been detected in electron micrographs of aqueous solutions.[15]

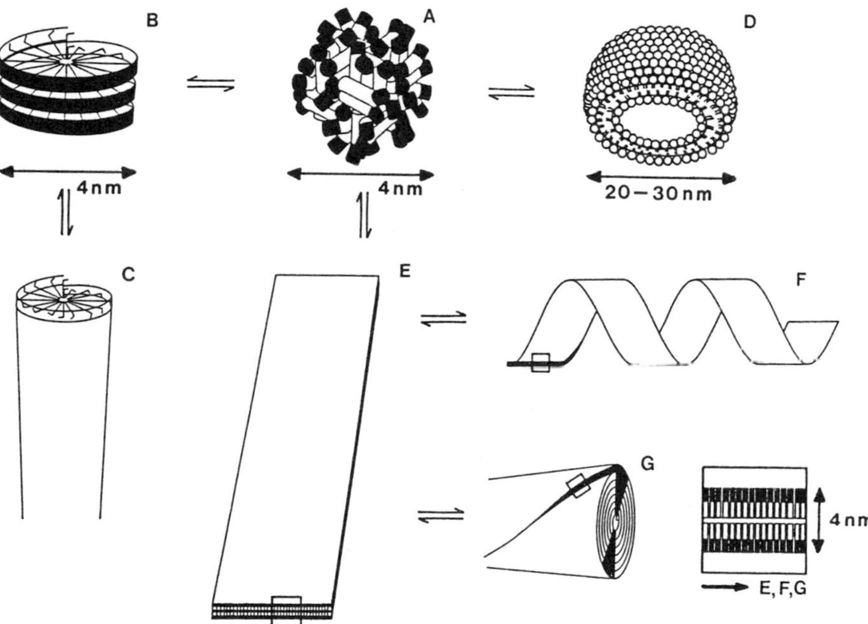

Figure 1: The five types of micellar aggregates, which have been characterized so far by electron microscopy.

All solid-like aldonamide fibre aggregates are quantitatively reconverted to small micelles above the melting point. They are in a temperature-dependent phase equilibrium down to room temperature.[6] The micelles have been characterized in hot solutions by titrations in the presence of brilliant blue G 250.[8] For D- and L-*N*-octyl-gluconamides a critical micellar concentration (cmc) of 2.77 x 10^{-2} mol dm^{-3} at 80 °C was determined. The ^1H NMR spectrum of the hot (>70 °C) micellar solution was identical to the spectrum in monomolecular dimethyl sulphoxide solution, since the formation of small aggregates does not blur the structure of NMR signals. On cooling, most of the signal intensity disappeared, but about 15% of the original signal intensity remained at 10 °C below the solidification point. Details of the ^1H NMR spectra of the micelles at temperatures above and below the gelation point give indications for an intermediate: head group signals are still well-resolved below the solidification point, but the hydrophobic chains produce much broadened singlets. This observation could indicate micelles with partly hydrophobic, immobile surfaces floating within the gel. At room temperature only signals of the order of 1% of the original intensity were found. Deuterated compounds with a CD_2-group next to the amide nitrogen were prepared and examined by the ^2H solid echo pulse technique (see Figure 2). The isotropic micellar solution

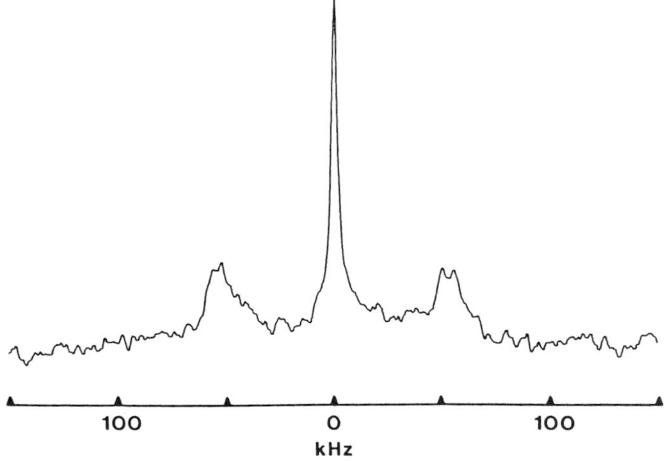

Figure 2: ^2H solid echo pulse NMR spectrum of **8a** gel containing 2% phosphotungstate in water at 60 °C; **8a** was di-deuterated at the methylene group next to nitrogen.

produced a singlet corresponding to the ^1H signal at δ 3.08 ppm which split into a symmetric doublet with a quadrupole splitting of 120 ± 2 kHz for the anisotropic fibre structures. This is in agreement with observations on vesicle membranes.[9] The typical powder spectrum of a solid is found. From the relative intensities of the signals one may deduce an approximate ratio of molecules in fibres and micelles of about 7:1 at 10 °C below the melting point, and of more than 100:1 at 7 °C. The clear, slightly opaque gels become turbid within a few hours. Crystals appear within a few days. It is very likely that the stepwise rearrangement from thin fibres to large multiple fibres and finally crystals, occurs by material transport with small micelles and not by direct combination of fibres.

The 'hydrophobic' micelles mentioned above should be responsible for the rapid growth of the primary fibres. On cooling, the spherical micelles are probably converted into ellipsoid bilayers by the cooperative formation of amide hydrogen bond chains. The ellipsoid platelets could then combine by end-on growth processes. The crystals grow only in one direction, because the association of the micellar platelets occurs only to the amphiphilic areas at the top of the columns and not to the hydrophilic head groups at the edges of the 'disk'. The same is true for the reverse process: micelles split preferably from the end of the fibres. Fibre growth appears as a one-dimensional crystallization process.[10] The process is also analogous to the assembly of microtubules *in vitro*.

The solubility of the aldonamides is also highly dependent on the stereochemistry of the head groups. Open-chain polyols with no 2,4- or 3,5-*syn*-hydroxyl group pair are the least soluble and they form stable gels at low concentration, (galacton, mannon; 0.3-0.5% w/v). If there is a 2,4-*syn*-hydroxyl pair, then the solubility rises by a factor of more than ten and the aldonamide still aggregates to fibres and forms gels (glucon, talon; solubility 0.5-50% w/v). The primary ultrathin helices with extremely high surface energy will again be in equilibrium with small micelles, which will speed up the growth of multiple fibres and crystals. Polyol head groups with 3,5-*syn*-hydroxyl pairs (idon **16**, allon **15**, altron **14**, gulon **13**) are more water-soluble and do *not* aggregate to form stable fibres or gels. Disturbed planarity of the bilayers at the interface with water is obviously not tolerated. *N*-Octylpentonamides with a 2,4-*syn*-

hydroxyl pair (ribon **3**, xylon **4**) are also too water-soluble to form gels, whereas the arabinon and lyxon analogues **1** and **2** form reasonably clear gels of short lifetime.[12] Racemic aldonamides **7** + **10** and **8** + **11** do not form gels,[2,3] although the racemates are less soluble that the pure enantiomers. Only the galactonamide racemate **6** + **9** forms stable clear gels at very low concentration (0.2% w/w, see Section 4).

To obtain micellar or hollow fibres with high length to diameter ratios and clear structural details, as described in the next Section, it is necessary to design simple experimental conditions of fibre formation. Reproducible results were obtained by any one of the following three procedures:

Distilled water was used as solvent and heated to 95°C until a clear solution was obtained. This solution was then cooled to room temperature in less than one minute. The ultrathin, primarily formed fibres, *eg* the bulgy helix of Figure 4, were stabilized by addition of 2% phosphotungstate. This presumably binds to the fibres by strong hydrophilic interactions and thereby prevents further aggregation and rearrangement processes.

Another very useful technique is rapid injection of an ethanol solution of the aldonamide into a 2% phosphotungstate solution at room temperature. Quantitative yields of just one type of fibre aggregate as judged by electron microscopy, have thus been obtained.

Finally one may dissolve water-insoluble aldonamides in hot lauryl sulphate micellar solution. On cooling, fibres will be extruded from the micelles (see Figure 8).

3 HEAD GROUP STEREOCHEMISTRY AND FIBRE STRUCTURES

The few aldonamide crystal structures known, correspond to gluconamides with various alkyl substituents on the amide nitrogen.[13,14] In these crystals the polyol chain is in the *all-anti* conformation. A slight bend at the amide bond is observed and the molecular sheets are arranged in a head-to-tail fashion. No hydration water is bound to the crystals. The corresponding esters have very similar conformations.[15] Both structural features, the straight polyol-chain and the enantiopolar arrangement of the amphiphiles, are not very likely to occur in aqueous gels. The steric interaction of the 2,4-*syn*-hydroxyl groups will certainly perturb the linear chain of the hydrated head groups and the hydrophobic effect will enforce the tail-to-tail arrangement of water-insoluble amphiphiles. A more realistic picture of the head group conformation of gluconamides in fibres is probably provided by the *X*-ray structure of glucitol.[16] Here no long-chain substituent enforces a linear

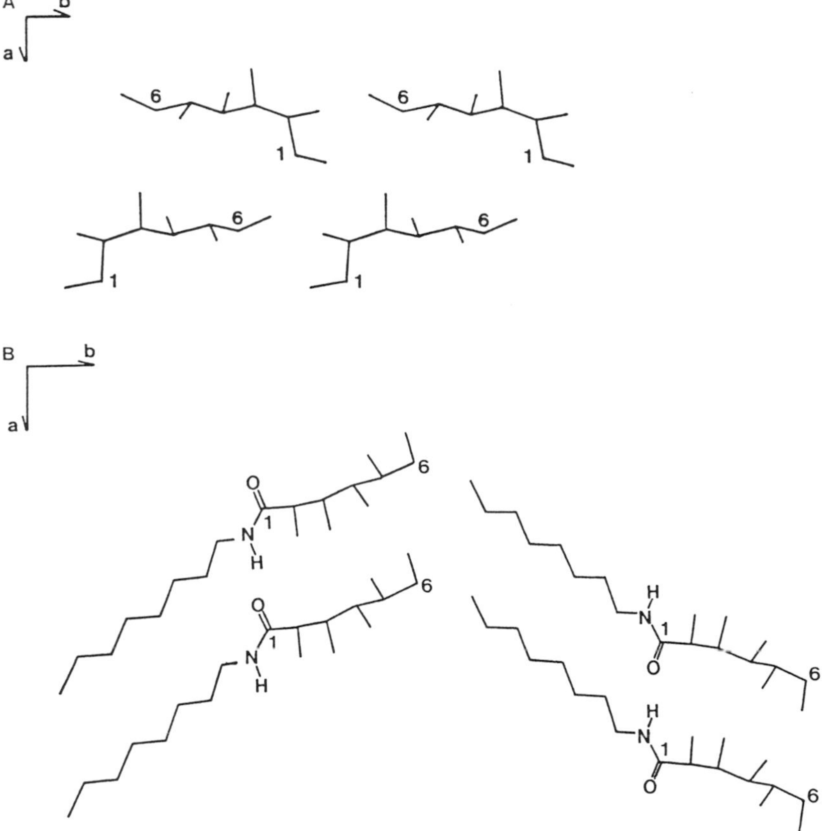

Figure 3: a) Partial crystal structure of D-glucitol: neighbouring sheets run antiparallel; the bend at C-2 occurs because 2,4-*syn*-hydroxyl groups repel each other in the *all-anti* conformation;

b) In the crystal structure of amphiphile **8a** no such bend is observed.

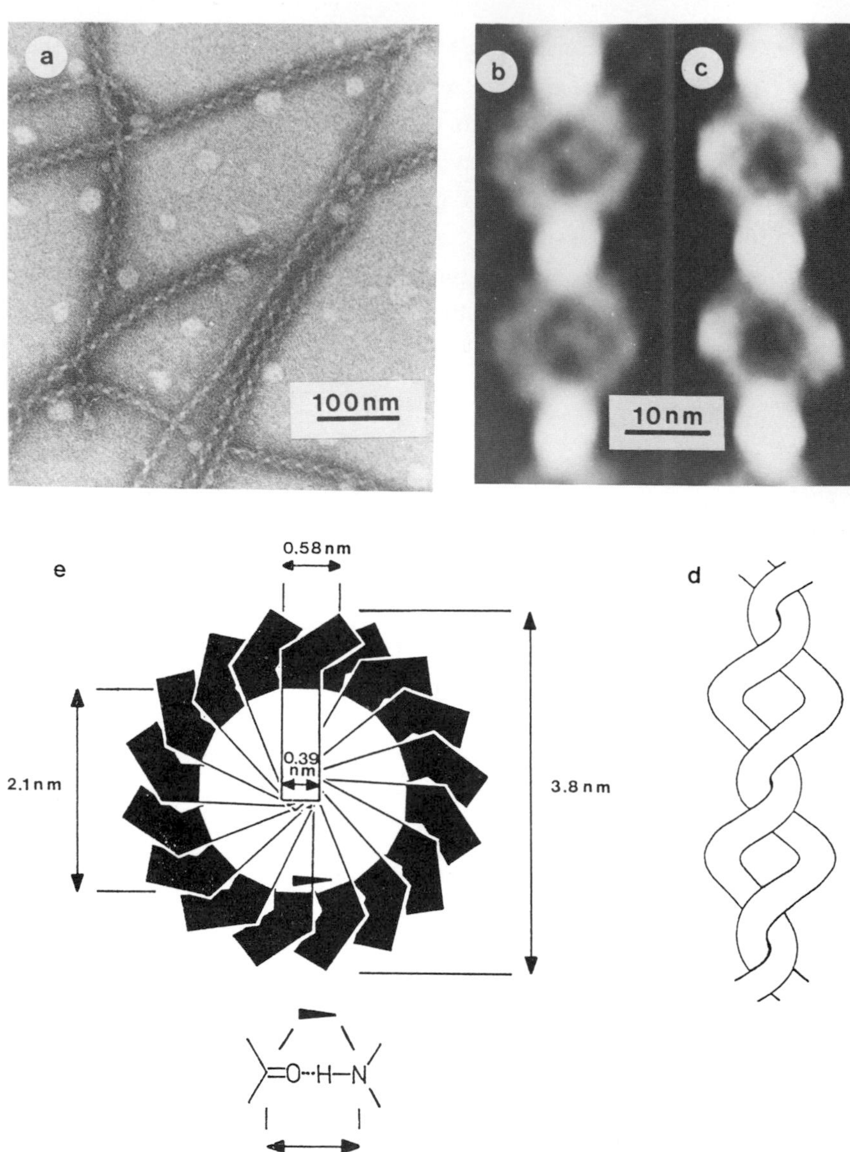

Figure 4: a) b) c) Electron micrograph of computer images of the electron micrographs of the bulgy helices made of **8a** (a, b) and **8b** (c); **d)** model of the bulgy helix and **e)** a cut through the model of one fibre.

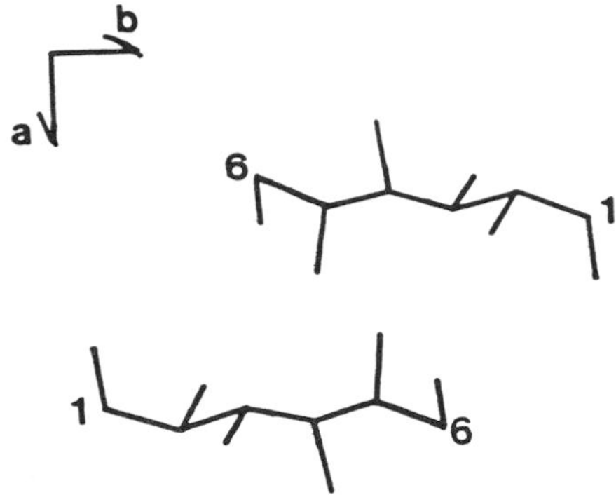

Figure 5: Partial crystal structure of β-D-mannitol: neighbouring sheets are displaced by about half a molecular length and run antiparallel; no bend is observed (compare Figure 3).

conformation and the molecular bend in the polyol is adjusted by: (i) molecular displacements, and (ii) alternating 6 ⟶ 1 and 1 ⟶ 6 arrangements of the molecules within the crystal. The bend in the gluconamide chain causes a broadening of the head group relative to the alkyl chain. There is now enough room to envelop a water molecule and a *micellar aggregate with high curvature* is formed.[4] Since the polyol is chiral the bending will only occur in one direction and will cause a chiral superstructure, namely a uniformly handed double helix with bulges and knots.[2] The single strains of the bulges are two molecular lengths or 4 nm thick; the knots have diameters of 8 nm.

Since there are no 2,4- or 3,5-*syn*-hydroxyl groups present the molecular conformations of mannitol[17] and galactitol show no bendings. The alternation of sheets with 1 ⟶ 6 and 6 ⟶ 1 arrangements is, however, still found. The corresponding mannon- and galacton-*N*-octylamides aggregate in water to form bilayer sheets. Galactonamide **6** sheets are much narrower than mannonamide **7** sheets, because galactonamide is less soluble. Less material is therefore available to form the bilayer sheets in rapid non-equilibrium processes. Galactonamide sheets appear as flat ribbons, twisted ribbons or tubes. The mannonamide molecular bilayer sheets usually form multilayered tubes, which look very similar to graphite whiskers.[18] Polymerizable lecithins form similar tubes,[19] but with relatively low length : diameter ratios. Very long ribbons with and without regular twists have been described by Kunitake,[20] who worked with double chain L-glutamic acid amphiphiles.

Amphiphiles with shorter head groups than in hexonamides may also form fibres and gels, but they have a much higher tendency to crystallize. The gel from *N*-octyl-D-ribonamide, for example, shows short tubules, which readily rearrange to multilayered crystals.[12] These tubules look similar to those obtained by Yager from lecithin type amphiphiles containing diacetylene moieties.[19] The arabinonamide **2** forms macroscopic whiskers. Similar observations have been made with the glyceronamide **1** and the lyxonamide **4**. Xylonamide **5** is too water-soluble to form gels. Soft tubes which aggregate to cloth-like structures have been obtained from negatively charged tartaric acid amides.[21]

The aggregate structure also depends on the lengths of the hydrocarbon chains. In the range between seven and twelve methylene groups aggregate structures are very similar. If one shortens the alkyl chain below six CH_2-groups the aldonamides become water-soluble. More than sixteen methylene groups favour ill-defined thick tubules and multihelices. Stereochemical details and well-defined edges often disappear. Diacetylene groups within an oligomethylene chain favour the formation of monolayered narrow tubes instead of cigar-like structures of low curvature.

Finally, the fine structure of a whisker-type aggregate is highly dependent on growth conditions. Fast growth favours large surface areas and complex structures. If the water-insoluble mannonamide **7b** with an octadecyl chain is, for example, dissolved in lauryl sulphate solutions, mannonamide helices separate from the micellar whiskers instead of the usual rolled-up sheets. The screw-sense of the whiskers, however, is statistical. This is a clear-cut case of helix growth by screw dislocation, as observed with metal whiskers.[22] We assume that under these conditions the whisker growth is so fast, that no dehydration of the intermediate mannonamide micelles, which is necessary in the formation of planar sheets, is possible. The micellar curvature is therefore retained in the aggregate.[6]

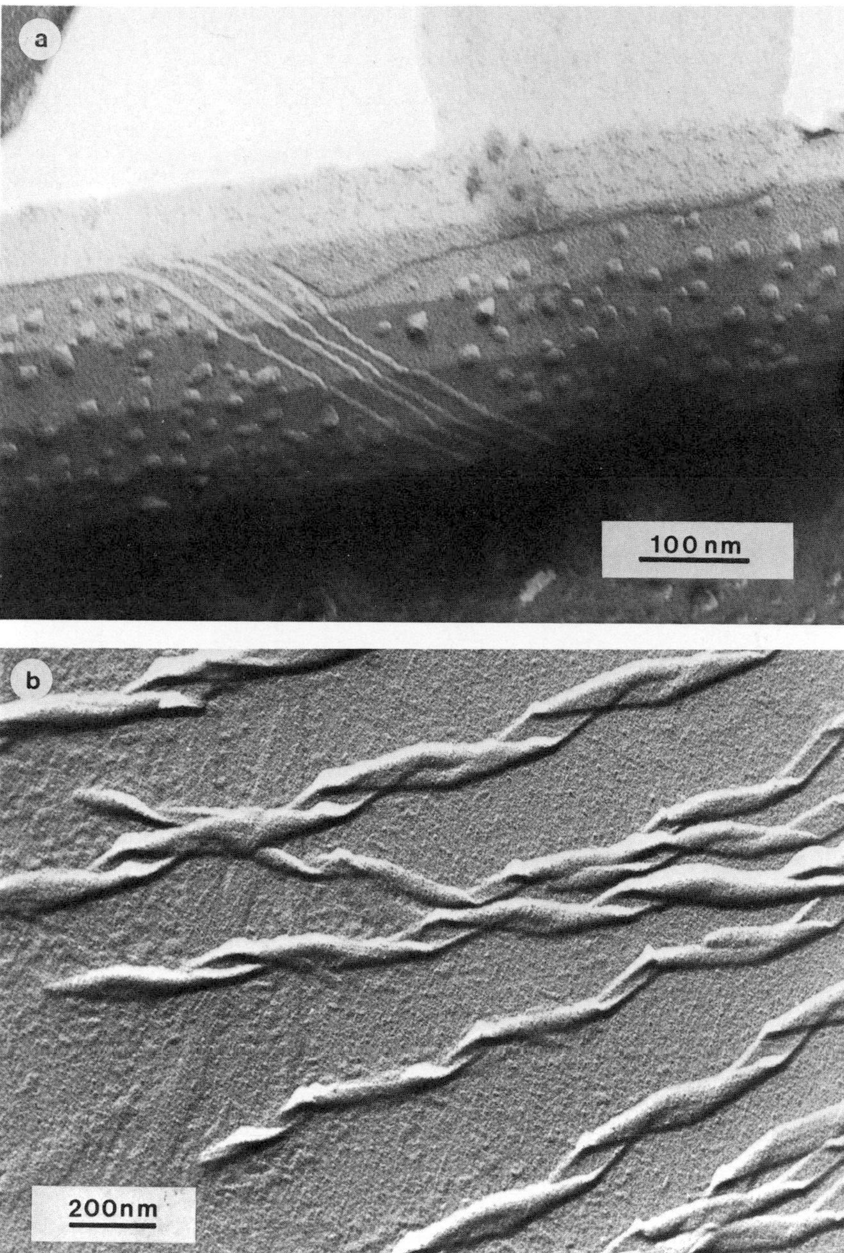

Figure 6: Electron micrographs of (a) rolled-up sheets of mannonamide **7a** with the vesicles absorbed to the surface and, (b) of twisted ribbons of galactonamide **6a**.

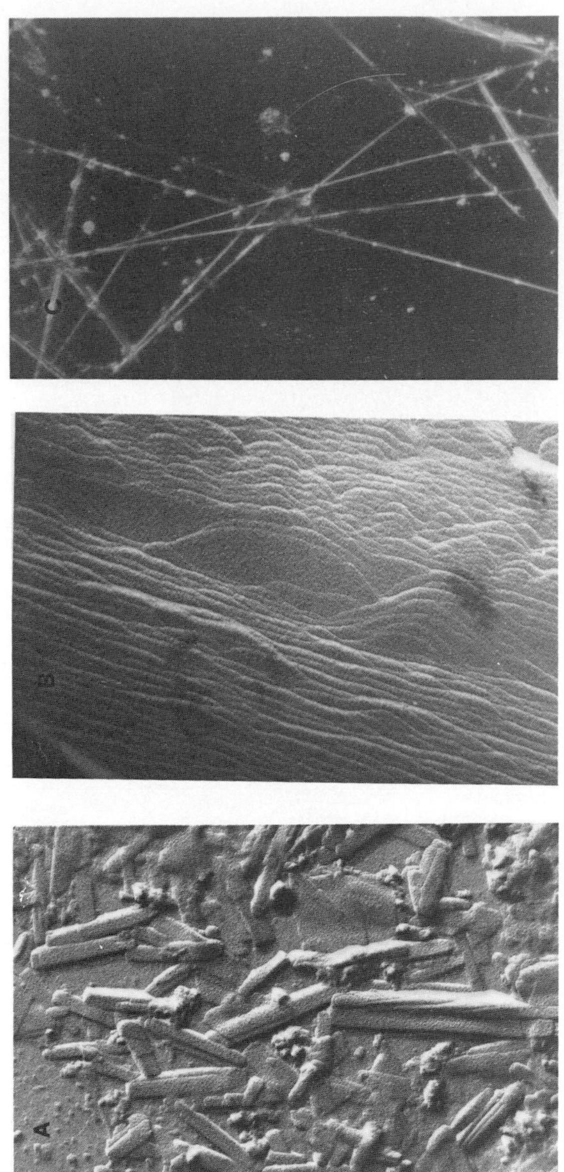

Figure 7: Electron micrographs of (a) ill-defined tubules or ribbons from *N*-octylribonamide **4**, (b) multilayered crystal plates from **4**, (c) whiskers from arabinonamide **2**.

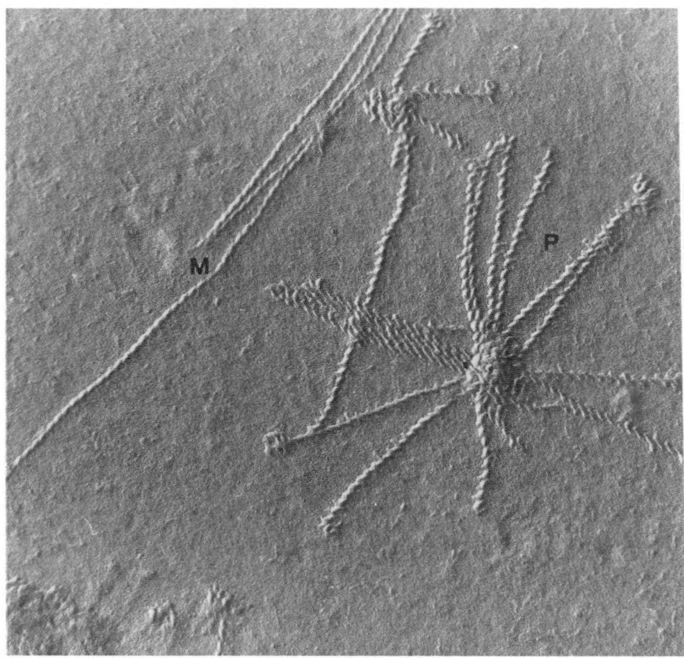

Figure 8: M– and P–helices from micellar solutions of octadecyl–L–mannonamide **10b** in lauryl sulphate.

4 THE CHIRAL BILAYER EFFECT

Once the bilayer fibre aggregate is formed, it is prone to reversible dissociation into small micelles and final irreversible precipitation as a three–dimensional crystal. Crystals of aldonamides and aldonesters are, however, enantiomer polar: the chiral head groups do not allow the tail–to–tail arrangement of the hydrophobic bilayer in regular crystal planes and the long apolar methylene chains do not allow the alternating 1 $\longrightarrow$ 6 and 6 $\longrightarrow$ 1 arrangement of the polyols. The rearrangement from micellar bilayers to crystalline monolayer sheets is therefore connected with a high loss of entropy and it is also thermodynamically unfavourable in water, because the hydration spheres of the head groups have to be removed from the micelles. Fibres are therefore stabilized toward crystallization by the chirality of the head groups ('chiral bilayer effect').[2]

The rearrangement of aldonamide fibres to crystals occurs much more rapidly with racemates. The racemic glucon- and mannon-amides **7+10** and **8+11** do not form fibres at all. They precipitate from solution in the form of multilayered platelets. It is assumed that these crystals are formed by symmetric tail-to-tail bilayers of the racemic compounds. An interdigitated racemic bilayer has been demonstrated in a crystal structure of a monoacetate of racemic tartaric acid octylamide.[23]

5 MULTIFIBRE SYSTEMS

Aldonamides of very low solubility can be used for the synthesis of gels with two types of fibres, say A and B, in one gel. Preliminary experiments show, for example, that a mixture of *N*-octyl-D-gluconamide **8a** and *N*-dodecyl-L-mannonamide **10c** clearly separates into helices, presumably made of **8a**, and thick bilayer rolls made of **10c**. The evidence for this conclusion comes from comparisons with electron micrographs of the pure compounds' aggregates. More quantitative investigations are currently being undertaken with radioactive materials and electron microscopy of autoradiographs.[24]

If the long term separation of ultrathin fibres in one gel is successful, then one could use fibre A to fix an electron donor, fibre B to isolate an acceptor, and the solvent phase to dissolve a redox active dye. Efficient systems for light induced charge separation may thus become accessible. The fibres and their redox systems could also be stabilized by fibre metallization and mineralization. Work along these lines has already been reported by O'Brien[25] and Yager.[19] They succeeded in electroless plating or metallization of lipid membranes. Our experience with phosphotungstates shows that this inorganic material can be efficiently deposited on carbohydrate surfaces.[2] The first purpose of these inorganic coatings should be to prevent dissociation of micelles from the fibres. Their second task would then be to help in the build-up of redox potential on the fibre surfaces and to catalyse the light-induced water splitting into hydrogen and oxygen. Platinum and manganese oxides are obvious materials for coating the fibres.

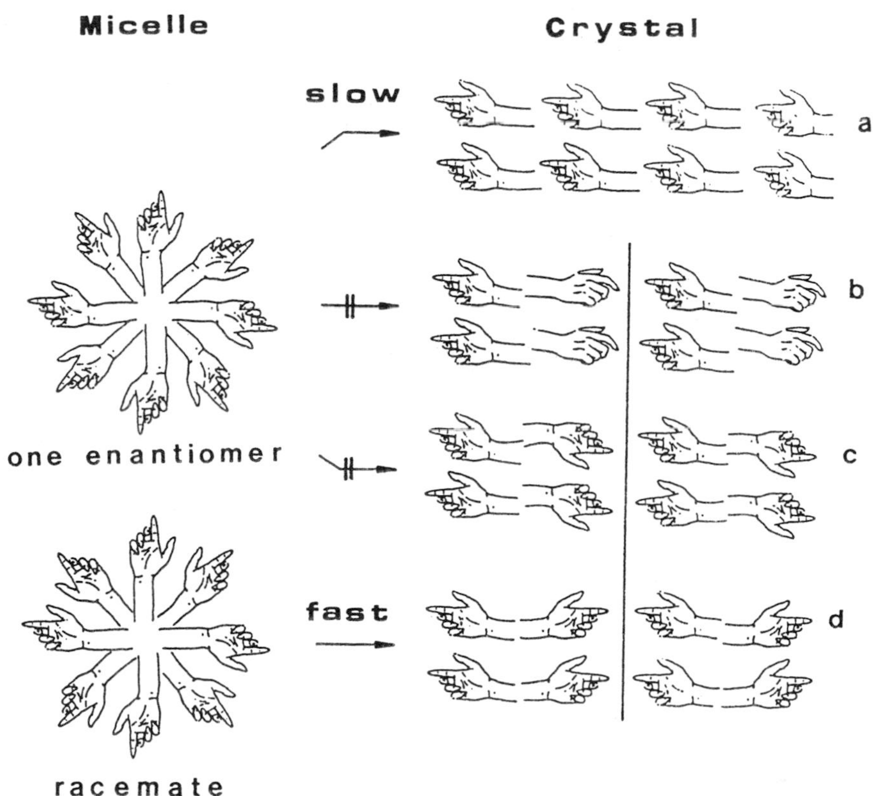

Figure 9: a) The rearrangement of hydrophobic bilayers of chiral molecules to enantiopolar crystals is not favourable in aqueous media and so it occurs very slowly; fibres with high surface areas are relatively stable.
b) c) Crystal planes with chiral groups in different orientations (*eg* palms up *and* down or thumbs up *and* down). The enantio non-polar tail-to-tail arrangement does not allow crystallization.
d) Racemic mixtures, on the other hand, crystallize rapidly ('chiral bilayer effect').

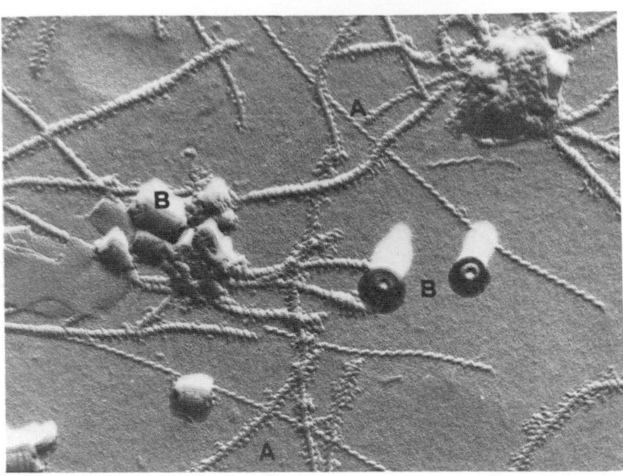

Figure 10: Electron micrograph of a mixed gel showing separate fibres of gluconamide **6a** (A) and mannonamide **10c** (B).

6 OUTLOOK

The development of hydrophobic fibres in an aqueous enviroment is very recent, but the available structural types are already established in our opinion: depending on stereochemistry, solubility and formation process, one obtains helical ropes, rolled-up sheets or twisted ribbons. We do not expect any other types of aggregates, but each may appear with very different length : diameter ratios. In terms of new developments we are most interested in the synthesis of hybrid membrane structures *eg* a vesicle membrane from which tubules or collapsed vesicles grow into the inner volume (mitochondrion model) or a tubule with rectangular branchings (glycoprotein model). Furthermore, the fibre aggregates may be good starting materials for the synthesis of high molecular weight polymers with a hydrophobic core and strictly uniform stereochemistry. We are also attempting to establish charge-separation systems and to apply redox-active and light-sensitive gel layers as substitutes for gelatine-based surface layers. Many more complex systems composed of the fundamental building blocks of biological organisates, namely soft, deformable bilayer sheets and rigid fibres can be envisaged.

SUMMARY

The combined effects of hydrophobicity and linear amide hydrogen bond chains in *N*-alkylaldonamides lead to the formation of fibrous aggregates in water. Rolled-up or twisted bilayer sheets and helical rods are observed. The formation of ultrathin (diameter ≃ 4 nm) micellar helices is dependent on the chirality of the head groups and eventual bends induced by stereochemical interactions. The micellar fibres are in equilibrium with small, probably flattened micelles. Crystallization, on the other hand, is usually a slow and irreversible process. Only racemic mixtures tend to crystallize rapidly to form platelets.

REFERENCES

1 B Pfannemüller and W Welte, *Chem Phys Lipids*, 1985, **37**, 227.
2 J-H Fuhrhop, P Schnieder, J Rosenberg, and E Boekema, *J Am Chem Soc*, 1987, **109**, 3387.
3 J-H Fuhrhop, P Schnieder, E Boekema, and W Helfrich, *J Am Chem Soc*, 1988, **110**, 2861.
4 D F Evans, *Langmuir,* 1988, **4,** 3 and references cited therein.
5 J-H Fendler, 'Membrane Mimetic Chemistry', Wiley, New York, 1982, p 25ff.
6 J-H Fuhrhop, S Svenson, C Boettcher, H-M Vieth, and E Rössler, submitted for publication.
7 J W Goodby, *Mol Cryst Liq Cryst,* 1984, **110**, 205.
8 K S Rosenthal and F Koussale, *Anal Chem,* 1983, **55**, 1115.
9 R Ebelhäuser and H W Spiess, *Ber Bunsenges Phys Chem,* 1985, **89**, 1208.
10 J J Gilman, ed, 'The Art and Science of Growing Crystals', Wiley, New York, 1963, p 42ff.
11 L O Bergen and G O Borisy, *Cell Biol*, 1980, **84**, 141.
12 R Bach, Diplomarbeit 1988, FU Berlin; J-H Fuhrhop and R Bach, submitted for publication.
13 V Zabel, A Müller-Fahrnow, R Hilgenfeld, W Saenger, B Pfannemüller, V Enkelmann, and W Welte, *Chem Phys Lipids*, 1986, **39**, 313.

14 A Müller-Fahrnow, V Zabel, M Steifan and R Hilgenfeld, *J Chem Soc Chem Commun*, 1986, 1573.
15 S Bhattacharjee, G Jeffrey, and J W Goodby, *Mol Cryst Liq Cryst*, 1985, **131**, 245.
16 N Arzania, G A Jeffrey, and M S Shen, *Acta Crystallogr, Sect B: Struct Crystallogr Cryst Chem*, 1972, **B28**, 1007.
17 H M Berman, G A Jeffrey, and R D Rosenstein, *Acta Crystallogr Sect B: Struct Crystallogr Cryst Chem*, 1968, **B24**, 442.
18 R Bacon, *J Appl Phys*, 1960, **31**, 283.
19 J H Georger, A Singh, R R Price, J M Schnur, P Yager, and P E Schoen, *J Am Chem Soc*, 1987, **109**, 6169.
20 N Nakashima, S Asakuma, and T Kunitake, *J Am Chem Soc*, 1985, **107**, 509.
21 J-H Fuhrhop, C Demoulin, J Rosenberg, and C Boettcher, submitted for publication.
22 S S Brenner, *Acta Met*, 1956, **4**, 62.
23 P Luger, C Lehmann, C Demoulin, and J-H Fuhrhop, submitted for publication.
24 C Boettcher, E Boekema and J-H Fuhrhop, submitted for publication.
25 W T Ferrar, D F O'Brien, A Warshawsky, and C L Voycheck, *J Am Chem Soc*, 1988, **110**, 288.

THE MOLECULAR MECHANISMS OF SENSITIVITY TO ANTICANCER DRUGS

Adrian L Harris and Ian D Hickson

ICRF Clinical Oncology Unit, Institute of Molecular Medicine, John Radcliffe Hospital, Headington, Oxford OX3 9DU, UK

1 INTRODUCTION: CLINICAL RESISTANCE TO CHEMOTHERAPY

One in five people dies from cancer. A major obstacle to reducing deaths from the disease is resistance to chemotherapy. In two thirds of cancer cases, death is due to metastases and in one third of cases to progression of local disease after failure of surgery and radiotherapy (for a detailed review see ref 1). Both groups of patients may receive chemotherapy.

Tumours can be classified into groups based on their sensitivity to chemotherapy (Table 1).[2] Most of the common solid neoplasms are not curable with chemotherapy. Although tumours may respond for periods of 2 to 3 years, usually responses are of shorter duration (*eg* the median response duration in colon cancer is 4 months; breast cancer, 10 months; and oat cell lung cancer, 1 year). These responses are worthwhile to individual patients and are important in showing that a differential effect in neoplastic tissue can be obtained. In most cases of moderately chemosensitive tumours, when resistance develops to one agent, there is cross-resistance to other drugs. In contrast, those tumours which are most chemosensitive often respond to second or even third-line therapies.

In the chemosensitive tumours, *eg* Hodgkin's disease, oat cell lung cancer, follicular lymphoma, and acute myeloid leukemia, patients may relapse a year or more after discontinuing chemotherapy. This is a different situation compared to relapse while on chemotherapy or tumour progression on therapy. Retreatment of patients who relapse off therapy can produce about 50% response rate when using the same drugs with which they were initially treated. The response of late relapsers to the same chemotherapy suggests that in many tumours, rather than eradicating stem cells, more sensitive daughter cells are killed, as in normal tissues. When the stem cells proliferate and the tumour recurs, the daughter cells may then remain sensitive. Thus, a relatively sensitive subpopulation may be responsible for tumour regression.

The chemoresistant group of tumours is usually resistant to any first-line agent, and only 20% of tumours show responsiveness. There are often only one or two types of drugs to which responses occur, *eg* colorectal cancer: 5-fluorouracil, mitomycin C; melanoma: 4-(3,3-dimethyltriazeno)imidazole-5-carboxamide (DTIC). Many tumours are intrinsically resistant (primary resistance) and may have many different mechanisms of resistance, since they are usually

Table 1: Chemosensitivity of tumours

Chemosensitive and curable	**Tumour type**
High complete response rate, combinations much better than single agents, sensitive to many different agents, chemotherapy essential in management	Hodgkin's disease, non-Hodgkin's lymphoma, acute leukaemias, choriocarcinoma, teratoma, Burkitt's lymphoma, paediatric tumours, oat cell lung cancer, ovarian carcinoma (10% long-term, 5 year survival)
Moderately chemosensitive	
50% response rate (± 15%) complete response rate (10%), combinations marginally better than single agents	Breast cancer (myeloma, CML) bladder cancer
Chemoresistant	
20% response rate, combinations not better than single agents, often only responsive to one drug type	Colon, rectal cancer, malignant melanoma, gliomas

cross-resistant to a wide range of drugs. It is intriguing that such tumours often arise in organs highly exposed to environmental carcinogens and toxins, *eg* gastrointestinal tumours, hypernephroma, hepatoma and melanoma. Their resistance mechanisms may involve inactivation mechanisms such as scavenging of the drug by glutathione transferases. Alternatively, there may be drug extrusion mechanisms such as that involving the P-glycoprotein, which can occur in normal tissue protection mechanisms.

The moderately chemosensitive tumours usually show only partial response to therapy and it is worth quantitating this effect when considering the proportion of resistant cells. A mass decreasing from maximum diameters perpendicular to each other of 6 x 6 to 1 x 1 cm would be considered a partial response. In volume terms, this represents approximately a 2-log reduction. Nevertheless, killing 99% of cells only produces partial response. Regrowth of the residual cells will produce relapse, *ie* drug resistance. A key question is whether the 'resistant' population is present from the beginning of therapy or induced by repeated exposure (acquired or secondary resistance). This has important implications for the duration of therapy. If a resistant population is present early on, prolonged chemotherapy is unjustified once a maximal response is achieved (usually four to six courses of treatment). Only a mutant drug-sensitive subpopulation may be responding.

There are very few randomized trials of duration of chemotherapy in common solid tumours. However, two trials in oat cell lung cancer and one in breast cancer showed no significant difference in survival comparing short courses of therapy with therapy until relapse.

Responses in tumours sensitive to chemotherapy may represent elimination of a subpopulation of sensitive daughter cells. However, the lack of cure is associated with failure to eliminate a resistant parent or stem cell population. Also, partial responses may represent elimination of one sensitive subclone of daughter cells. If this is the case, continued therapy after maximal response is not worthwhile.

Tumours relapsing early (within one year) after stopping chemotherapy may represent progression of the resistant population and are usually not responsive to second-line therapy. Tumours relapsing late are often sensitive to drugs used initially to induce remission. This has been shown in breast cancer and myeloma. Repopulation with the initially sensitive daughter cells must have occurred.

These different clinical patterns of behaviour suggest that different resistance mechanisms may be relevant. In the very chemosensitive tumours (*eg* teratoma) it is clear that there is more rapid proliferation and higher growth fraction than in the more chemoresistant tumours. However, the normal tissues from which the chemosensitive tumours arise are also relatively the most sensitive to chemotherapy. This particularly applies to hemopoietic lymphoid and testicular neoplasms and suggests that there are mechanisms related to differential sensitivities of normal tissues that are inherited by cancers arising from those tissues. This has been shown most convincingly for human bladder and teratoma cell lines, where the latter are much more sensitive to radiation, adriamycin and *cis*-platin.

As most antitumour agents are thought to exert their toxicity principally *via* an interaction with DNA, it is important to understand both the mechanisms of action of these agents and the DNA repair mechanisms present in normal cells to counteract drug-induced killing. It is presumably the resistance of normal tissues that allows the use of chemotherapy or radiotherapy.

In our studies of drug resistance we have sought to define normal basal mechanisms by which drugs kill cells or cells resist drugs. We have used mutant Chinese hamster ovary (CHO) cell lines that can ultimately be analysed at the molecular level. We have developed mutants that are hypersensitive to various classes of cytotoxic drugs and also mutants defective in drug activation or very effective at drug inactivation.

The major mechanisms regulating the basal sensitivity of cells to cytotoxic agents are DNA repair, interaction with topoisomerase II and the operation of multidrug resistance genes, each of which is discussed below.

2 DNA REPAIR

DNA repair is at the interface of many biological processes, including mutagenesis, carcinogenesis and free radical toxicity, and is involved in certain cancer-prone human syndromes and in the development by cancer cells of resistance to therapy. The major repair pathways in *Escherichia coli* and mammalian cells are summarized below (for a review, see ref 3):

DNA Repair Pathways

Excision repair. This involves incision on both sides of a DNA adduct, filling in of the gap with polymerase(s), and rejoining with ligase(s). The types of lesions recognized by excision repair enzymes are caused by UV light, numerous carcinogens, and DNA cross-linking alkylating agents such as mitomycin C and *cis*-platin.

Glycosylases. These recognize specific abnormal bases (*eg* alkylated adenine and guanine residues). They excise the base to leave an apurinic site which is further processed by endonucleases. The lesions repaired by glycosylases are caused by methylating agents such as methyl nitrosourea (MNU) and methyl methanesulphonate (MMS)

Error-free repair mechanisms. A specific suicide protein, which dealkylates the O^6 position of guanine residues in DNA, exists in both bacteria and eukaryotes. Photolyases reverse pyrimidine dimer formation induced by UV light.

Recombination repair. This involves strand exchange between chromosomes and is implicated in double-strand break repair. This type of damage is considered to be the main cytotoxic lesion induced by *X*-rays and bleomycin. Recombination is also important in DNA cross-link repair in *E coli*.

Mismatch repair. This removes mispaired bases and preferentially repairs GT in favour of GC. This may protect against miscoding of deaminated 5-methylcytosine residues. Defects in this system are associated with sensitivity to *cis*-platin and methylating agents in *E coli*.

Much of the knowledge of mechanisms of DNA repair has been derived from an analysis of *E coli* mutants, of which there are over 60 showing an alteration in some aspect of DNA repair. The existence of this range of mutants has facilitated the cloning of most of the relevant genes and, consequently, the purification of the gene products and the reconstitution of repair activities *in vitro* with defined DNA substrates.

To analyse DNA repair in mammalian systems in the same detail it is essential to have available a similar range of repair-defective mutants. Several groups, therefore, have sought to isolate mutants hypersensitive to DNA-damaging agents, with the expectation that these cells would possess an abnormality in DNA processing. Consequently, there has been a rapid increase in the numbers of such mutants reported in the past five years.

Isolation of Mutants

The general approach to the isolation of mutants is to mutagenize the parent population with radiation or with an alkylating agent, and then to screen for clonal isolates hypersensitive to a particular DNA-damaging agent (for a review, see ref 4). In most cases, a replica plating technique has been used. This ensures that clones killed on a test plate by the selective agent (at a dose non-toxic to normal cells) can be isolated from a master plate, on which the cells are not treated.

The cell lines most commonly used are from the Chinese hamster, although mutant murine lines have also been reported. The hamster lines grow well, have a high plating efficiency and, of particular importance when considering their use as hosts for DNA cloning, generally have a high uptake of exogenous DNA compared with human cell lines. Also, large areas of the hamster genome appear to be functionally haploid, which greatly increases the chance of isolating mutants bearing recessive mutations. Most groups report a mutant isolation frequency of 1 in 1000-2000, far higher than expected for mutations at both *loci* of a diploid genome. Recently, even higher frequencies (around 1 in 200) have been reported for hamster V79 cells.[5] Even with functional hemizygosity, this is a remarkably high frequency, suggesting that there may be 100-200 genes involved in DNA repair and the related processes in mammalian cells. Considering the number of repair genes in *E coli*, which has no nucleus or histones, and also the number of human hereditary repair defects compatible with life (>20), the number may be even greater.

Agents used for Mutant Selection

The mechanisms of action of DNA-damaging agents vary considerably. Simple monofunctional alkylating agents, such as methyl methanesulphonate (MMS) and methyl nitrosourea (MNU), interact directly with cellular DNA causing the formation of covalent adducts. Several other drugs interact with DNA after activation, which can be either spontaneous (*eg* the nitrogen mustards, that yield reactive aziridinium ions as a prelude to DNA cross-linking) or metabolically induced (*eg* the mitomycins, that can be reduced to a reactive quinol form). Other agents, including ionizing radiation and bleomycin, generate highly reactive free radicals, which directly damage DNA bases or sugars and lead to the formation of DNA strand breaks. An additional group of drugs, including intercalating agents and *epi*-podophyllotoxins, indirectly produce single- and double-stranded breaks in DNA *via* an interaction with the nuclear enzyme topoisomerase II (see below).

Because each agent generates a characteristic spectrum of DNA lesions, which are in turn recognized by specific repair enzymes, the choice of selective agent greatly influences the range of mutants that can be isolated, such as mitomycin C for DNA cross-link repair-defective mutants, ultraviolet light (UV) for excision repair, bleomycin and *X*-rays for DNA strand break repair and recombination. Most of these agents are used in cancer therapy, and are mutagens and carcinogens.

The most commonly used selective agent has been radiation (both UV and ionizing). However, our approach has been to concentrate on DNA-damaging drugs which are used extensively in antitumour therapy. We have isolated mutants hypersensitive not only to agents that damage DNA directly, but also to those drugs whose cytotoxicity is mediated *via* topoisomerase II.

Classification of Mutants

A number of the Chinese hamster mutants isolated so far can be correlated with a defined abnormality in DNA processing, but others have defects that are undefined. Although one aim of isolating mutants is to define DNA repair pathways, there may be several other basal mechanisms involved in drug and radiation resistance/sensitivity. Mutants are obviously useful for defining these alternative pathways.

Defects in Incision and Cross-link Repair

References 4–30 describe the mutants of Chinese hamster cells with a defined defect in DNA damage induction and/or repair. The best characterized of these are the UV-sensitive CHO mutants representing UV complementation groups 1 to 5. These mutants possess a similar phenotype both to one another and to *Xeroderma pigmentosum* cell lines (a human syndrome characterized by extreme sensitivity to sunlight and a predisposition to skin cancers), in being defective in the incision step of excision repair.[8]

Mutants from groups 1 to 5 are also characterized by their reduced capacity for unscheduled DNA synthesis (UDS),[6] otherwise known as repair synthesis, which reflects the filling-in of DNA patches from which chemical adducts have been excised by repair enzymes. Another UV-sensitive mutant, UV61,[22] falls into a different complementation group and shows normal rates of UDS after irradiation,[11] presumably reflecting an ability to excise UV adducts successfully.

Defects in DNA Strand Break Repair

The second major class of repair mutants isolated comprises mutants defective in the repair of DNA strand breaks. Such damage is induced by a variety of agents, including ionizing radiation, and the radiomimetic and alkylating agents. Four CHO mutants with a defined defect in strand break repair have been identified, two of which (XR-1 and *xrs*-1) were isolated as *X*-ray sensitive, one (EM9) as ethyl methanesulphonate (EMS) sensitive, and one (BLM-2) as bleomycin sensitive.[10,15,17,20] These mutants tend to be sensitive to the same classes of DNA-damaging agent, although to differing degrees. Their repair characteristics, however, are not identical. EM9 and BLM-2 are defective in the repair of both single- and double-strand breaks[12] (see also C N Robson *et al*, submitted), while XR-1 and *xrs*-1 show a significant defect in double strand break repair only.[16,18] XR-1 displays its repair defect only during the G1 phase of the cell cycle (and is similarly only *X*-ray sensitive during G1 phase).

Consistent with these phenotypic differences, it has recently been shown that the repair-defective mutants represent four different genetic complementation groups.[20] Mutant EM9 has been characterized in considerable detail and displays a high baseline frequency of sister-chromatid exchanges.[12] This phenotype resembles that of cell lines from patients with Bloom's syndrome, a hereditary disorder involving chromosome instability and a predisposition to acute leukaemia. Bloom's lines have recently been shown to have a defect in DNA ligase I,[26] although EM9 and Bloom's cells are genetically distinct and EM9 cells have no apparent alteration in ligase activity.[27]

Undefined Mechanisms

There are also CHO and Chinese hamster V79 mutants hypersensitive to DNA-damaging agents, but which have no defined defect in DNA repair.

Cross-sensitivity Profiles

For mutants isolated on the basis of hypersensitivity to a single class of agents, the various profiles of cross-sensitivities to different drugs provide evidence for different biochemical defects. This can also provide valuable information when testing for correction of defects by DNA transfection - if all hypersensitivities are corrected, it is highly likely that they were due to one defect. The profiles are also interesting because they point to unexpected interrelationships among different types of drugs and provide an insight into which type of assays to use in defining the defect.

Thus, of four MMC-sensitive mutants, only MMC-2 is UV sensitive, suggesting a defect in excision repair. Similarly, only two mutants, MMC-1 and MMC-2, are *cis*-platin sensitive, suggesting a problem in the repair of bulky adducts or DNA cross-links.

BLM-1 and -2, which are sensitive to an agent considered to kill by producing double-strand breaks in DNA, are cross-sensitive to topoisomerase II inhibitors. Similarly, ADR-1, isolated on the basis of hypersensitivity to topoisomerase II inhibitors, is cross-sensitive to bleomycin. This raises the question, previously unaddressed, of a role for topoisomerase II in the repair of double-strand breaks.

Only three of the six MMS-sensitive mutants are sensitive to another methylating agent, *N*-methyl-*N*'-nitro-*N*-nitrosoguanidine (MNNG), which produces a higher proportion of O^6-methylguanine adducts than does MMS. Since CHO cells do not possess detectable amounts of the repair protein for this lesion, the implication is that there must be another repair process involved. Without the comparison of different classes of methylating agent, these leads to further investigation would not be apparent.

Hydrogen peroxide is toxic to three different classes of mutants (BLM-1, ADR-1, MMS-2) which are sensitive to strand breakage, topoisomerase II inhibitors and methylating agents, respectively. This points to complexity in the processes involved in handling free radical damage. Finally, some mutants are sensitive to only one class of cross-linkers (MMC-3 and MMC-4), others to mustards and *cis*-platin as well as mitomycin C (MMC-1, MMC-2). Different cross-links may therefore require different repair processes.

3 INTERACTION WITH TOPOISOMERASE II

There are two major classes of drugs whose cytotoxicity appears to be mediated *via* an interaction with the nuclear enzyme topoisomerase II (for a review see ref 31). These are the intercalating agents (*eg* adriamycin and actinomycin D) and the non-intercalative *epi*-podophyllotoxins, such as etoposide (VP16) and teniposide (VM26). A common feature of their action is the formation of protein-associated single- and double-strand breaks in cellular DNA. Topoisomerase II cleaves both strands of the DNA and binds covalently to the 5' termini of the breaks. These normally transient protein-DNA complexes are apparently stabilized *in vivo* by the topoisomerase II-interactive drugs, with the result that the normal progress of the enzyme reaction (*ie* strand passage and ligation) is halted.[32,33] Although toxicity correlates with the formation of these complexes, it is not clear why these protein-concealed lesions are cytotoxic. The amount of complex formed also depends on the availability of topoisomerase II (since it is bound covalently) and therefore factors regulating the amount of topoisomerase II are probably important in determining the effectiveness of these drugs. Thus, the role of topoisomerase II as a positive effector in this reaction contrasts markedly with that of most intracellular targets for cytotoxic drugs, such as dihydrofolate reductase, in that overexpression of the target does not negate the action of the drug.

To study the role of topoisomerase II in drug-induced cell killing, we have isolated six mutants of CHO cells (designated ADR-1 to -6) on the basis of hypersensitivity to adriamycin or actinomycin D. One of these, ADR-1, has been characterized in detail.[21] The cross-sensitivity patterns of these mutants are such that in all cases, an approximately equivalent degree of sensitivity to the intercalating agents adriamycin, *m*-AMSA (amsacrine), mitoxantrone and actinomycin D is evident, as is sensitivity to the *epi*-podophyllotoxins, VP16 and VM26. This confirms that the cytotoxicity of this range of structurally distinct compounds is mediated *via* one or more common pathways. A clear feature of these mutants is their lack of cross-sensitivity to most classes of DNA-damaging agents, including radiation and mono- and bi-functional alkylating agents.

We have shown that ADR-1 cells exhibit higher topoisomerase II activity than do CHO-K1 cells, and that this results from an overproduction of topoisomerase II protein in the mutant.[39-41] There is a good correlation between the degree of drug hypersensitivity and the level of drug-induced DNA strand breaks in ADR-1 cells. This provides strong evidence that these topoisomerase II-dependent strand scissions lead directly or indirectly to cell death, and that the higher topoisomerase II level in the ADR-1 cells underlies its drug-hypersensitive phenotype.

4 MULTIDRUG RESISTANCE GENE(S)

Mutants ADR-2 and ADR-6, although isolated on the basis of hypersensitivity to intercalating agents, show extreme sensitivity to both vincristine and colchicine. These latter two drugs are not known as DNA-damaging agents at all; instead, they block cell division by interfering with the formation of the mitotic spindle. ADR-2 and ADR-6 appear to have a reciprocal phenotype to that of the so-called multidrug-resistant (MDR) cell lines (for a review, see ref 34), which have been extensively studied as a model system for how tumour cells acquire resistance to antitumour drugs. The pleiotropic phenotype of MDR cells is associated with the overproduction of a 170 kDa membrane glycoprotein (P-glycoprotein) which appears to act by promoting drug export. The protein has two ATP-binding sites and two regions that are homologous to each other, each consisting of three membrane-spanning loops. P-glycoprotein mediates the active export of a wide range of antitumour antibiotics. Inhibition of this mechanism has potential for overcoming anticancer drug resistance.

The possibility exists that our mutants are mutated in the *mdr1* gene, leading to a 'multidrug sensitive' phenotype. It is consistent with this suggestion, as we have already shown, that both these mutants accumulate elevated levels of radiolabelled drug. Studies are in progress to assign the defect positively to the *mdr1 gene*. If this is confirmed, these mutants should facilitate identification of the normal role of P-glycoprotein in mammalian cells, and the identification of protein domains crucial for drug export. These domains may subsequently allow effective drug targeting in patients with highly resistant tumours.

5 USE OF DRUG-HYPERSENSITIVE MUTANTS

The following key points should be noted:

1 The major purpose is to clone human DNA repair genes. Genomic DNA or an expressing cDNA library can be transfected into hamster or mouse cells with high efficiency, in the hope of complementing the defect in repair. Identification of mutants

complemented in this way is by virtue of the restoration of normal cellular resistance to a cytotoxic agent. This approach has led to the first successful cloning of a human DNA repair gene (see below).

2 The function of repair genes can be readily analysed in the mutants, without which the phenotype associated with a gene would be unknown.

3 Comparison of cross-sensitivity to other agents may show interactions between types of DNA-damaging agents that would not be detected in wild-type cells, and may suggest alternative modes of drug action *etc.*

4 It is possible to elucidate the most important biochemical targets mediating cell death.

5 Using a bank of mutants with different cross-sensitivities may be a more relevant way to screen for environmental toxins than using bacterial or yeast systems.

6 Screening for novel cytotoxic drugs for use in cancer therapy could be enhanced by selecting those that were *not* more toxic to the mutant panel than they were to wild-type cells. This would exclude cytotoxic drugs that are mechanistically similar to ones that already exist. The need is for novel compounds rather than 'me too' drugs.

7 Extracts from the mutants can provide the basis for purifying repair proteins by *in vitro* complementation in repair assays. Similarly, microinjection of extracts from normal cells into mutants could be used to follow purification of normal repair proteins.

8 Genes complementing human hereditary DNA repair defects could be cloned, if by genetic analysis a Chinese hamster or murine mutant can be shown to have the same defect as one of the human syndromes. All the processes involved in cloning the human genes should be easier in the mutant rodent cells.

The First Cloned Human Repair Gene (ERCC 1)

The *ERCC 1* gene (excision repair, complementing Chinese hamster cells, see ref 35 for a review) was cloned by transfecting human genomic DNA into the UV-sensitive CHO mutant 43-3B.[9] The gene is 15-17 kb long with 10 exons, one of which can be alternatively spliced. Only the larger transcript is functional (M_r 32562). Clones expressing the smaller transcript fail to complement

43–3B cells. The complete sequence of the *ERCC 1* cDNA has been determined.

An extensive homology has been found between the deduced amino acid sequence of ERCC 1 protein and that of the protein encoded by the previously cloned yeast repair gene *RAD10,* with 34% identity in the region of highest homology. There is an additional homology with the UvrA protein (31%), an *E coli* repair enzyme. It is interesting that most of this homology is at the carboxy-terminal end of ERCC 1, beyond the region homologous with RAD10. The regions of RAD10 and ERCC 1 with the highest homology are putative DNA-binding domains.

A potential ADP ribosylation site is also present in the predicted ERCC 1 sequence, at the border of exon 7 and the alternatively spliced exon 8. Thus, proteins lacking the sequence coded by this exon would lack an ADP ribosylation site. Although much can be deduced from the sequence, as yet studies with purified protein have not been reported. The conservation of repair genes from *E coli* through yeast to man is perhaps not surprising, in view of the fundamental processes involved, but it is tempting to speculate that the *ERCC 1* gene has evolved through the fusion of several functional domains from different proteins. Several groups have now reported successful transfection of human DNA into mutants, with correction of the hypersensitivity,[36–38] and it is anticipated that the sequences of several human DNA repair genes will be available in the near future.

Future Studies with Cloned Human DNA Repair Genes

The availability of probes for human DNA repair genes will have important implications in most areas of biology and medicine. Gene amplification is common in cancers, but not in normal tissues, and occurs *via* genetic recombination, among other mechanisms. It will be possible to investigate the basis for genetic instability in cancers. Other key areas include the generation of diversity in immunoglobulin genes, DNA rearrangements in development, and alterations in the ability of cells to repair DNA damage with increasing age. Considering the large number of genes that must be involved, it is likely that there will be polymorphisms that could affect cancer susceptibility or the organ specificity of carcinogens. Endogenous toxins such as free radicals are one of the major agents that damage DNA and thus may be responsible for many degenerative diseases. Therefore, individual differences in susceptibility to non-carcinogenic DNA damage may also be studied using such probes. X-rays and cytotoxic drugs can cure some patients with cancer, but most develop resistance, although their normal tissues do not. The contribution of DNA repair to these phenomena is suspected, but a molecular approach is needed to identify specific mechanisms and, hopefully, to open up new therapeutic possibilities.

ACKNOWLEDGEMENTS

We would like to acknowledge the considerable contributions made to the work described here by members of the Cancer Research Unit, University of Newcastle upon Tyne: in particular, Craig Robson, Paul Hoban, Janice Reid, Stella Davis, and Sally Davies.

REFERENCES

1 V T deVita, S Hellman, and S A Rosenberg, 'Cancer: Principles and Practice of Oncology', Lippincott Publishers, Philadelphia, 1985.
2 H M Pinedo, D L Longo, and B A Chabner, eds, 'Cancer Chemotherapy and Biological Response Modifiers', Annual 10, Elsevier, Amsterdam, 1988.
3 T Lindahl, *Annu Rev Biochem*, 1982, **51**, 61.
4 L H Thompson in 'Molecular Cell Genetics', M Gottesman, ed, John Wiley & Sons, 1985, p 641.
5 M Z Zdzienicka and J W I M Simons, *Mutat Res*, 1987, **178**, 235.
6 L H Thompson *et al*, *Somat Cell Genet*, 1980, **6**, 391.
7 L H Thompson *et al*, *Proc Natl Acad Sci USA*, 1981, **78**, 3734.
8 L H Thompson *et al*, *Somat Cell Genet*, 1982, **8**, 759.
9 R D Wood and H J Burki, *Mutat Res*, 1982, **95**, 505.
10 C N Robson, A L Harris, and I D Hickson, *Cancer Res*, 1985, **45**, 5303.
11 L H Thompson *et al*, *J Cell Sci*, 1987, **6** (Suppl), 97.

12 L H Thompson *et al, Mutat Res,* 1982, **95**, 427.
13 T D Stamato and C A Waldren, *Somat Cell Genet*, 1977, **3**, 431.
14 T D Stamato, L Hinkle, A R S Collins, and C A Waldren, *Somat Cell Genet,* 1981, **7**, 307.
15 T D Stamato, R Weinstein, A Giaccia, and L Mackenzie, *Somat Cell Genet,* 1983, **9**, 165.
16 A Giaccia, R Weinstein, J Hu, and T D Stamato, *Somat Cell Genet,* 1985, **11**, 485.
17 P A Jeggo and L M Kemp, *Mutat Res,* 1983, **112**, 313.
18 L M Kemp, S G Sedgwick, and P A Jeggo, *Mutat Res,* 1984, **132**, 189.
19 C–C Chang *et al, Somat Cell Genet,* 1981, **7**, 235.
20 C N Robson, A L Harris, and I D Hickson, *Mutat Res,* 1989, **217**, 93.
21 C N Robson, P R Hoban, A L Harris, and I D Hickson, *Cancer Res*, 1987, **47**, 1560.
22 D B Busch, J E Cleaver, and D A Glaser, *Somat Cell Genet,* 1980, **6**, 407.
23 C N Robson and I D Hickson, *Mutat Res,* 1986, **163**, 201.
24 M Stefanini, A Reuser, and D Bootsma, *Somat Cell Genet,* 1982, **8**, 635.
25 C N Robson and I D Hickson, *Carcinogenesis,* 1987, **8**, 601.
26 A E Willis and T Lindahl, *Nature,* 1987, **325**, 355.
27 J Y H Chan, L H Thompson, and F F Becker, *Mutat Res,* 1984, **131**, 209.
28 R A Schultz, J E Trosko, and C–C Chang, *Environ Mutagen,* 1981, **3**, 53.
29 R I Pinto *et al, Tsitologia,* 1980, **22**, 1085.
30 N J Jones, R Cox, and J Thacker, *Mutat Res,* 1987, **183**, 279.
31 B S Glisson and W E Ross, *Pharmac Ther,* 1987, **32**, 89.
32 K M Tewey, G L Chen, E M Nelson, and L F Liu, *J Biol Chem,* 1984, **259**, 9182.
33 G L Chen *et al, J Biol Chem*, 1984, **259**, 13560.
34 G F–L Ames, *Cell*, 1986, **47**, 323.
35 J H J Hoeijmakers, *J Cell Sci,* 1987, **6** (Suppl), 111.
36 M A McInnes, J M Bingham, L H Thompson, and G F Strniste, *Mol Cell Biol,* 1984, **4**, 1152.

37 I J Spiro, L R Barrows, K R Kennedy, and C C Ling, *Radiat Res,* 1986, **108**, 146.

38 L H Thompson *et al, Mol Cell Biol,* 1985, **5**, 881.

39 S M Davies, C N Robson, S L Davies, and I D Hickson, *J Biol Chem*, 1988, **263**, 17724.

40 S M Davies, S L Davies, A L Harris, and I D Hickson, *Cancer Res*, 1989, **49**, 4526.

41 S M Davies, A L Harris, and I D Hickson, *Nucleic Acids Res*, 1989, **17**, 1337.

ON THE MECHANISM OF BLEOMYCIN ACTIVATION AND POLYNUCLEOTIDE STRAND SCISSION

Sidney M Hecht, Eric C Long, Reuel B Van Atta, Erik de Vroom, and Barbara J Carter

Departments of Chemistry and Biology, University of Virginia, Charlottesville, Virginia 22901, USA

1 INTRODUCTION

The bleomycins (BLM's, see Figure 1 for a representative structure) are glycopeptide-derived antitumour antibiotics that are believed to exert their therapeutic effects at the level of DNA degradation.[1] Bleomycin-mediated DNA degradation is an oxidative process that requires oxygen and one of several redox-active metal ions.[2,3] By the use of an oligonucleotide that undergoes efficient, site-selective modification by bleomycin, two sets of products have been shown to result from this process (Scheme 1). One of these, believed to derive from an intermediate C-4' hydroperoxynucleoside (i), involves DNA strand scission with the formation of a 'base propenal', a 5'-fragment terminating with a nucleoside 3'-phosphoroglycolate, and a

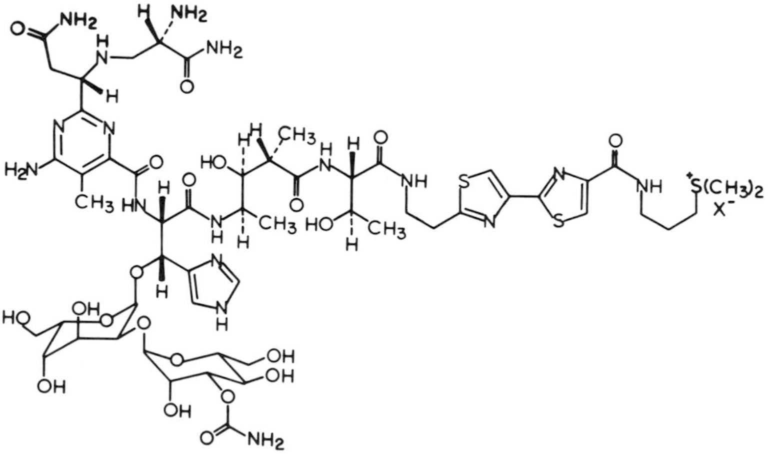

Figure 1: Structure of bleomycin A_2

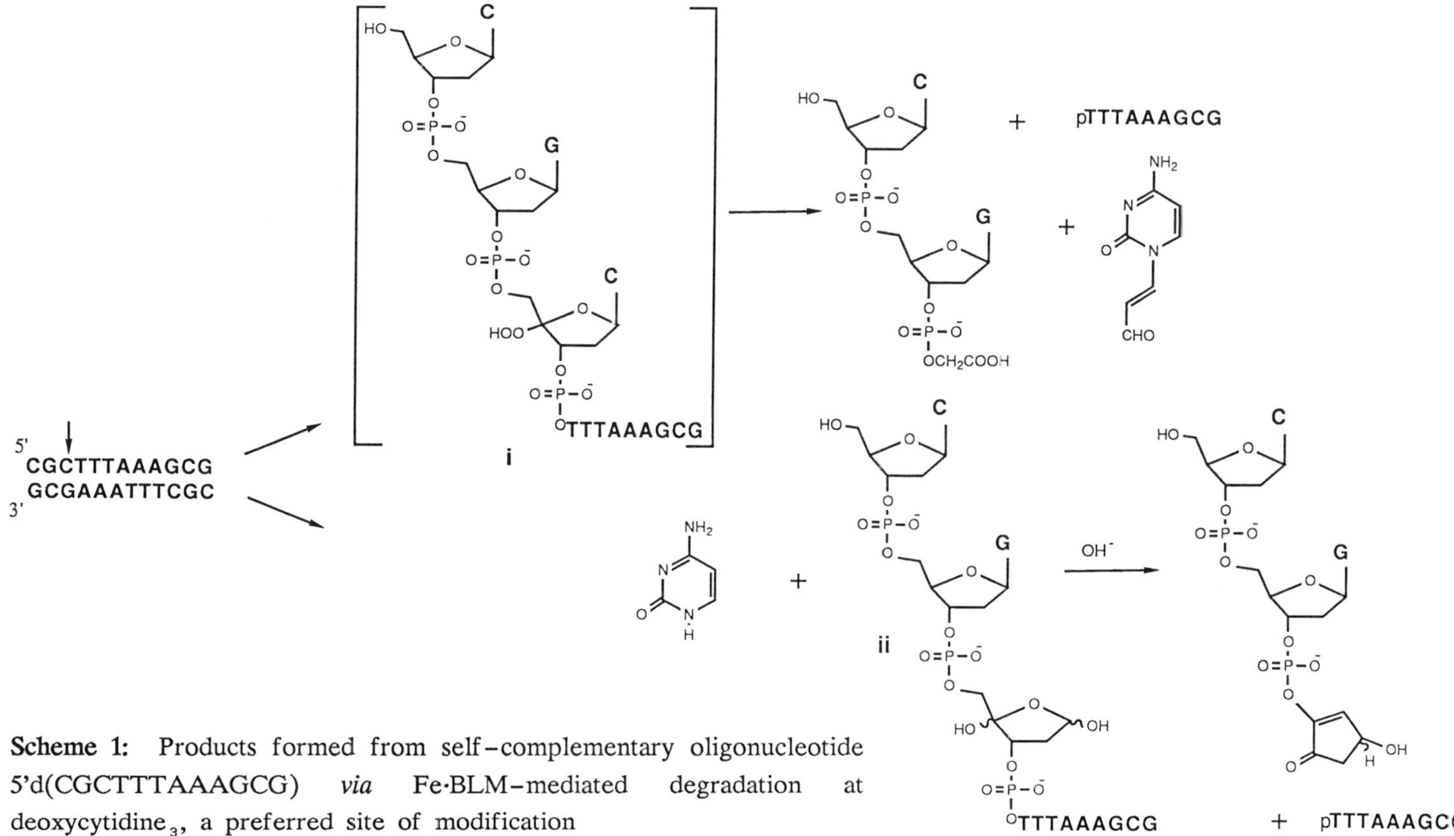

Scheme 1: Products formed from self-complementary oligonucleotide 5'd(CGCTTTAAAGCG) *via* Fe·BLM-mediated degradation at deoxycytidine$_3$, a preferred site of modification

3'-fragment terminating with a nucleoside 5'-phosphate.[2-5] The other set of products, apparently formed by C-4' hydroxylation of a deoxyribose moiety in DNA, includes a free base and a C-4' hydroxyapurinic acid (ii).[6,7] Although this pathway does not lead directly to DNA strand scission, the initially formed product can react chemically with a number of reagents (*eg* alkali, alkylamines and hydrazine) affording DNA strand scission at the site of the initially formed lesion.

It has been shown that cobalt bleomycin can be activated in an alternative fashion involving light but not dioxygen.[8] A recent publication by Saito *et al*[9] has shown that the DNA degradation products resulting from this process are limited to free bases and alkali-labile lesions; these authors have also proposed a mechanistic scheme that accommodates the observed behaviour of this metallobleomycin.

Our ongoing studies of the mechanism of BLM activation and subsequent DNA degradation have attempted to define better certain aspects of this process. Presently we describe the electrochemical activation of Fe·BLM at a glassy carbon electrode, and compare the derived activated species with activated Fe·BLM formed upon admixture of Fe(II) + BLM in aerated aqueous solution. Also described is the attempted degradation of RNA by bleomycin, and an analysis of the factors that limit degradation of this species by activated Fe·BLM.

2 ELECTROCHEMICAL ACTIVATION OF FE·BLM

On the basis of oxygen consumption data,[10] spectroscopic analysis of reactive intermediates,[11,12] and quantification of products formed from DNA by activated Fe·BLM,[2,5] it has been proposed that the aerobic activation of Fe(II)·BLM requires an additional electron. This could be obtained by disproportionation of two Fe(II)·BLM's in the presence of O_2, or by provision of an external reducing agent such as dithiothreitol or ascorbic acid. In order to facilitate the study of the mechanism of activation of Fe·BLM, and the behaviour of the resulting activated complex, we have developed an efficient electrochemical system for bleomycin activation.[13]

Cyclic voltammetry at a glassy carbon electrode (Figure 2) permitted determination of the electrochemical potential of Fe(III)·BLM A_2. The cyclic voltammogram contained a wave corresponding to a one-electron metal-centred reduction of Fe(III)·BLM at −0.13 V and a wave at −0.03 V that represents Fe(II)·BLM reoxidation; $E_{1/2}$ = −0.08 V *vs* Ag/AgCl ($i_{p,c}/i_{p,a}$ ~ 1, ΔEp = 0.10 V) corresponding to a nearly reversible, Fe(III)/Fe(II) redox couple. This $E_{1/2}$ value was consistent with the redox potential of Fe·BLM measured by microcoulometric and optical absorption techniques.[14]

While multiple scans could be conducted anaerobically without change of current amplitude, suggesting that neither the

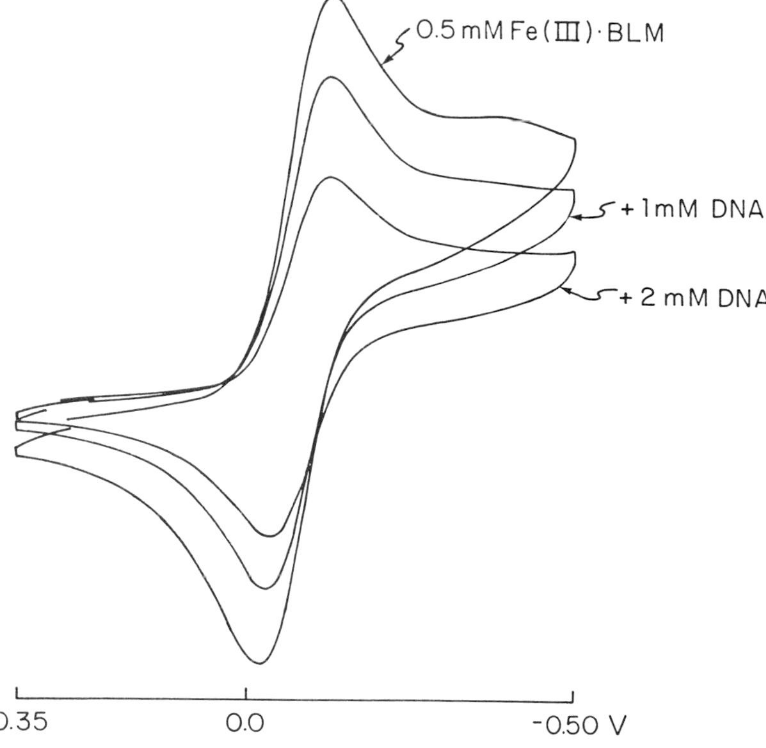

Figure 2: Cyclic voltammograms of Fe(III)·BLM (anaerobic conditions) in the absence of DNA, and in the presence of 1 mM and 2 mM calf thymus DNA. The experiments employed 0.5 mM Fe·BLM in 50 mM Na cacodylate, pH 7.2, at a scan rate of 20 mV/sec; $E_{1/2}$ = −0.08 V *vs* Ag/AgCl reference (Pt auxiliary electrode).

metallobleomycin nor electrode surface was altered during this process, admission of oxygen during the reductive sweep of the cyclic voltammogram resulted in increased current for the reduction wave and the appearance of a multielectron reduction wave at more negative potentials. We suggest that the increased current for the first reduction wave corresponds to further reduction of the Fe(II)·BLM formed following combination with O_2. Activated Fe·BLM should undergo facile reduction;[11,12,15] the new multielectron reduction wave at more negative potential presumably reflects this process, as does the absence of a reoxidation wave in the oxidative sweep of the cyclic voltammogram.

In order to determine the relationship between electrochemically reduced, oxygenated Fe·BLM and activated bleomycin formed by admixture of Fe(II)·BLM + O_2 in solution, the former was used for the degradation of DNA oligonucleotides that have been shown to act as substrates for activated Fe·BLM (*cf* Scheme 1). Solutions of Fe(III)·BLM were electrolysed under aerobic conditions at −0.22 V in the presence of an excess of (self-complementary) d(CGCTTTAAAGCG) or d(CGCTAGCG). As shown in Table 1, electrochemically activated Fe·BLM degraded both of these oligonucleotides, producing the same products in the same relative amounts obtained upon admixture of Fe(II)·BLM + O_2 in the presence of these oligonucleotides.[5,16] The availability of reducing potential at the electrode surface should logically result in the production of multiple lesions for each Fe(III)·BLM present. As shown in Table 1, as many as 12.9 lesions were produced for each Fe·BLM; further, the amounts of products produced electrochemically were comparable to those obtained by chemical activation of Fe·BLM in the presence of sodium ascorbate. We have shown previously that the ratio of lesions produced from d(CGCT$_3$A$_3$GCG) at deoxycytidine$_3$ *versus* deoxycytidine$_{11}$ (*ie* at the two major sites of BLM-mediated degradation) is a strong function of the specific structural nature of individual activated bleomycins.[16] Therefore, the finding that electrochemically and chemically activated Fe·BLM's produced damage at C_3 and C_{11} in very similar ratios (8/92 and 7/93, respectively) constituted further support for the thesis that these activated species were the same.

Table 1: Products resulting from degradation of DNA oligonucleotides by electrochemically and chemically activated Fe·BLM[a]

Oligonucleotide	Treatment	C Propenal + Cytosine (μM)	Total DNA Lesions (μM)	DNA Lesions/ Fe·BLM
d(CGCTAGCG)	-0.22 V, 6 h	580	580	12.9
d(CGCT$_3$A$_3$GCG)	-0.22 V, 4 h	170	188	4.2
d(CGCT$_3$A$_3$GCG)	-0.22 V, 22 h	338	371	8.2
d(CGCT$_3$A$_3$GCG)	Na ascorbate, 0.25h	177	199	10.0

[a] Reaction mixtures (200 μl total volume) contained 45 μM Fe(III)·BLM and either 0.9 mM d(CGCTAGCG) or 0.4 mM d(CGCT$_3$A$_3$GCG) in 50 mM Na cacodylate, pH 7.0. The reaction mixtures were electrolysed as indicated above in a continuously aerated cell containing a 1.1 cm^2 polished glassy carbon plate electrode. Product analysis was carried out by C$_{18}$ reverse phase HPLC. Chemically activated Fe·BLM (20 μM) was run in comparison as described.[16]

Mechanistically, it seemed possible that the electrochemical activation of Fe(III)·BLM might involve Fe·BLM catalysed two-electron reduction of dioxygen to peroxide, the latter of which is known to combine with Fe(III)·BLM to form 'activated Fe bleomycin'. In order to test this possiblity, we included high concentrations of catalase in the electrochemical cell during activation. The lack of any effect observed for catalase suggests that H_2O_2 is not a reaction intermediate, a conclusion reinforced by the finding that intentional admixture of H_2O_2 and Fe(III)·BLM to d(CGCT$_3$A$_3$GCG) under conditions comparable to those in our electrochemical activation experiments produced lesions much less efficiently than did the electrochemical system.

Another interesting finding resulted from the measurement of current and total charge as a function of electrolysis time. As shown in Figure 3, aerobic electrolysis of Fe(III)·BLM resulted in a steady decrease in current; the current was < 20% of the initial value after 10.2 electron equivalents. That this was due to self-inactivation of activated Fe·BLM was determined directly by recovery of the electrolysed BLM and demonstration that this material functioned very poorly in DNA degradation, even when fresh Fe(II) was provided. Further, when the experiment outlined in Figure 3 was repeated in the presence of 630 μM (dCGCT$_3$A$_3$GCG), the current had diminished only to 23.5 μA after passage of 14.4 electron equivalents, demonstrating protection of the activated Fe·BLM against self-inactivation by added DNA.

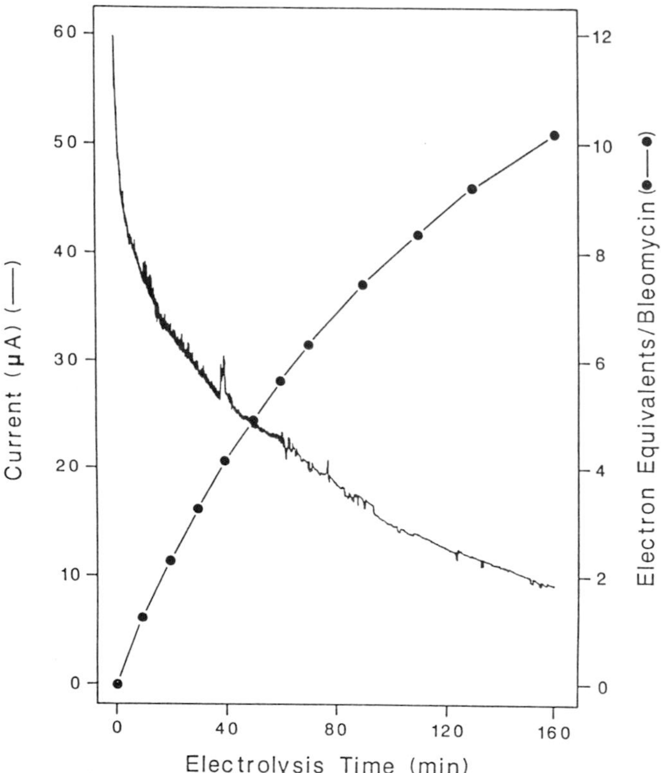

Figure 3: Plot of current (solid line) and total charge (dashed line) for the electrolysis of Fe·BLM A_2 under ambient O_2. Electrolysis was carried out using 0.20 mM Fe(III)·BLM A_2 in 1 ml of 50 mM Na cacodylate, pH 7.2, in a cell containing a 3.14 cm² glassy carbon plate electrode. The background current was negligible (< 2% of the initial current).

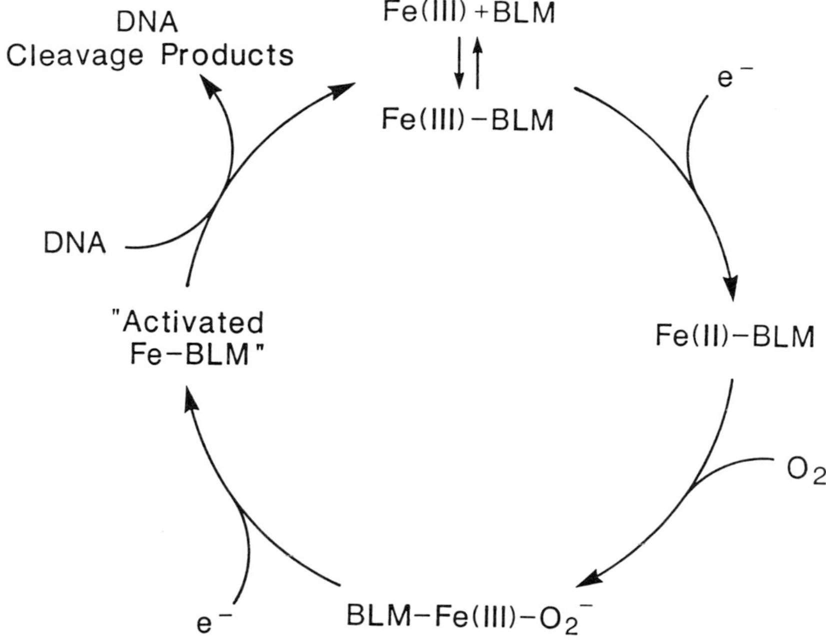

Scheme 2: Proposed catalytic cycle for Fe·BLM activation

During the electrochemical studies that employed DNA (oligonucleotide) substrates, it was noted that the presence of a large excess of such substrates actually diminished the observed current, and the reduction of Fe(III)·BLM to Fe(II)·BLM. This is illustrated in Figure 2 for the cyclic voltammogram of Fe(III)·BLM obtained under anaerobic conditions in the presence of 1 mM and 2 mM (nucleotide concentrations of) calf thymus DNA. This suggested that Fe(III)·BLM was largely bound to DNA under these experimental conditions, and that the bound Fe(III)·BLM was reduced poorly, if at all. Parallel observations have been made in other BLM activating systems,[17–19] suggesting that this was not due simply to physical exclusion of the Fe(III)·BLM from the electrode surface. More surprising was the additional finding that high concentrations of DNA also inhibited the reductive activation of preformed Fe(II)·BLM under aerobic conditions.

The accumulated data are consistent with the catalytic cycle for BLM activation outlined in Scheme 2. This scheme posits a one-electron reduction of Fe(III)·BLM to afford Fe(II)·BLM, followed by O_2 binding and further reduction of the initially formed ternary complex. The foregoing data, which demonstrate inhibition of activation at two stages by exogenous DNA, argue that Fe·BLM activation must precede DNA binding. Clearly, this facet of the scheme places important mechanistic constraints on other BLM mediated processes, *eg* the way in which Fe·BLM produces double-strand DNA lesions with high efficiency.

3 DEGRADATION OF RNA BY FE·BLM

Although several metallobleomycins mediate reasonably efficient degradation of double-stranded B-form DNA, a few different lines of evidence have suggested that RNA is not a good substrate for degradation. These include an early study in which tRNA was shown to afford poor protection to SV40 DNA against cleavage by Fe·BLM.[20] The same authors reported that both single-stranded and double-stranded RNA homopolymers afforded only limited quenching of the bithiazole moiety upon admixture of Cu·BLM A_5, although a recent report indicated that tRNA did quench bithiazole fluoresence at concentrations not dramatically different from DNA.[21] Two laboratories have reported that BLM cleaves only the DNA strand of DNA-RNA hybrids.[22,23] Finally, a recent study[21] demonstrated limited degradation of tRNA's at very high concentrations of added Fe·BLM; the detection of free adenine and uracil was also indicated, although no quantification was reported.

In addition to the foregoing data which suggests that BLM may bind to RNA somewhat less effectively than DNA, or perhaps in an orientation not conducive to RNA degradation, there is also a potential mechanistic dilemma on one pathway leading to RNA strand scission. Shown in Scheme 3 is the presumptive intermediate formed by Criegee-type rearrangement[24] of hydroperoxynucleoside intermediate **i** (*cf* Scheme 1). For the Fe·BLM-mediated degradation of DNA, it has been shown by the use of radiolabelled substrates that the H lost during the collapse of this intermediate derives exclusively from the C-2' αH.[25] Further, the finding[2,26] that the *trans* base propenals are the primary products of DNA degradation by bleomycin

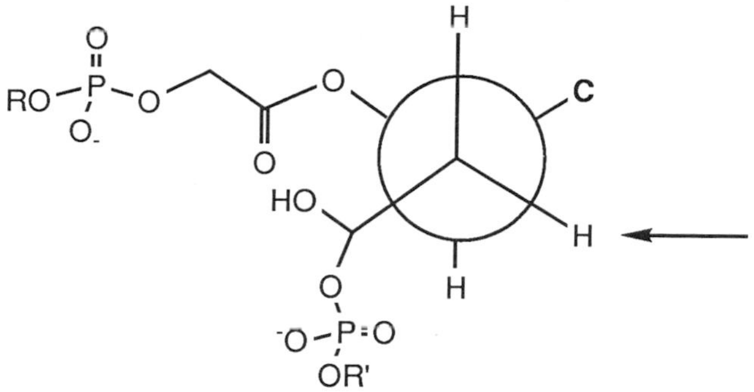

Scheme 3: Formation of a nucleoside 3'-phosphoroglycolate and *trans*-3-(cytosin-1'-yl) propenal *via* a putative *anti* elimination.

argues for the abstraction of C-2' αH in an *anti* elimination process. Were the same mechanistic pathway to obtain in the degradation of RNA by Fe·BLM, the intermediate analogous to the one shown in Scheme 3 could not collapse by the same mechanism as there is no C-2' αH in RNA.

In order to investigate the mechanism of RNA cleavage by Fe·BLM, we initially studied the effects of activated Fe·BLM on a 5'-[^{32}P] labelled *Bacillus subtilis* precursor tRNA^{His}, produced as an RNA transcript from a DNA expression plasmid. Because this molecule was ^{32}P-labelled, alterations of tRNA primary structure involving micromolar concentrations of tRNA's would be readily detectable (whereas damage to unlabelled DNA or RNA duplexes can be detected readily only in the millimolar range). As shown in Figure 4, this tRNA precursor underwent strand scission in a highly selective fashion; nucleotidc sequence analysis (not shown) indicated that scission had occurred at the base of the acceptor stem. As is evident from the Figure, the tRNA was actually a better substrate under these conditions than a ^{32}P-end labelled DNA fragment. Lanes 13 and 14 illustrate the Fe·BLM-mediated cleavage of a mixture of DNA and RNA. In the absence of 'carrier' polynucleotide, the DNA was completely degraded and the RNA gave the same pattern of degradation as in lane 6. Interestingly, admixture of carrier DNA (lane 13) eliminated the observed cleavage of RNA, but still gave some DNA degradation, suggesting that the binding of RNA by Fe·BLM may limit degradation by this agent.

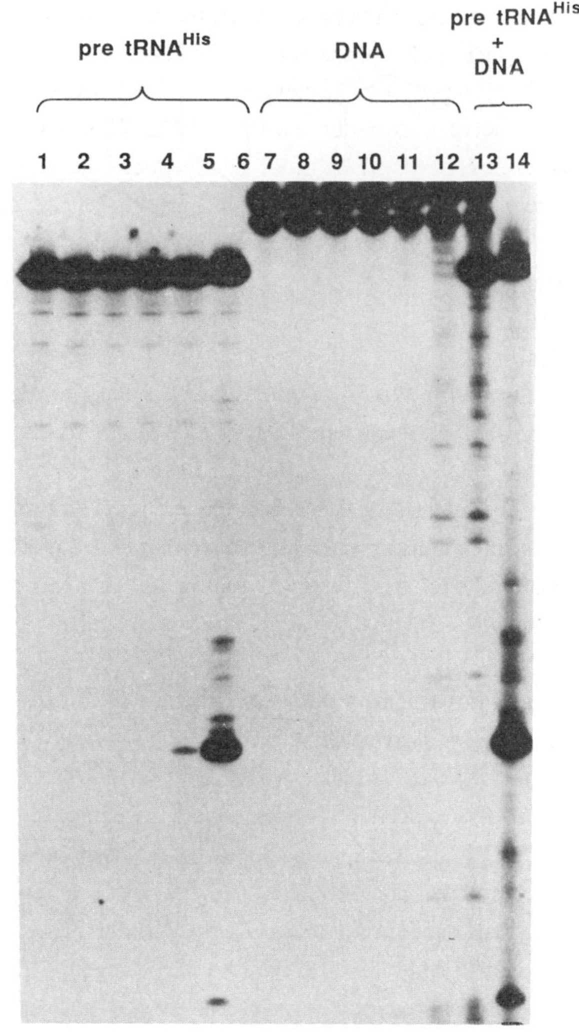

Figure 4: Treatment of (~2.5 μM) precursor tRNA^{His} with Fe(II)·BLM + O_2 in the presence of 3 mM carrier tRNA. The reaction was carried out in 5 mM phosphate buffer, pH 7.0, at 20°C for 1 h using 3 μM (lane 4), 30 μM (lane 5), or 300 μM (lane 6) Fe(II)·BLM A_2. Lanes 10–12 represent the same treatment of a ^{32}P–end labelled DNA containing 3 mM unlabelled carrier DNA. Lanes 13 and 14 contained both RNA and DNA treated with 300 μM Fe(II)·BLM A_2; lane 13 contained 3 mM unlabelled DNA as well. Lanes 1 and 7 contained untreated tRNA and DNA, respectively. Lanes 2 and 8 contained 300 μM BLM A_2; lanes 3 and 9 contained 300 μM Fe(II).

To define the actual chemistry of RNA degradation by Fe·BLM, we tested two specifically prepared heteroduplexes as substrates for activated Fe·BLM. As shown in Scheme 4, these self-complementary DNA duplexes were modelled after d(CGCTAGCG), a highly efficient substrate for Fe·BLM that undergoes degradation almost exclusively at deoxycytidine_3 and deoxycytidine_7. In the present case, deoxycytidine_3 was replaced by *ribo*-cytidine in one octanucleotide, and by *ara*-cytidine in the other. Each of the three substrates in Scheme 4 was treated with Fe(II)·BLM + O_2; the results are quantified in Table 2.

As shown in Table 2, all three of these oligonucleotides were substrates for degradation by Fe·BLM. All three gave quantities of 2'-deoxycytidylyl(3'→5')[2'-deoxyguanosine 3'-(phosphoro-2"-O-glycolate)] ($CpGpCH_2COOH$, *cf* Scheme 1). Thus, Fe·BLM is capable of mediating oxidative destruction of an RNA nucleotide, at least when this nucleotide is a constituent of a DNA molecule known

Table 2: Products resulting from treatment of three oligonucleotides with Fe(II)·BLM[a]

Oligonucleotide	$CpGpCH_2COOH$ (µM)	C propenal (µM)	Cytosine (µM)	5'-GMP (µM)
d(CGCTAGCG)	5	50	24	49
C_3-*ribo*(CGCTAGCG)	2	51	23	54
C_3-*ara*(CGCTAGCG)	30	32	45	40

[a] Reaction mixtures (50 µl total volume) contained oligonucleotide (1.0 mM final nucleotide concentration), 50 mM Na cacodylate, pH 7.2, and 0.2 mM Fe(II)·BLM A_2. The reactions were run at 0 °C for 60 min, and then analysed by C_{18} reverse phase HPLC.

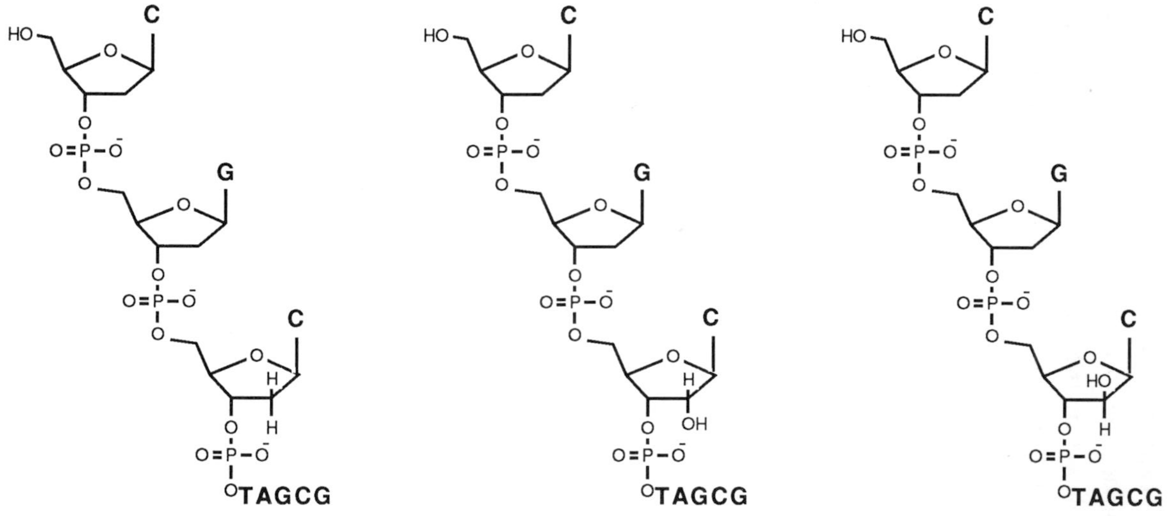

deoxy **CGCTAGCG** ***C_3 - ribo*** **CGCTAGCG** ***C_3 - ara*** **CGCTAGCG**

Scheme 4: Three self–complementary octanucleotides used as substrates for degradation by Fe·BLM

to be a good substrate for Fe·BLM–mediated DNA degradation. The data in the table make it clear that all three oligonucleotides gave approximately the same amount of total products, but that the oligonucleotide containing *ribo*–cytidine at position 3 underwent much greater damage at deoxycytidine$_7$ than at cytidine$_3$. Although the mechanism of degradation of this oligonucleotide at cytidine$_3$ is unclear at present, the fact that the same product is formed from all three oligonucleotides makes it attractive to think that all three utilize species **i** (Scheme 1) as an intermediate. Presumably, collapse of this intermediate in the case of the *ribo*–cytidine–containing oligonucleotide involves either abstraction of C–2' βH, or else direct solvolysis of the C–4' carboxylate ester (*cf* Scheme 3).

REFERENCES

1 H Umezawa, 'Bleomycin: Current Status and New Developments', S K Carter, S T Crooke, and H Umezawa, eds, Academic Press, New York, 1978, p 15ff.

2 S M Hecht, *Acc Chem Res*, 1986, **19**, 383.

3 J Stubbe and J W Kozarich, *Chem Rev*, 1987, **87**, 1107.

4 S Uesugi, T Shida, M Ikehara, Y Kobayashi, and Y Kyogoku, *Nucleic Acids Res*, 1984, **12**, 1581.

5 H Sugiyama, R E Kilkuskie, S M Hecht, G A van der Marel, and J H van Boom, *J Am Chem Soc*, 1985, **107**, 7765.

6 H Sugiyama, C Xu, N Murugesan, and S M Hecht, *J Am Chem Soc*, 1985, **107**, 4104.

7 H Sugiyama, C Xu, N Murugesan, S M Hecht, G A van der Marel, and J H van Boom, *Biochemistry*, 1988, **27**, 58.

8 C–H Chang and C F Meares, *Biochemistry*, 1984, **23**, 2268.

9 I Saito, T Morii, H Sugiyama, T Matsuura, C F Meares, and S M Hecht, *J Am Chem Soc*, 1989, **111**, 2307.

10 H Kuramochi, K Takahashi, T Takita, and H Umezawa, *J Antibiot*, 1981, **34**, 576.

11 R M Burger, J Peisach, and S B Horwitz, *J Biol Chem*, 1981, **256**, 11636.

12 R M Burger, J S Blanchard, S B Horwitz, and J Peisach, *J Biol Chem*, 1985, **260**, 15406.

13 R B Van Atta, E C Long, S M Hecht, G A van der Marel, and J H van Boom, *J Am Chem Soc*, 1989, **111**, 2722.
14 D L Melnyk, S B Horwitz, and J. Peisach, *Biochemistry*, 1981, **20**, 5327.
15 K Miyoshi, T Kikuchi, T Takita, S Murato, and K Ishizu, *Inorg Chim Acta*, 1988, **151**, 45.
16 H Sugiyama, R E Kilkuskie, L-H Chang, L-T Ma, S M Hecht, G A van der Marel, and J H van Boom, *J Am Chem Soc*, 1986, **108**, 3852.
17 R M Burger, J Peisach, W E Blumberg, and S B Horwitz, *J Biol Chem*, 1979, **254**, 10906.
18 J-P Albertini, A Garnier-Suillerot, and L Tosi, *Biochem Biophys Res Commun*, 1982, **104**, 557.
19 M R Ciriolo, R S Magliozzo, and J Peisach, *J Biol Chem*, 1987, **262**, 6290.
20 M Hori, 'Bleomycin: Chemical, Biochemical, and Biological Aspects', S M Hecht, ed, Springer-Verlag, New York, 1979, p 195ff.
21 R S Magliozzo, J Peisach and M R Ciriolo, *Mol Pharmacol*, 1989, **35**, 428.
22 C W Haidle and J Bearden, *Biochem Biophys Res Commun*, 1975, **65**, 815.
23 C R Krishnamoorthy, D E Vanderwall, J W Kozarich, and J Stubbe, *J Am Chem Soc*, 1988, **110**, 2008.
24 I Saito, T Morii, and T Matsuura, *Nucleic Acid Symp Ser*, 1983, **12**, 95.
25 J C Wu, J W Kozarich, and J Stubbe, *J Biol Chem*, 1983, **258**, 4694.
26 N Murugesan, C Xu, G M Ehrenfeld, H Sugiyama, R E Kilkuskie, L O Rodriguez, L-H Chang, and S M Hecht, *Biochemistry*, 1985, **24**, 5735.

THE INFLUENCE OF BASE-PAIR MISMATCHES AND MODIFIED BASES ON THE STRUCTURE AND STABILITY OF THE DNA DUPLEX

Tom Brown, Ewan D Booth, and Gordon A Leonard

Department of Chemistry, University of Edinburgh, Edinburgh EH9 3JJ, UK

1 INTRODUCTION

The accurate transmission of genetic information is essential in maintaining the integrity of living organisms. Point mutations can arise in DNA as a result of base mispairing during replication and a complex group of enzymes has evolved to recognise and remove these mistakes.[1] These proofreading and repair enzymes detect some base-pair mismatches more efficiently than others and certain modified bases give rise to incorrect pairs which are particularly evasive. Such mutagens, which are produced by irradiation or chemical damage to the four natural bases, may give rise to genetic defects and cancer. A number of theories have been proposed to explain the observed frequency of substitution mutations,[2] but until recently no detailed information was available on the precise conformations of mismatched base pairs in DNA. In the late 1970s the development of reliable methods for the synthesis and purification of deoxyoligonuclcotides made it possible to contemplate structural studies on these materials by *X*-ray crystallography and high-field NMR methods. In this chapter we discuss some recent results obtained in this field in our laboratory.

2 THE A.C MISMATCH

An *X*-ray crystallographic analysis of the A.C mismatch in the self-complementary B-DNA dodecamer, d(CGC*A*AATT*C*GCG) at 2.5 A resolution had previously shown the presence of a wobble base

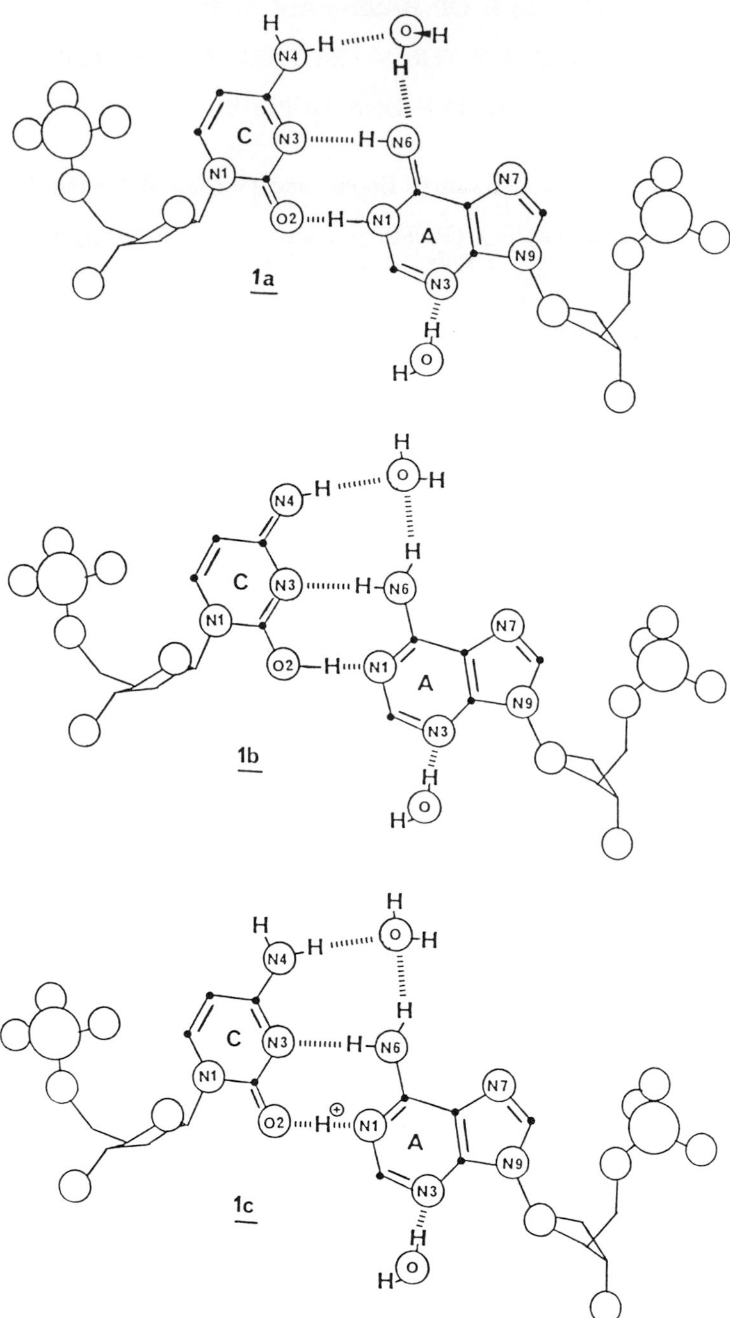

Figure 1: Possible forms of the A.C mismatch: (**1a**) C.A(imino); (**1b**) C(enol, imino).A; (**1c**) $C.A^{+}$.

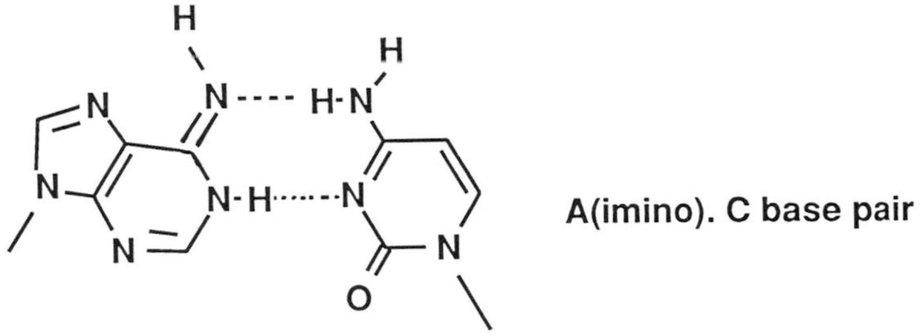

A(imino). C base pair

Figure 2: A minor tautomer C.A mismatch isomorphous with a Watson–Crick base pair.

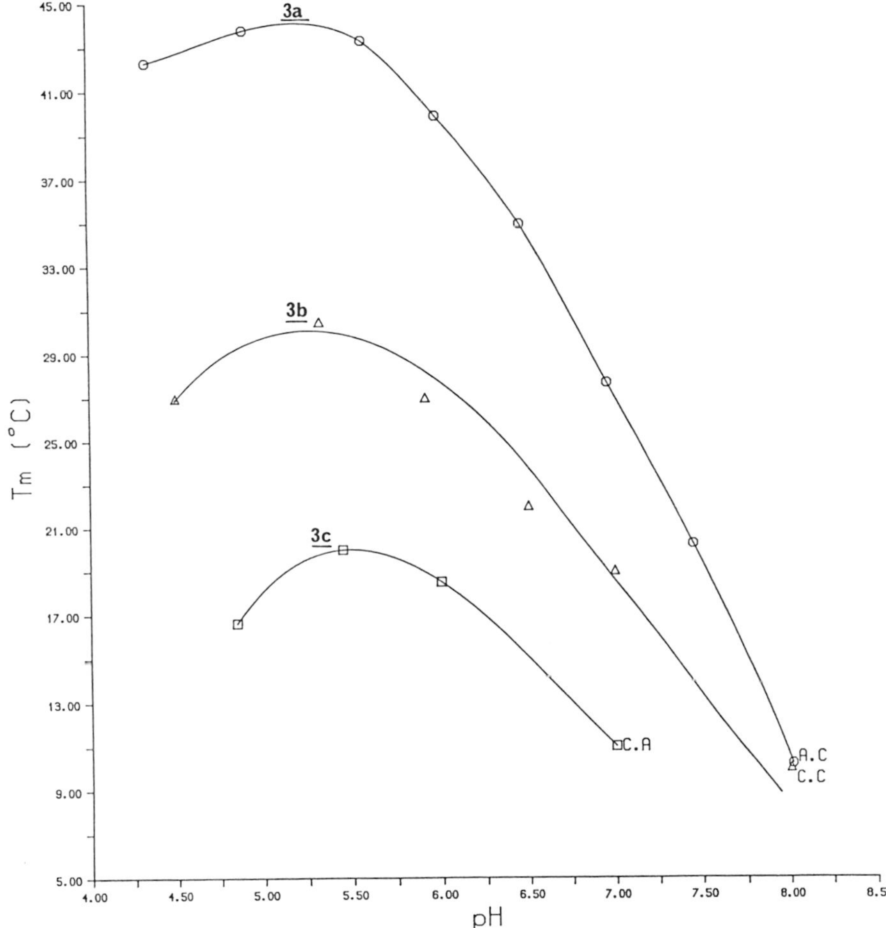

Figure 3: pH–dependence of melting temperature in 0.1M phosphate buffer: (**3a**) d(CGC*A*AATT*C*GCG); (**3b**) d(CGC*C*AATT*C*GCG); (**3c**) d(CGC*C*AATT*A*GCG).

pair.[3] At this resolution it is not possible to identify hydrogen atoms, so the crystal structure is consistent with any of the three A.C mispairs shown in Figure 1. The base pairs (1a) and (1b) seem unlikely as in each case one base is present as a rare tautomer. In particular, the occurrence of the imino tautomer of adenine (1a) might be expected to produce a Watson–Crick base pair (Figure 2) rather than a wobble base pair. The existence of the protonated base pair (1c) is also uncertain as the crystals were grown in sodium cacodylate buffer at pH 7, far from the pK_a for protonation of adenine N1 (*ca* 4.1). In order to resolve the situation we measured the melting temperature of the A.C dodecamer as a function of pH (Figure 3a). The results clearly show a correlation between duplex stability and hydrogen ion concentration, strongly supporting base pair (1c). At high pH the duplex is very unstable and in the absence of a high proton concentration it is likely that A.C base pairs do not form. Interestingly, the presence of a protonated base does not seem to destabilise the duplex and the A.C dodecamer is more stable than the native sequence around pH 5.

The correlation between pH and duplex stability for the sequence d(CGC*C*AATT*A*GCG) was also examined. This sequence relates to the previous one by inverting the A.C mismatches at positions 4 and 9, and this changes the base stacking environment. Again, the duplex becomes more stable as the pH is lowered until maximum stability is reached just above pH 5. Beyond this point protonation of cytosine N3 and adenine N1 atoms destabilises all Watson–Crick base pairs in the usual way (Figure 3c). The lower melting temperature of the C.A dodecamer relative to the A.C dodecamer might be due to the occurrence of an unfavourable dipole moment contribution to the CpC base stacking step in the former, compared to a favourable CpG step in the latter. This is speculative in the absence of accurate theoretical calculations.

3 THE G.A MISMATCH

The G.A mismatch is of particular interest as experiments with DNA polymerase III have shown that about 10% of these mismatches avoid detection, whereas only 0.5% of G.T mismatches escape.[4] The G.A mismatch has been shown to be conformationally variable[5–8] and both G(*anti*).A(*syn*) and G(*anti*).A(*anti*) base pairs have been identified in different DNA sequences at neutral pH (Figure 4b, 4a). Theoretical calculations suggest that both have similar stability when incorporated in a DNA duplex although the G(*anti*).A(*anti*) base pair causes greater distortion to the sugar–phosphate backbone.[9,10] It has recently been shown by nuclear magnetic resonance that the G.A mismatch in the sequence d(CGGGAATTCACG) displays conformational flexibility as a function of pH.[11] The G(*anti*).A(*anti*) base pair (**4a**) is observed at neutral pH, whereas the G(*syn*).A(*anti*) base pair (**4c**) predominates below pH 5.5. In the latter case the adenine base is protonated.

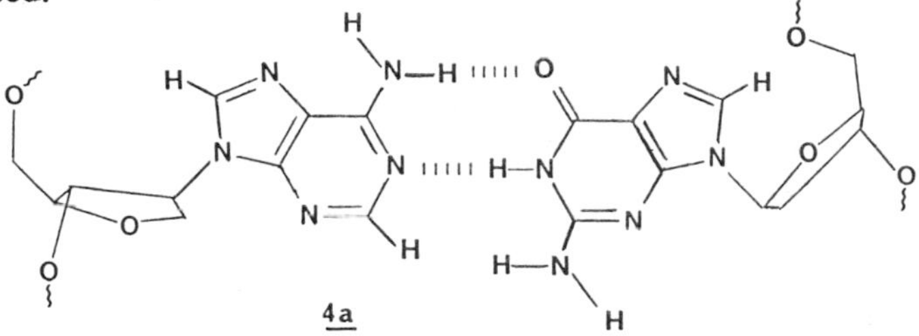

4a

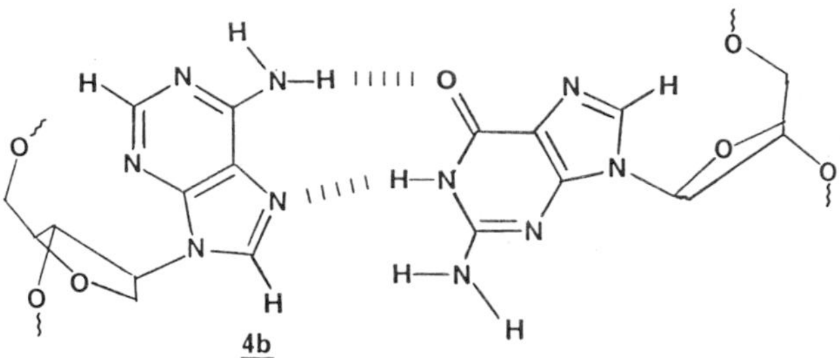

4b

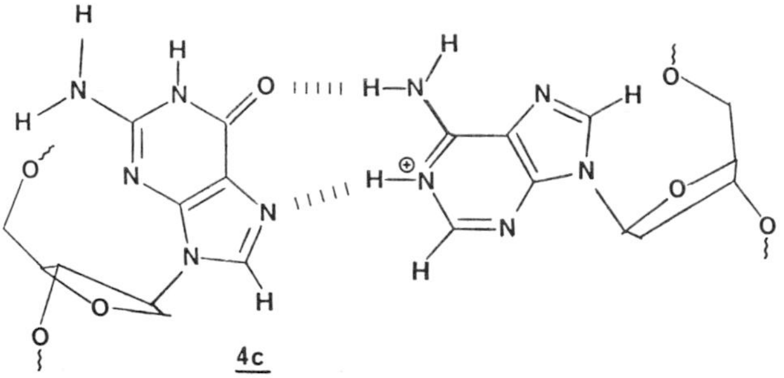

4c

Figure 4: Possible forms of the G.A mismatch:
(**4a**) A(*anti*).G(*anti*); (**4b**) A(*syn*).G(*anti*); (**4c**) G^{+}(*syn*).A(*anti*).

Our earlier crystallographic analysis of a G.A mismatch in the sequence d(CGC*G*AATT*A*GCG) showed the presence of a G(*anti*).A(*syn*) pairing (Figure 4b).[7] Very recently we have solved the structure of the dodecamer d(CGC*A*AATT*G*GCG) which is related to the previous sequence by interchanging the adenine and guanine bases of the G.A mismatches.[12] As a consequence of this change the mismatches are in a different base stacking environment. Clearly base stacking interactions must be influential in determining the form of conformationally flexible mismatches. Crystals were grown at pH 6.6 and *X*-ray data were measured at 4°C on a Stoe-Siemens AED2 four circle diffractometer equipped with a long arm and a helium path. Intensities were corrected for Lorenz and polarisation factors, absorption and time dependent decay. A total of 2262 unique reflections with F≥2 sigma(F) in the range 8A to 2.25A were used in the refinement. Starting co-ordinates were those of the native dodecamer d(CGCGAATTCGCG) and the model was refined using the restrained least squares methods[13,14] with periodic examination of F_o-F_c maps on an Evans and Sutherland PS300 graphics system. During these calculations the mismatches were allowed to move freely relative to each other. It was clear from the fragment F_o-F_c electron density maps calculated with the mismatched base pairs omitted, that G(*syn*).A(*anti*) gave a much better electron density fit than either G(*anti*).A(*anti*) or G(*anti*).A(*syn*) [Figure 5].

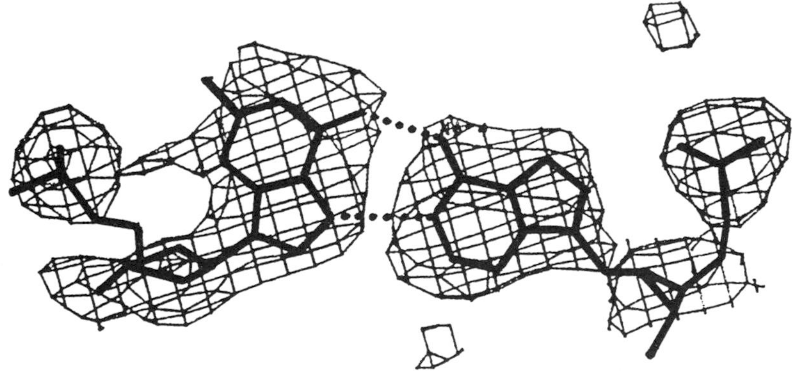

Figure 5: Fragment F_o-F_c electron density map of the G(*syn*).A(*anti*) base pair in the dodecanucleotide duplex d(GCG*A*AATT*G*GCG). All atoms have been omitted from the structure factor calculations to give an unbiased representation of the true electron density.

The refinement converged at R = 0.16 with the location of 94 solvent molecules. Figure 4c shows the G(*syn*).A(*anti*) base pair schematically. The N6 atom of adenine and the O6 atom of guanine are 2.6Å apart and the N1 atom of adenine and the N7 atom of guanine are separated by 2.8Å. These distances are indicative of strong hydrogen bonds, but with both bases in their major tautomeric forms a proton is required to enable the N1-adenine N7-guanine hydrogen bond to form. There is good evidence from thermal denaturation studies that the mismatch base pair is indeed protonated. Ultraviolet melting studies in the pH range 8.0 to 4.5 on the A.G dodecamer d(CGC*A*AATT*G*GCG) [Figure 6b] and the native control sequence d(CGCGAATTCGCG) [Figure 9a] show that the behaviour of the two duplexes in aqueous media is strikingly different (Figure 6). The melting temperature of the native sequence decreases sharply below pH 6.5, whereas that of the A.G dodecamer reaches a maximum around pH 5.0 before falling away at lower pH. These results suggest that the G(*syn*).A(*anti*) base pair is present in solution over a wide pH range. It can, however, be concluded that at high pH other species are present as the melting temperature does not change between pH 7 and 8.

In the G(*syn*).A(*anti*) base pair the 2-amino group of guanine lies in the major groove where it is free to hydrogen bond to two water molecules. This is not the case for G(*anti*).A(*syn*) and G(*anti*).A(*anti*) base pairs (Figure 4b and 4a, respectively) in which the 2-amino group is sterically crowded in the minor groove and can only interact with a single water molecule. This will lead to a loss of hydrogen bonding when the fully hydrated single strands come together to form a duplex. The disposition of major groove hydrogen bond acceptors and donors (heteroatoms) in the A(*anti*).G(*syn*) base pair is completely different from that of other forms of the G.A mismatch although in general shape it resembles the A(*syn*).G(*anti*) mismatch. These observations may be important when considering the recognition of mismatches by proofreading and repair enzymes. In the sequences d(CGC*A*AATT*G*GCG) and d(CG*G*GAATTC*A*CG), both of which have been shown to accommodate G(*syn*).A(*anti*) base pairs, the guanine base of the mismatch is flanked by the guanine of a G.C base pair. Guanine has a large dipole moment (7.5D)[15] and a GpG step in B-DNA involving G(*anti*) bases would be expected to have an

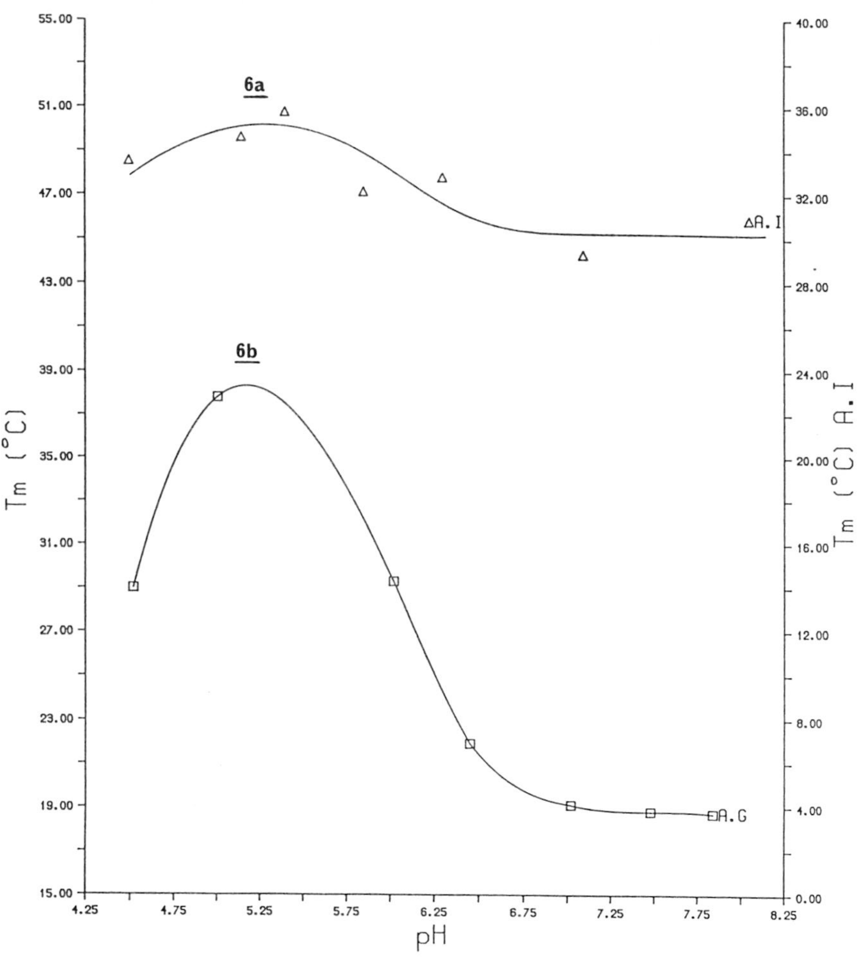

Figure 6: pH–dependence of duplex melting in 0.1M phosphate buffer. **(6a)**: d(CGC***A***AATT***I***GCG), temperature scale on right hand side. **(6b)**: d(CGC***A***AATT***G***GCG), temperature scale on left hand side.

unfavourable dipole–dipole contribution as the dipole moments of the two bases would be unfavourably aligned, with a relative rotation of only *ca* 33°. However, with the guanine base of the G.A mismatch in the *syn*-conformation, the static dipole–dipole interactions are likely to be much more favourable. In the sequences d(CG*A*GAATTC*G*CG) and d(CGC*G*AATT*A*GCG) in which the guanine bases of the G.A mismatches are *anti* [G(*anti*).A(*anti*) and

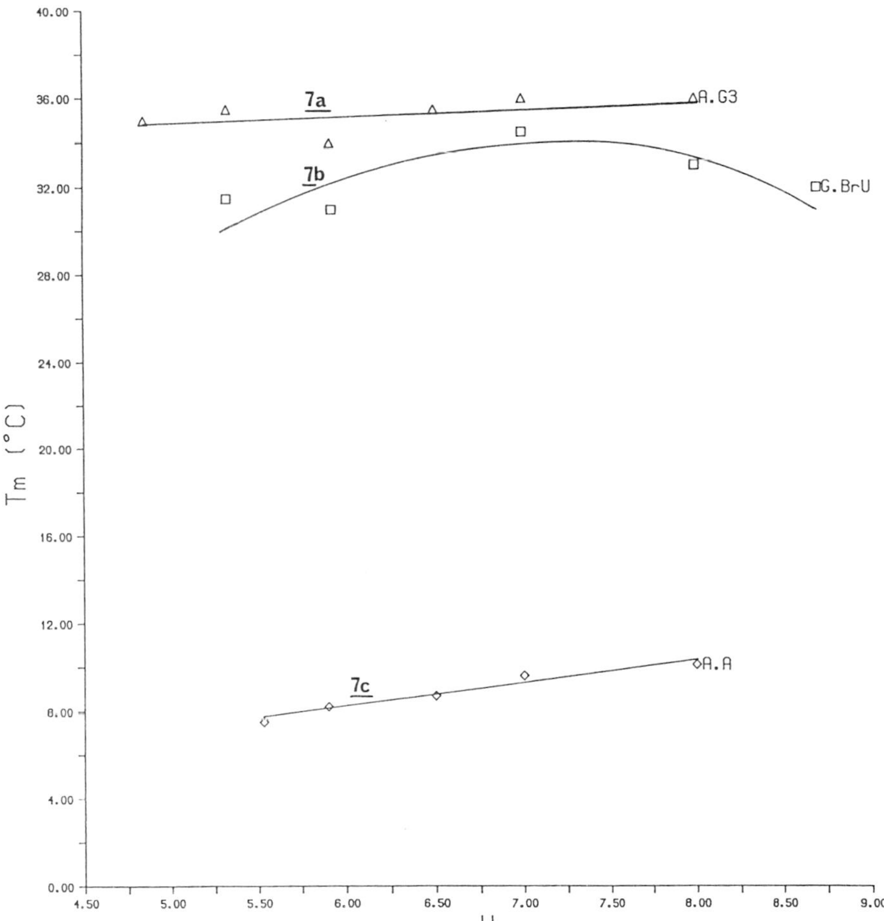

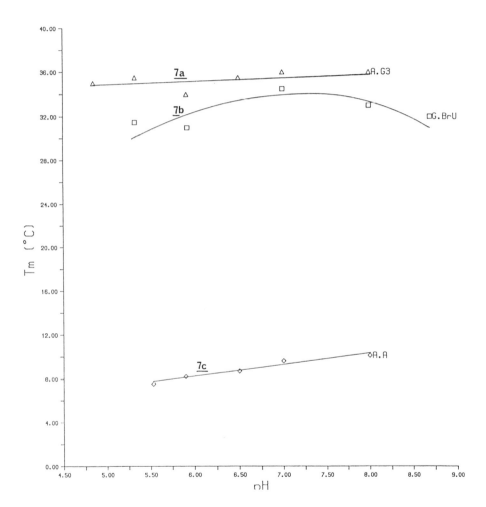

Figure 7: pH dependence of duplex melting in 0.1M phosphate buffer. (**7a**) d(CG*A*GAATTC*G*CG); (**7b**) d(CGC***G***AATT***BrU***GCG); (**7c**) d(CGC*A*AATT*A*GCG).

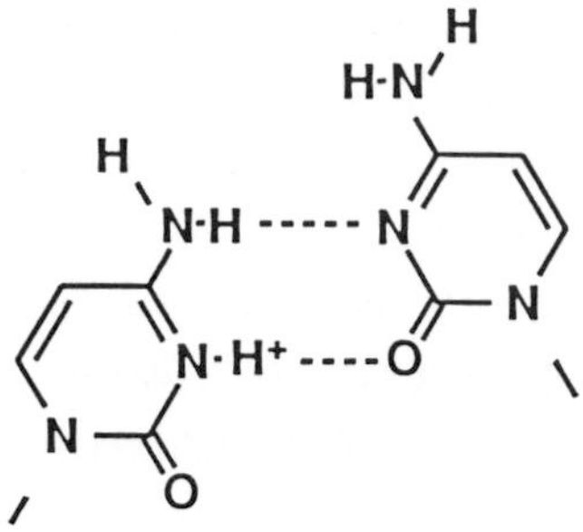

8a $C^+.C$ mismatch.

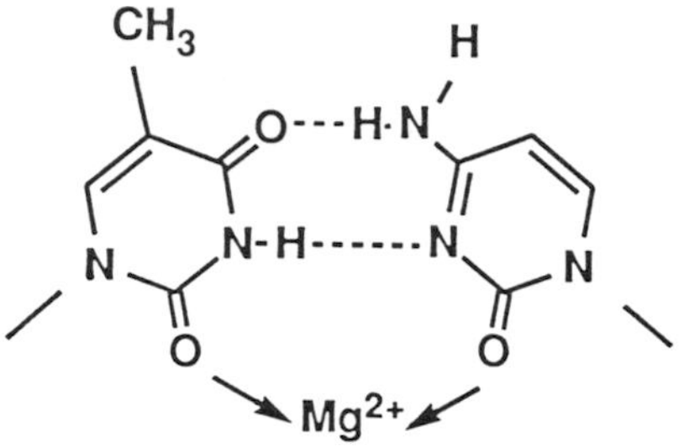

8b T.C mismatch?

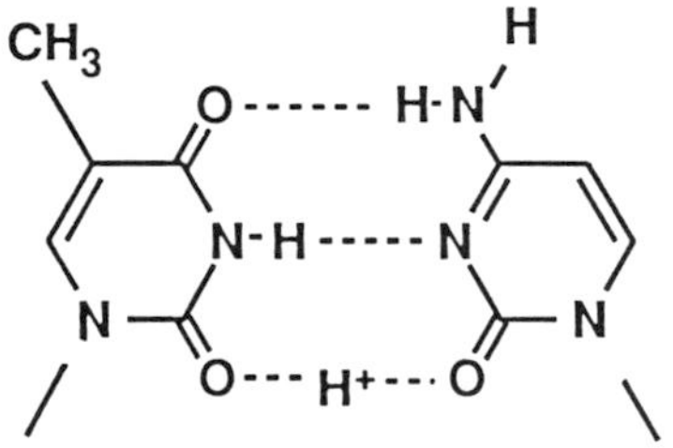

8c T.C mismatch?

Figure 8: (**8a**): Putative $C^+.C$ mismatch; (**8b**) and (**8c**): Possible forms of the T.C mismatch.

G(*anti*).A(*syn*), respectively] the guanines are involved in CpG stacking interactions. The dipole moment of cytosine is almost identical to that of guanine (7.6D) and when on the 5'–side of guanine in B–DNA it lies in almost exactly the opposite direction. Thus, the CpG(*anti*) base stack will have a favourable static dipole contribution. The dipole moment of adenine is relatively small so its influence on the G.A mismatch will be less than that of guanine. Dipole moments may be particularly important in the two *single* strands as they come together and intrastrand stacking interactions can be optimised, locking the G.A mismatch into a specific conformation as the duplex is formed. Once the duplex *is* formed, any change in the conformation of the G.A mismatch will require rotation of a purine base around its glycosidic bond, an operation that would considerably disrupt the surrounding Watson–Crick base pairs. It is well known that poly d(CG) forms a more stable duplex than poly dG.poly dC[16] and that the CpG base stacking step is the most stable in B–DNA,[17] so our results with mismatches are consistent with these observations. Thermal denaturation studies on the dodecamer d(CG*A*GAATTC*G*CG) in our laboratory show no indication of increased stability with decreasing pH (Figure 7a), suggestion that the guanine base of the G.A mismatch has no tendency to adopt the *syn*–conformation in this sequence. This observation is consistent with the prediction that the *anti*–conformation should be strongly preferred for the mismatched guanine base when it is on the 3'–side of a cytosine.

4 OTHER MISMATCHES

Any mismatch containing either adenine or cytosine bases can in principle be stabilised by protonation near physiological pH as the N1–atom of adenine and the N3–atom of cytosine both have pK_a values around 4. Thermal denaturation studies on the dodecamer d(CGC*A*AATT*A*GCG) indicate that in this sequence the A.A. mismatch is not stabilised by protonation (Figure 7c). Due to a lack of high resolution crystallographic data we are unable to speculate on the precise form of the A.A mismatch. So far we have been unable to grow suitable crystals of a DNA duplex containing pyrimidine–pyrimidine base pairs and no crystallographic information is available. It is likely that these mismatches will not span the same distance across the duplex as Watson–Crick base pairs so their incorporation into B–DNA will lead to some distortion of the sugar–phosphate backbone. In general pyrimidine–pyrimidine base pairs substantially destabilise the duplex and the strong pH dependence of the stability of the C.C mismatch in the sequence d(CGC*C*AATT*C*GCG) (Figure 3b) is consistent with the protonated base pair in Figure 8a. The T.C mismatch in the dodecamer d(CGC*T*AATT*C*GCG) shows no such pH dependent stability (Figure 9b) so the base pair (8c) is unlikely to predominate. An alternative base pair that is not stabilised by protonation is illustrated in Figure 8b.

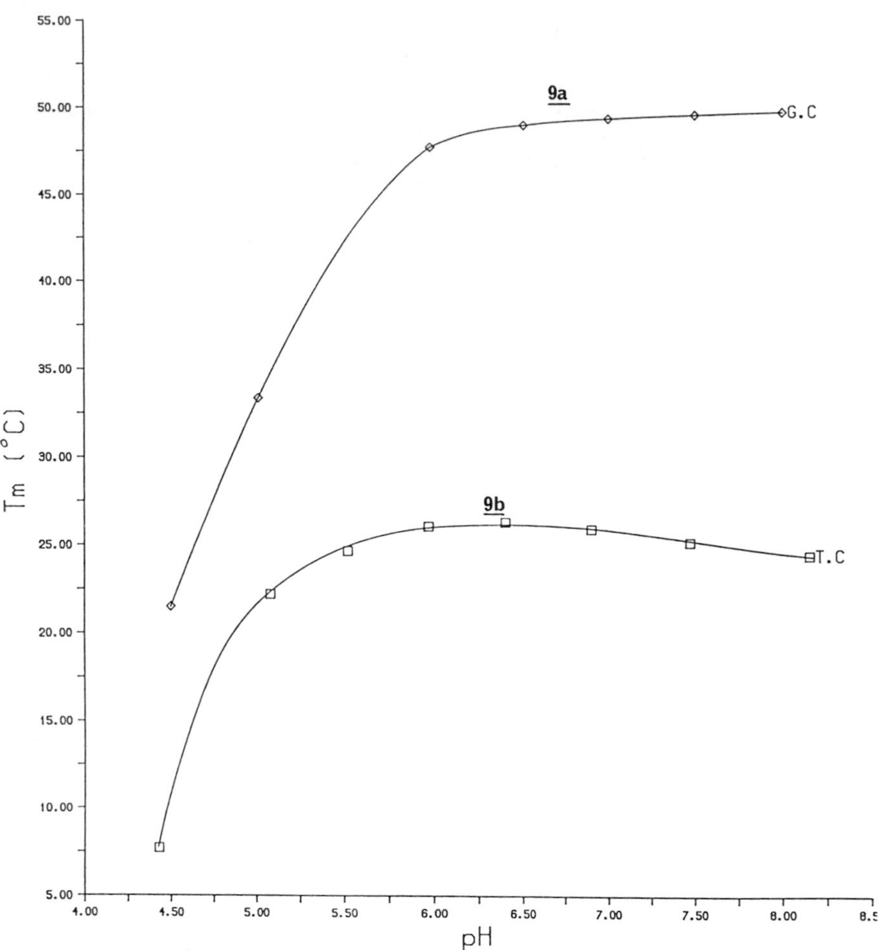

Figure 9: pH-dependence of duplex melting in 0.1M phosphate buffer: **(9a)** d(CGCGAATTCGCG); **(9b)** d(CGC*T*AATT*C*GCG).

5 5-BROMO-URACIL

5-Bromo-uracil is an analogue of thymine that has an enhanced tendency to form mismatches with guanine.[18] The stability of the G.BrU mismatch could be due to any of the following:

(i) A stable wobble base pair **(10a)** with the bromine atom involved in favourable base stacking interactions.

(ii) The formation of **(10b)**, a Watson–Crick–like base pair with the deprotonated N3–atom of 5–bromo–uracil acting as a hydrogen bond acceptor.

(iii) The occurrence of **(10c)**, a Watson–Crick–like base pair involving a rare enol tautomer of 5–bromo–uracil.

Although **(10b)** and **(10c)** would be indistinguishable at the limit of resolution of a typical oligonucleotide crystal structure, **(10c)** is unlikely to occur, as there are no grounds for proposing a very high ratio of the enol form of 5–bromo–uracil. However, as the pK_a for deprotonation of the N3 position of 5–bromo–uracil is 8.5, compared to a pK_a of 10 for thymine, base pair **(10b)** might well occur. In order to attempt to identify this species we crystallised the hexamer d(***BrU***GCGC***G***) at pH 8.5 where 5–bromo–uracil should be 50% ionised. The hexamer crystallised in the Z–form and the structure was solved to a resolution of 2.2Å.[19] The Z–DNA conformation was deliberately chosen because of the high diffracting power of crystals

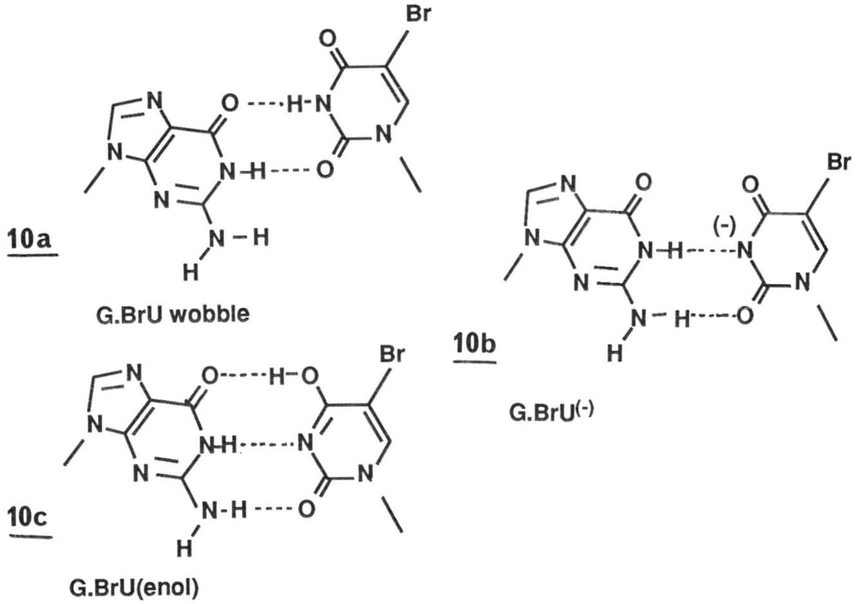

Figure 10: Possible G.BrU mismatches: **(10a)** G.BrU(wobble); **(10b)** G.BrU$^-$; **(10c)** G(enol).BrU or G.BrU(enol).

of this form. Crystallisation of oligonucleotides is impeded at high pH so we were forced to work with small crystals. Analysis of the structure revealed that the wobble base pair **(10a)** was the only detectable species. Ultraviolet melting studies (Figure 7b) of the sequence d(CGC***G***AATT***BrU***GCG), a self-complementary dodecanucleotide containing two G.BrU mismatches were consistent with this finding. There was no significant change in the melting temperature of the duplex between pH 5.5 and 8.5 even though the formation of **(10b)** might be expected to be more favourable at high pH. If this base pair *were* to form it would be destabilised by the close contact between the O6 atom of guanine and the O4 atom of BrU. This unfavourable interaction could only be stabilised at *very low* pH by protonation of the carbonyl oxygen of either base. Therefore conditions suitable for the stabilisation of the guanine O6-bromo-uracil O4 hydrogen bond are totally unsuitable for the formation of the guanine N1-bromo-uracil N3 hydrogen bond. It has been shown by NMR that the presence of the electron withdrawing bromine atom in the 5-position of uracil strengthens the hydrogen bonding capacity of the O2-atom and weakens the hydrogen bonding capacity of the O4-atom.[20] Clearly this would stabilise the G.BrU wobble base pair relative to the A.BrU 'Watson-Crick' base pair and produce a mutagenic effect.

6 DEOXYINOSINE

Inosine is an analogue of guanosine which lacks the 2-amino group. It occurs in ribonucleic acids, especially in the wobble position in some transfer RNA anticodons, where it is known to base pair with A, C or U at the messenger RNA codon.[21] It is also able to occupy the middle position of the anticodon and pair with A.[22] The occurrence of deoxyinosine in genomic DNA is much less common, although it is occasionally produced by the deamination of deoxyadenosine. As it is potentially mutagenic[23] it is efficiently removed by the enzyme hypoxanthine DNA glycosylase.[24–26] Because of its tendency to form stable mismatches, deoxyinosine has been used as a universal base in synthetic hybridisation probes.[27–29]

In order to investigate the precise nature of the I.A pairing in a

DNA helix we carried out an X-ray structural analysis on the synthetic dodecamer d(CGC*I*AATT*A*GCG) to 2.5Å resolution.[30] This analysis established that an I(*anti*).A(*syn*) base pair (Figure 11c) can be accommodated in the B-DNA duplex without any major distortion of the local or global conformation. The structure is essentially isomorphous with that of d(CGC*G*AATT*A*GCG) which contains two G(*anti*).A(*syn*) mispairs (Figure 11d), but because inosine lacks a 2-amino group there is no possibility of a steric clash in the minor groove. This is likely to be the origin of the greater stability of I.A mismatches and it can be seen from Figure 6 that A.I mismatches are significantly more stable than G.A mismatches around neutral pH. We are currently crystallising the dodecamer d(CGC*A*AATT*I*GCG).

An analogous comparison was made between the I.T and G.T mismatches in the A-DNA octamers d(GG*G*GC*T*CC)[31] and d(GG*I*GC*T*CC),[32] respectively. A high resolution X-ray analysis[32] on the I.T octamer (1.7Å) showed the presence of wobble base pairs (Figure 11a) similar to those found previously (Figure 11b) for the G.T mismatches.[31,33,34] As the 2-amino group of guanine in the G.T mismatch is not involved in any unfavourable steric clashes it is not surprising that the G.T and I.T mismatches have almost identical stability.

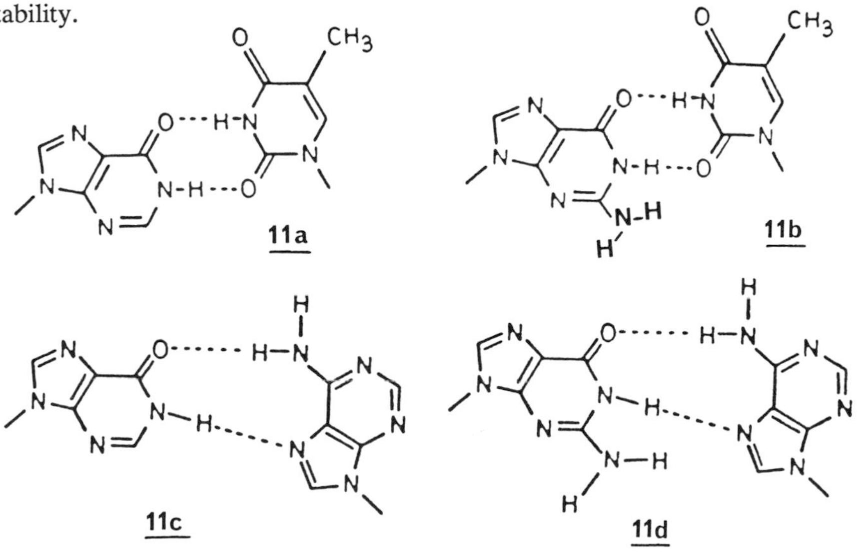

Figure 11: A comparison of mismatches of guanine and of hypoxanthine: (**11a**): I.T(wobble); (**11b**): G.T(wobble); (**11c**): I.A(*syn*); (**11d**): G.A(*syn*).

7 SUMMARY

A combination of X-ray crystallography and ultraviolet melting studies have been used to examine the structure and stability of DNA duplexes containing mismatched base pairs and modified bases. Base stacking interactions clearly have an influence on mismatch conformation and we are currently measuring thermodynamic parameters of a wide range of duplexes in order to gain a clearer insight into the factors influencing the stability of abnormal base pairs in DNA.

ACKNOWLEDGEMENTS

We are grateful to Olga Kennard, Bill Hunter and Geoff Kneale for helpful discussions. This work was funded by SERC Molecular Recognition grants, and Awards from the Research Corporation Trust and the Nuffield Foundation to TB.

REFERENCES

1 E C Friedberg, 'DNA Repair', W H Freeman and Co, New York, 1985.
2 M D Topal and J R Fresco, *Nature*, 1976, **263**, 285.
3 W N Hunter, T Brown, N N Anand, and O Kennard, *Nature* 1986, **320**, 552.
4 A R Fersht, J W Knill-Jones, and W C Tsui, *J Mol Biol*, 1982, **156**, 37.
5 L-S Kan, S Chandrasegaran, S M Pulford, and P S Miller, *Proc Nat Acad Sci USA*, 1983, **80**, 4263.
6 D J Patel, S A Koslowski, S Ikuta, and K Itakura, *Biochemistry*, 1984, **23**, 3207
7 T Brown, W N Hunter, G Kneale and O Kennard, *Proc Nat Acad Sci USA*, 1986, **83**, 2402.
8 G G Prive, U Heinemann, L-S Kan, S Chandrasegaran, and R E Dickerson, *Science*, 1987, **238**, 498.
9 V P Chuprina and V I Poltev, *Nucleic Acids Res,* 1983, **11**, 5205.

10 J W Keepers, P Schmidt, T L James, and P A Kolman, *Biopolymers*, 1984, **23**, 2901.
11 X Gao and D J Patel, *J Am Chem Soc*, 1988, **110**, 5178.
12 T Brown, G A Leonard, E D Booth, and J Chambers, *J Mol Biol*, 1989, **207**, 455.
13 W A Hendrickson and J H Konnert, 'Biomolecular Structure, Conformation, Function and Evolution', Plenum Press, 1981, pp43–57.
14 E Westhof, P Dumas, and D Moras, *J Mol Biol*, 1985, **184**, 119.
15 B Pullman and A Pullman, *Adv Het Chem*, 1971, **13**, 77
16 O Gotoh and Y Takashira, *Biopolymers*, 1981, **20**, 1033.
17 R L Ornstein, R Rein, D L Breen, and R D MacElroy, *Biopolymers*, 1978, **17**, 2341.
18 E Freeze, *J Mol Biol*, 1959, **1**, 87.
19 T Brown, G Neale, W N Hunter, and O Kennard, *Nucleic Acids Res*, 1986, **14**, 1801.
20 H Iwahashi and Y Kyogoku, *J Am Chem Soc*, 1977, **99**, 7761.
21 F H C Crick, *J Mol Biol*, 1966, **19**, 548.
22 B D Davis, P Anderson, and P F Sparking, *J Mol Biol*, 1973, **76**, 223.
23 T Lindahl, *Prog Nucleic Acids Res Mol Biol*, 1979, **22**, 135.
24 P Karran and T Lindahl, *J Biol Chem*, 1978, **253**, 5877.
25 B Myrnes, P–H Guddal, and H Krokan, *Nucleic Acids Res*, 1982, **10**, 3693.
26 T Lindahl and P Karran, *Biochemistry*, 1980, **19**, 6005.
27 Y Takahashi, K Kato, Y Hayashizaka, T Wakabayashi, E Ohtsuka, S Matsuki, I Ikehara, and K Matsubara, *Proc Nat Acad Sci USA*, 1985, **82**, 1931.
28 E Ohtsuka, S Matsuki, M Ikehara, Y Takahashi, and K Matsubara, *J Biol Chem*, 1985, **260**, 2605.
29 F H Martin, M M Castro, F Aboul–ela, and I Tinoko, *Nucleic Acids Res*, 1985, **13**, 8927.
30 P W R Corfield, W N Hunter, T Brown, P Robinson, and O Kennard, *Nucleic Acids Res*, 1987, **15**, 7935.
31 T Brown, O Kennard, G Kneale, and D Rabinovich, *Nature*, 1985, **315**, 604.
32 W B T Cruse, W Aymani, O Kennard, T Brown, A G C Jack, and G A Leonard, *Nucleic Acids Res*, 1989, **17**, 55.

33 G Kneale, T Brown, O Kennard, and D Rabinovich, *J Mol Biol*, 1985, **186**, 805.

34 W N Hunter, T Brown, G Kneale, N N Anand, D Rabinovich, and O Kennard, *J Biol Chem*, 1987, **262**, 9962.

HISTONE–DNA INTERACTIONS IN CHROMATIN – SOME INSIGHTS FROM A STUDY OF HISTONE VARIANTS

Jean O Thomas

Department of Biochemistry, University of Cambridge, Tennis Court Road, Cambridge CB2 1QW, UK

1 INTRODUCTION

A Hierarchy of Folding: The Nucleosome, the 10 nm Filament and the 30 nm Filament

Essentially all the DNA in the eukaryotic nucleus is complexed with a roughly equal weight of five small basic proteins termed histones (M_r ~11 000 – 22 000); the complex, which may also be associated with small amounts of non–histone proteins, is termed chromatin. It serves to package the DNA such that about 2 metres are contained within a nucleus 5–10 μm in diameter. The first stage in packaging produces a filament about 10 nm in diameter – about five times as thick as naked DNA. The beaded appearance of the filament in the electron microscope arises from wrapping of the DNA in a periodic manner around a succession of disk–shaped histone octamers with the composition $(H3)_2(H4)_2(H2A)_2(H2B)_2$, each comprising a central tetramer $(H3)_2(H4)_2$ of the arginine–rich histones, H3 and H4, and two flanking heterodimers of the lysine–rich histones, H2A and H2B. Two turns of a DNA superhelix are constrained on the surface of each octamer, largely by electrostatic interaction between the polycationic histones and the polyanionic DNA. One molecule of a fifth histone, H1, serves to seal the two turns.

The structural repeat in chromatin is the nucleosome, which is composed of the histone octamer, two turns (~ 166 bp) of DNA and (usually) one molecule of H1, together with a variable length of linker DNA. The nucleosomal DNA repeat length is commonly about

~ 200 bp, the range being ~165 - 240 bp; the origin of linker length variation is unknown but some signals at least reside in the histones (see below). DNA extracted from nuclei incubated with a double-stranded endonuclease such as micrococcal nuclease and analysed according to size by agarose gel electrophoresis gives a regular 'ladder' of bands, each a multiple of a unit size which is the nucleosomal DNA repeat length. Lengths of chromatin (oligonucleosomes) for biophysical and other studies may be generated by brief digestion of nuclei with micrococcal nuclease; more extensive digestion releases individual nucleosomes, and the linker DNA is ultimately trimmed from these by the exonuclease action of the enzyme, giving first a two-turn particle (the chromatosome) containing the octamer, ~166 bp of DNA and H1, and finally a ~146 bp nucleosome core particle by removal of 10 bp at the two ends of the chromatosome DNA, with loss of the H1 that was bound to them and which sealed the two turns.[1] Core particles have been crystallized and their structure is known to 7Å resolution.[2] Core particles, chromatosomes and nucleosomes have recently been reviewed.[3]

The compaction of DNA achieved in the nucleosome filament (about 5–6 fold) is compounded in the nucleus by further coiling into the 30 nm filament.[1,3] *In vitro* the 10 nm to 30 nm transition may be brought about by monovalent cations (*eg* 60 mM NaCl and above) or lower concentrations of divalent cations (*eg* submillimolar concentrations of Mg^{2+} ions) and is H1-dependent. Probably the most widely accepted model for the structure of the 30 nm filament, which is supported by *X*-ray diffraction on oriented fibres,[4] is the solenoid.[5,6] This is a simple one-start contact helix formed by coiling of the nucleosome filament with about 6 nucleosomes per turn and achieves a roughly 40-fold compaction of the DNA. Other models have, however, been proposed, and have been reviewed,[3,7] and the issue is essentially unresolved. However, there is general agreement on a radial disposition of nucleosomes in the 30 nm filament, the faces of the disk-shaped nucleosomes being roughly parallel to the filament axis (rather than perpendicular to it). The location of H1 in the filament is still uncertain but salt-dependent formation of a helix of H1, which then lines the centre of the solenoid, has been suggested to drive chromatin folding,[6] and indeed H1–H1 contacts in both the 10 nm and the 30 nm filaments have been demonstrated by chemical cross-linking.[8,9] Folding of the 10 nm to the 30 nm filament brought about by cooperative H1–H1 interactions is an attractive possibility (discussed further in Section 2).

Most of the DNA in the nucleus (except for the small proportion that is transcriptionally active in a given cell type) exists in the 30 nm filament form. Further compaction is achieved not by further coiling into successively thicker fibres, but by looping of the 30 nm filament on to proteins of the nuclear matrix. This is an attractive situation because decondensation of the chromatin in particular loops (~5–100 kilobase pairs of DNA), from the 30 nm to the 10 nm form, in principle provides a means whereby the transcriptional status of individual genes or groups of genes could be independently regulated[10] (see Section 2).

The Histones: Distinct Structural Domains

The histones as a class are amongst the most evolutionarily conserved proteins, H3 and particularly H4 being the most highly conserved, consistent with their (literally) central role in the nucleosome,[11] and H1 the most variable. In all four core histones (H3, H4, H2A and H2B) the N-terminal 10–30 residues, which are 'unstructured' in the free histone octamer, are characterized by a high concentration of basic residues and an unusually low content of hydrophobic residues. These basic regions ('tails'), which can be proteolytically cleaved from the octamer in solution or in chromatin (*eg* using trypsin), presumably interact with DNA in chromatin. Sequence variations between histones from different species (which are greatest in H2A and H2B) occur in these regions and could well modulate DNA binding. The N-terminal regions of H3 and H4 are relatively constant in terms of amino acid sequence but are subject to multiple enzymic acetylations at up to four lysine ϵ-amino groups and this could also modulate interactions with DNA. Experiments in which the core histone tails were removed by controlled proteolysis suggested that they may stabilize the nucleosome core rather than playing a major role in its structural organization, and that they may also play some role in the formation of higher-order structures (the core histone tails are discussed further elsewhere[12,3]). More precise information on their roles is, however, lacking. In addition to the species-specific variants of the core histones, more subtle variations may occur within a cell type,[13] and the structural significance of these is even less well understood. Histone sequences have been conveniently collated for comparison.[14,15]

Histone H1 and its variants have a tripartite structure with a central globular domain which is folded in solution at high ionic strength (*eg* 0.5 M NaCl) and resistant to tryptic digestion, flanked by basic tails which are readily removed by trypsin.[16] These appear to be largely random coil in solution but it seems inherently unlikely that they are 'structureless' in chromatin (see below). Within a cell type there may be several H1 subtypes[13,17] which perhaps differ in only a few amino acid residues. However, there are also extreme variants in certain cell types (notably H5 in nucleated erythrocytes (red blood cells) and spH1 in sea urchin sperm) whose appearance in the chromatin during cell differentiation correlates with shut-down of transcription. The greatest similarity between H1 and extreme variants such as H5 and sperm H1 is in the globular domain, which is sufficient to seal two turns of DNA around the octamer,[18] although even here there are substantial sequence differences that could in principle contribute to differences in the affinity of different H1 variants for chromatin. The main differences between the H1 molecules from different sources are in the length and composition of the basic tails. The N-terminal tail[19] (whose role is unclear) is generally relatively short (~ 20-40 residues), whereas the C-terminal tail (which serves to condense linker DNA[20] is about 100 residues long; both tails are rich in lysine, alanine and proline. The C-terminal tails of H5 and spH1, in contrast to that of H1, also contain several arginine residues. This might well be significant, since arginine residues probably form stronger (bidentate) linkages with DNA phosphates than the simple salt-linkages formed by lysine ϵ-amino groups.[21]

Constancy and Variation

Although the organization of chromatin is broadly the same in all species and cell types, largely due to the evolutionary conservation of the histones, there are characteristic variations on the theme which presumably underlie functional differences. Many of these are subtle and ill-understood. However others are more clear-cut and offer some insights into chromatin structure–function relationships. Some of these will be discussed here.

The range of structural variation in chromatin is encapsulated in Table 1. There is a broad correlation between increasing linker length and decreasing transcriptional activity. However, in addition to linker length changes, the type and even content of H1 may also vary in different chromatins, and the altered transcriptional status is probably a composite of these changes which are probably closely linked. If one of the functions of H1 is to neutralize the charge on the linker DNA and allow the formation of higher-order structures, the apparent absence of H1 in yeast, and in the case of cortical neurons[22] a lowering of the H1 content by about half relative to 'canonical' chromatin, would be consistent with the short linker length. At the other end of the scale, sea urchin sperm chromatin,

Table 1: The Range of Variation in Chromatin Structure

Source	Transcriptional state	H1 content (per nucleosome) and type	Repeat length (bp)	Linker length (bp)
Rabbit cerebral cortex neurons	Active	0.5 H1[22]	~166	~0
Rat liver	Moderately active	0.8 H1[23]	~200	~34
Chicken erythrocytes	Inactive permanently	0.9 H5 + 0.4 H1[23]	~212	~46
Sea urchin sperm	Inactive: reversed on fertilization	1.0 spH1[24]	~240	~74

which has the longest known repeat length (~ 240 bp; linker ~ 74 bp), contains a special H1 (one molecule per nucleosome[24]) which is larger and carries more positive charges than the normal H1, as might be expected if a longer linker has to be organized, as well as a special H2B. Both of these will be considered below.

Hydrodynamic studies of chromatin folding on the various chromatins shown in Table 1 have revealed similarities and differences in behaviour.[25,26] The similarities suggest a similar mode of structural organization, consistent with a solenoid with 6 nucleosomes per turn; the differences suggest a difference in stability of the 30 nm filament, with an increase in the series: rat liver< chicken erythrocyte< sea urchin sperm (the very short repeat chromatin was not studied in this respect). A stable structure such as that in sea urchin sperm is not designed to be easily unravelled, as it would have to be to allow access to RNA polymerase (a large enzyme) to permit transcription of the DNA into RNA, and indeed the 'function' of the sperm is efficient packing not transcription.

The purpose of this article is to show how natural variation in histone structure has provided insights into some important features of chromatin structure and folding, and will deal principally with recent work from this laboratory, and in particular with comparisons of a 'typical' H1 with the extreme variants H5 (from mature chicken erythrocytes) and spH1 (from sea urchin sperm) as free proteins in solution, complexed with DNA, and in chromatin. The appearance of these histones on the chromatin in their respective cell types [27,28] correlates with an increase in the nucleosome repeat length, shut-down of transcription, chromatin condensation and nuclear contraction. In the case of sea urchin sperm, these changes are accompanied also by the accumulation of a sperm-specific H2B (spH2B). The H1 and H2B variants found in sea urchin sperm are particularly distinctive. Both histones have extensions of a basic domain, spH2B of its N-terminal domain and spH1 of its C-terminal domain, and in both cases the extensions contain reiterated basic peptide motifs (see Section 5). It is natural to wonder whether the extensions to the tails are related to the extensions in the linker DNA. The basic C-terminal tail of H1 generally is assumed to promote chromatin condensation by association with the linker, but there is no direct information on the binding site for the N-terminal tail of H2B. The distinctive tail of spH2B makes the task of investigating this easier.

In the following sections some of the hitherto unresolved questions about chromatin structure and the roles of the various histones will be addressed. Does H1 bind cooperatively to DNA or in chromatin? Cooperative interactions would provide a possible explanation for chromatin folding and for stabilization of the folded state, and degrees of cooperativity might provide a mechanism for variation in chromatin stability. What is the structure of chromatin-bound H1? How does the basic C-terminal tail organize the linker? What is the role of the core histone tails? Do they bind to linker DNA? In particular, in sea urchin sperm, does the long N-terminal tail of spH2B interact with the unusually long linker DNA?

2 IS THE BINDING OF H1 COOPERATIVE?

Chromatin superstructure is dependent on the presence of H1, which not only seals the nucleosome but also promotes the coiling of the 10 nm filament into the 30 nm filament.[1,3] H1 molecules on successive nucleosomes form an array that runs along the length of the 10 nm filament, the molecules being in sufficiently close proximity to be cross-linked by bifunctional amino-group reagents such as bisimidoesters or bishydroxysuccinimide esters with a span of about 12Å.[29] Analysis of the cross-linked H1 dimers generated using the bifunctional active ester dithiobis(succinimidyl propionate) (Figure 1)

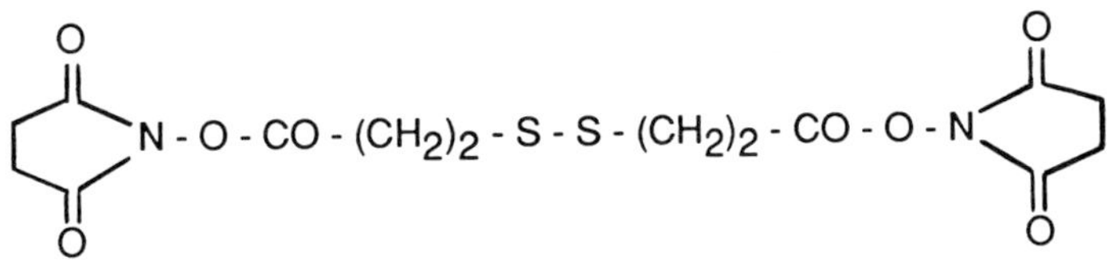

O O
N - O - CO - $(CH_2)_2$ - S - S - $(CH_2)_2$ - CO - O - N
O O

Figure 1: Dithiobis(succinimidyl propionate): an amino group cross-linking reagent.

showed that the H1 molecules were arranged in a polar manner, head-to-tail, along the nucleosome filament (such that the N-terminal domain of one molecule was close to the C-terminal domain of the next) and that additional contacts formed between C-terminal domains in the 30 nm filament.[9] Salt-dependent H1-H1 interactions are proposed to be the means whereby folding of the 10 nm filament into the solenoid occurs *in vitro*.[6] If the interactions were cooperative, and if this cooperativity existed *in vivo*, it is envisaged that the higher-order structure of chromatin could be altered over long distances by local disruption of H1-H1 interactions (*eg* by H1 loss or H1 modification) and cooperative transmission. Such long-range changes, possibly over entire loops of chromatin (several kilobases of DNA) could provide a means of regulating the accessibility of genes, or groups of genes, for transcription by converting chromatin into an 'enabling', looser, state. Further modifications (which are outside the scope of this article) would be required to make it fully transcriptionally competent and, beyond that, transcriptionally active.[30]

H1 cooperativity in chromatin is an attractive idea. It was therefore important to establish whether or not H1 molecules are capable of cooperative interactions and if so, whether there are differences between H1 variants that might be relevant to differences in the stability of the higher-order structures formed by different chromatins.[1,25,26] As determined by a variety of methods, H1 molecules free in solution show virtually no tendency to self-associate. There is some circumstantial evidence for cooperative binding of H5 to chromatin as shown by the preference of H5 (but not of H1) for long chromatin fragments when allowed to migrate between short and long fragments at NaCl concentrations of 30 - 70 mM.[31]

To gain further insights into H1 cooperativity, we carried out a systematic study of the binding of a typical H1 and the variants H5 and spH1 to DNA over a range of conditions. The choice of naked DNA to simplify the analysis is not without justification, since a face or cage of DNA is what the H1 'sees' in the nucleosome, where it could conceivably interact with as many as three DNA segments: the central turn of DNA on the dyad axis, and the entering and exiting segments of DNA that connect with the preceding and succeeding

DNA linkers.[11,18] As a simple test for cooperativity we added limited quantities of H1 to DNA of a particular (average) size and then asked whether the H1 was distributed over the entire population of DNA molecules or had concentrated on some molecules leaving the others free, consistent with cooperativity in H1 binding.[32] In the former case we would expect to see a single peak of sedimenting material containing both DNA and H1 in sucrose gradients, and in the latter two peaks: one of free DNA and a more rapidly sedimenting peak containing both DNA and H1. In fact the former result was seen at low ionic strength and the latter at higher ionic strength (the exact ionic strength for the change in behaviour depended on the DNA size, but was in the range ~25 mM for ~2 kbp DNA to ~40 mM for 0.2 kbp DNA). Evidently there is a salt-dependent change from non-cooperative to cooperative binding of H1 to DNA[32] in agreement with earlier findings.[33] When the behaviour of H5 and spH1 was examined[34] they were also found to bind to DNA cooperatively by this criterion at 'high' ionic strength (35 mM NaCl), but, in contrast to H1, they also bound cooperatively at low ionic strength (5 mM NaCl). The low ionic strength complexes, which sedimented very much more slowly than the high ionic strength complexes, appeared from electron microscopy to be thin filaments consisting of two DNA duplexes bridged by histone molecules. The high ionic strength complexes were thick filaments formed by association of the thin filaments (evidently formed transiently in the case of H1); above certain DNA lengths a proportion of the thick filaments circularized (see below). Full details of these studies and discussion of related work have been published.[32,34]

The sedimentation tests of cooperativity were complemented by the use of chemical cross-linking to investigate the proximity of neighbouring H1 molecules. At low ionic strength (5mM) there was no intermolecular cross-linking of a 'typical' H1 complexed with DNA, showing that the molecules are bound distributively over the DNA, whereas at 35 mM ionic strength the H1 molecules were clustered sufficiently to give large cross-linked polymers, consistent with cooperative binding.[32] In contrast, there is little difference between the cross-linking patterns for H5 or spH1 at low and high ionic strength, indicating cooperativity over the entire ionic strength range.[34]

Cooperativity in H1 binding to DNA under certain conditions has therefore been demonstrated, together with differences in cooperativity for different variants. In particular the two variants (H5 and spH1) that are involved in the formation of chromatin higher-order structures of increased stability relative to chromatin containing a 'typical' H1[25] show cooperativity at very low ionic strength. Exactly how the H5 or spH1 binds to the two DNA molecules in the thin filaments is unknown; one possibility is that it uses the binding sites on the globular domain which can probably accommodate up to three DNA segments (see above), and that the histone tails bind to the contiguous DNA as they would to linker DNA in chromatin. The cooperativity observed at low ionic strength for H5 and spH1 may therefore reflect the interactions between neighbouring molecules of these histones in the nucleosome filament. Formation of thick filaments from thin filaments as the ionic strength is raised presumably occurs because charge screening allows histone-DNA interactions between neighbouring filaments to dominate over negative charge repulsions.

Two intriguing features emerge from a comparison of the composition of the thick complexes formed between H1, H5 or spH1 and DNA. One is an apparent 'signal' in the histone structure that determines the histone packing density and hence the spacing of successive molecules along the DNA. The spacing deduced for thin filaments is proportional to the linker length of the chromatin which the histone normally stabilizes, which has obvious implications for nucleosome spacing in chromatin.[34] Chromatin reconstitution experiments have also suggested a role for H1-type in repeat length determination.[35,36]

The second interesting feature is that H1, H5 and spH1 differ in their ability to form circular thick filaments with linear DNA, suggesting that they differ in their ability to bend DNA.[32,34] The basis for bending is at present unclear (although it may be significant that the proportion of circular complexes is roughly proportional to the arginine:lysine ratio in the histone), but it is an attractive possibility that it is related to the bending of linker DNA that is probably required in H1-dependent higher-order structures.

3 WHAT IS THE STRUCTURE OF H1 IN CHROMATIN?

The existence of three distinct structural domains in histones of the H1 class is well established, as is the folding of the central domain at high ionic strength.[16] However, we know remarkably little in any detail about the structure of H1 in chromatin, particularly about the flanking basic domains ('tails'), and our understanding of how they function is therefore limited. Although the tails appear to be largely unstructured and flexible in the free histone in solution,[16] it is unlikely that this is the case when H1 is bound to DNA in chromatin.

```
    Chicken H1

         *           *   *  *    *      *         *     *           *        *
106 KLNKKPGETKEKATKKKPAAKPKKPAAKKPAAAAKKPKKAAAVKKSPKKAKKPAAAATKKAAKSPKKATKAGRP-
          *       *         *
    KKTAKSPAKAKAVKPKAAKSKAAKPKAAKAKKAATKKK 217


    Chicken H5

               *            *     *     *   *              *     *                    *
94  RLAKSDKAKRSPGKKKKAVRRSTSPKKAARPRKARSPAKKPKATARKARKKSRASPKKAKKPKTVKAKSRKASKAKKVKRSKP-
             *
    RAKSGARKSPKKK 189


    Sea urchin sperm H1

           *                                                         *          *    *
112 RVGAVAKPKKAKKTSAAAKAKKAKAAAAKKARRAKAAAKRKAALAKKKAAAAKRKAAAKAKKAKKPKKKAAKKAKKPAKKSP-
    ----------------------E--R-RAA-KK------------------------T-----------T-A..........
         *    *        *           *        *        *
    KKAKKPAKKSPKKKKAKRSPKKAKKAAGKRKPAAKKARRSPRKAGKRRSPKKARK 248
    .......................................................
```

Figure 2: Amino acid sequences of the C–terminal fragments of H1, H5 and spH1. Note the preponderance of Lys (K), Ala (A) and Pro (P), and to a lesser extent Arg (R); the Pro residues are marked with asterisks for emphasis. The complete sequence (top line) for spH1 is for *Parechinus angulosus*; the partial sequence shown beneath it is for *Echinus esculentus* used in our laboratory; note that they both have a 57–residue proline–free segment. The glutamic acid (E) at position 134 in *E Esculentus* is probably involved in an α–helix stabilizing salt–linkage (see text).
(– – – –), identical residues in the two sequences;
(.....), residues not determined. (From ref 37, which also gives sequence sources).

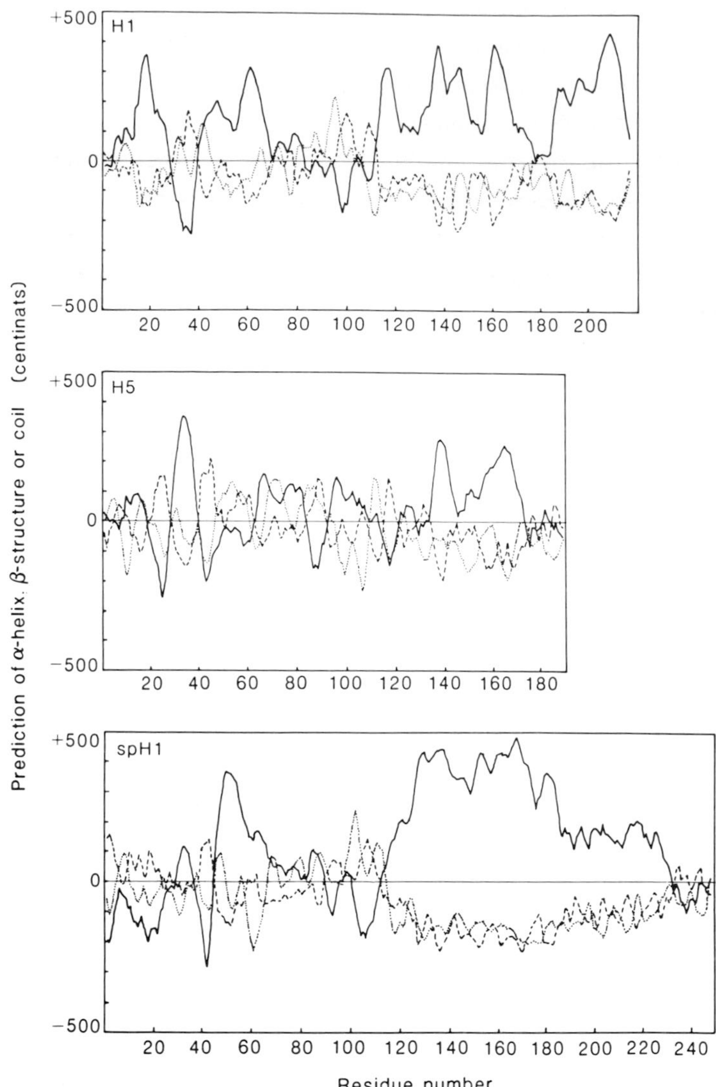

Figure 3: Secondary structure predictions for H1, H5 and spH1 using the Robson algorithm.[38]
(——), α–helix; (.....), β–structure; (------), coil. (From ref 37).

Because of the high lysine and alanine content of the C-terminal domains (Figure 2), secondary structure prediction algorithms predict a substantial amount of α-helix (Figure 3), especially in the C-terminal domain of spH1.[37] Not surprisingly, this is not realized in solution at low ionic strength, presumably due to the repulsive forces between protonated ε-amino side chains of the many lysine residues, as well as to the several proline residues dispersed throughout the domain. However, it seemed likely that when bound to DNA in chromatin the positive charges would be neutralized and α-helix might be stabilized. We therefore studied free H1, H5 and spH1 and their C-terminal domains (excised from the parent histones by chymotryptic cleavage at a unique phenylalanine residue) in solution under conditions that might be expected to stabilize α-helices, using the intensity of the negative circular dichroism (CD) band at 220 nm as a measure of α-helix. (We could not use difference spectra of peptide-DNA complexes *versus* DNA to test directly for α-helix in the peptide on binding to DNA due to mutual distortion of the peptide and DNA spectra.) Briefly, we found that 2,2,2-trifluoroethanol (65%, v/v) which might be expected to stabilize axial H-bonds in α-helices, indeed allowed formation of appreciable α-helix in the C-terminal tails of H1, H5 and particularly spH1, accounting for ~29%, ~39% and ~60%, respectively, of the residues in the tails. High concentrations (1 M) of NaCl or of $NaClO_4$, conditions which should also stabilize α-helices by screening of repulsive positive charges or formation of direct ion pairs, respectively, induced little α-helix in the C-terminal tails of H1 and H5. In contrast, about 58 residues in the C-terminal domain of spH1 were α-helical in 1M $NaClO_4$. Examination of the amino acid sequence of the domain (Figure 2) suggested that the helical segment corresponded to a proline-free region of 57 residues.[37] Remarkably, peptides representing much of this region were subsequently shown to be relatively resistant to tryptic digestion in the C-terminal domain of spH1 bound to DNA (but not free in solution, even in the presence of 1M NaCl), despite a large number of potential internal tryptic cleavage sites,[39] and likewise corresponding peptides fused to the adjacent globular domain survived tryptic digestion of chromatin. The most likely reason for these results is stabilization of the 57-residue proline-free segment in α-helix when the C-terminal domain of spH1 is bound to DNA.

Remarkably, however, the α–helical potential of this region is evidently manifested even in the absence of DNA, since a trypsin–resistant peptide corresponding to much of the proline–free region was ~70% helical in 1M $NaClO_4$ (despite likely end–effects *etc*) and was even partially helical in 1 mM sodium phosphate[39] (Figure 4). There are therefore strong reasons for believing that the 57–residue proline–free region in the C–terminal domain of spH1 is α–helical in chromatin.

About 57 residues (~16 turns) of helix is unusually long compared with the α–helices generally found in globular proteins [although exposed central helices of eight and nine turns, respectively, exist in troponin C and calmodulin (see ref 39 for references)], but the C–terminal tail of H1 is far from globular. Moreover, a peptide of ~ 80 residues isolated from troponin T is helical in aqueous solution and has been proposed to exist in the native protein as a long helix (of about 18 turns) that interacts in striated muscle with the coiled–coil of tropomyosin.[40]

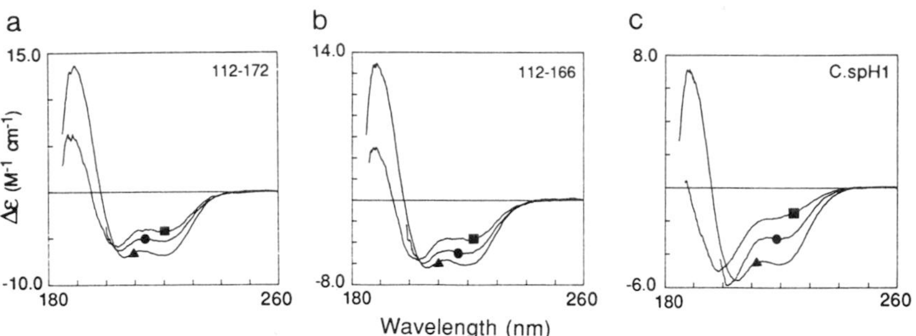

Figure 4: α–Helix in trypsin–resistant peptides from the C–terminal domain of spH1. CD spectra at 25°C of (a,b) peptides (residues 112–172 and 112–166) containing most of the proline–free segment and (c) the C–terminal chymotryptic fragment (residues 112–248):
(■) low salt buffer (1 mM Na phosphate, 0.2 mM Na_2EDTA, pH 7.4);
(●) in buffer containing 1M NaCl;
(▲) in buffer containing 1M $NaClO_4$. (From ref 39).

An important contributory feature to the partial stability of the helix in spH1 in aqueous solution at low ionic strength (1 mM sodium phosphate) in the absence of DNA, is probably a salt linkage between the side chains of lysine and glutamic acid four residues apart (positions 130 and 134 respectively in the amino acid sequence; Figure 2), a feature that has been convincingly demonstrated to contribute to α-helix stability,[41] and in addition a high alanine content.[42] Intrahelical salt bridges are known to stabilize calmodulin and troponin C, and are also invoked for the very long helix proposed for troponin T.[40] The 57-residue helix from spH1 viewed in projection as a 'helical wheel' (Figure 5) shows two features of note. First, the basic residues (lysine and arginine) are rather dispersed on the surface suggesting that DNA-binding is not confined to one side. Secondly, the alanine residues with small ($-CH_3$) side-chains are clustered on one side (bracketed in Figure 5) and would generate a hydrophobic furrow along the helix. The amphipathic nature of the helix means that it will probably be curved towards the hydrophobic side, and the small alanine side-chains would facilitate this.[39]

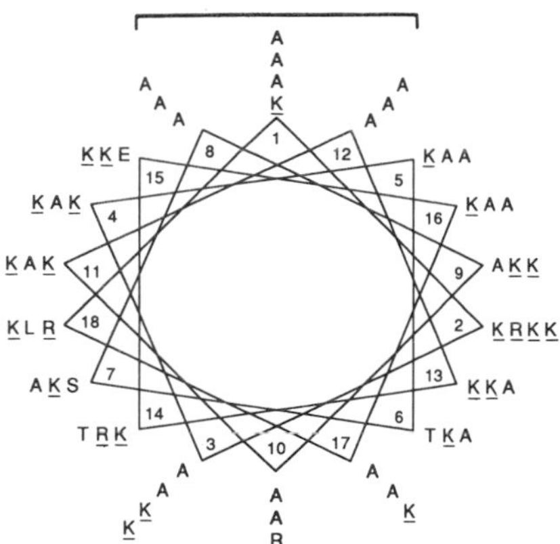

Figure 5: Helix projection (18-point 'helical wheel') for the proline-free segment, residues 120-176 of *E esculentus* sperm H1. Basic residues [Lys (K), Arg (R)] are underlined. The innermost letters in any set refer to amino acids 1-18 of the helix; the second set of letters refers to the next 18 residues *etc*. The bracket indicates the cluster of alanine (A) residues. (From ref 39).

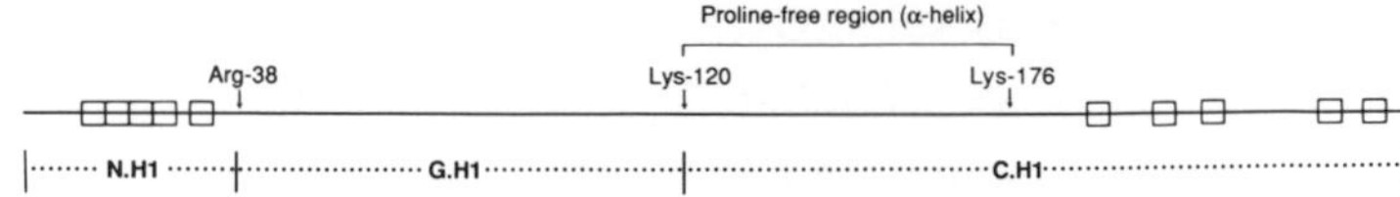

Figure 6: Some features of secondary structure and sequence organization in sea urchin sperm H1 (spH1). A linear representation (to scale) of the organization of sperm H1 into three domains (N-terminal (N.H1), central globular (G.H1), C-terminal (C.H1)), showing the location in the amino acid sequence of the 57-residue proline-free segment (α-helix) and the -Ser-Pro-Basic-Basic- motifs (□) that may form β-turns (see text). (From ref 39).

What might be the structural significance of the proline-free segment in sea urchin sperm H1? It is located immediately adjacent to the globular domain (Figure 6), which seals the two turns of DNA around the histone octamer. A rigid, possibly curved, structural element in this position would be ideally placed to guide the path of the linker DNA , to which the C-terminal domain of H1 is believed to bind, as the linker enters or leaves the nucleosome. At present we can only speculate about the mode of binding of the helix to DNA (*eg* whether or not it tracks a groove) and this is currently under investigation.

The C-terminal domains of somatic H1 and H5 do not contain a comparable long proline-free region and this is presumably why no α-helix is stabilized in these domains by NaCl or $NaClO_4$. However α-helix does form in trifluoroethanol, in accord with the secondary structure predictions (Figure 3) and, likewise, additional helix can form in the ~ 60% of the C-terminal tail of spH1 that is not contained in the proline-free segment[37] (Figure 6). If this helical potential is realized in chromatin, helical segments in these regions may be interspersed by kinks or bends at proline residues.[37] In the case of spH1, and to a lesser extent in H5 and H1, some of these proline residues occur in a distinct motif which may also have structural significance (see Section 5).

```
SpH2B         PSQKSPTKRSPTKRSPTKRSPQKGGKGGKGAKRGGKAGKRRRGVQVK
Chicken H2B                               PEPAKSAPAPKKGSKKAVTKTQKKGDK-

RRRRRRESYGIYIYKVLKQVHPDTGISSRAMSVMNSFVNDVFERIAAEAGRLTTYNRRS
-KKS-K---S--V---------------K--GS-------I-----GL-S--AH--K--

TVSSREVQTAVRLLLPGELAKHAVSEGTKAVTKYTTSR
-IT---I----------------------------S-K
```

Figure 7: Comparison of the amino acid sequences of sea urchin sperm H2B (spH2B) and a 'typical' chicken H2B.[14,15] spH2B (the variant spH2B$_1$ from *Parechinus angulosus*) has an extended N-terminal domain and contains repeats of –Ser–Pro–Thr–Lys– (underlined) which are sites of phosphorylation in spermatids (see text). Note that the sequence differences are greatest in the N-terminal regions; the globular remainders of the molecules are highly conserved.
Dashes (– – –) indicate residues identical with those in spH2B.

4 WHAT IS THE ROLE OF SPERM-SPECIFIC HISTONE H2B?

The highly inert and condensed chromatin of sea urchin sperm has two distinguishing components, spH2B and spH1, that are presumably jointly responsible for the unusually long repeat length and the remarkably tight packing of 30 nm filaments. spH2B has 10–20 (depending on the variant) more residues in the N-terminal tail than a 'typical' H2B; the globular domains are virtually identical (Figure 7). As a step towards resolving the various contributions of the two histones and their various domains we have asked a specific question: in the nucleosome filament in sea urchin sperm does the N-terminal tail of spH2B interact with the long linker?

To answer this question we used a chemical approach that we had used earlier to map DNA-binding sites on the surface of the histone octamer[43] and to identify the nucleosome-binding site on the globular domain of H5.[44] This relies on the fact that the ϵ-amino groups of lysine side-chains bound to DNA in chromatin are offered

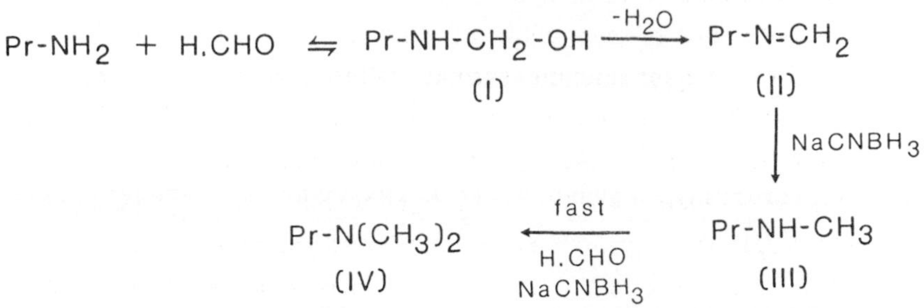

$$\text{Pr-NH}_2 + \text{H.CHO} \rightleftharpoons \underset{\text{(I)}}{\text{Pr-NH-CH}_2\text{-OH}} \xrightarrow{-\text{H}_2\text{O}} \underset{\text{(II)}}{\text{Pr-N=CH}_2}$$

$$\underset{\text{(II)}}{\text{Pr-N=CH}_2} \xrightarrow{\text{NaCNBH}_3} \underset{\text{(III)}}{\text{Pr-NH-CH}_3}$$

$$\underset{\text{(III)}}{\text{Pr-NH-CH}_3} \xrightarrow[\text{H.CHO, NaCNBH}_3]{\text{fast}} \underset{\text{(IV)}}{\text{Pr-N(CH}_3)_2}$$

Figure 8: Reductive methylation of lysine ϵ–amino groups in proteins. The imine (Schiff base) (II) produced by dehydration of the initial hydroxymethyl adduct (I) of an amino group on the protein (Pr) with formaldehyde is reduced with sodium cyanoborohydride to the ϵ–*N*–monomethyllysine derivative (III). Rapid reaction with a further equivalent of formaldehyde, followed by reduction, results in the ϵ–*N*,*N*–dimethyllysine derivative (IV). The pK_a values of the unmethylated and dimethylated ϵ–amino groups are essentially the same (~ 10.2).[46]

(partial) protection from chemical modification (specifically, in this case, with formaldehyde, as the first step in reductive methylation) compared with lysines that are fully exposed; they become fully accessible when the DNA–protein complex is dissociated and under these conditions may be selectively radiolabelled (using [^{3}H]HCHO). Any remaining lysines are reductively methylated under denaturing conditions to ensure chemical homogeneity with respect to the state of the lysine side–chains, and it is then a simple matter to digest the modified protein at arginyl peptide bonds only, using trypsin, and to identify radiolabelled lysines within the amino acid sequence by standard methods of protein chemistry to reveal DNA–binding regions. The chemistry of the modification[45] is outlined in Figure 8; the imine formed with formaldehyde at each stage is immediately reduced by sodium cyanoborohydride (added before the formaldehyde to prevent protein–protein crosslinking). Full details of the method and background have been published.[43,44,47]

To determine whether the N-terminal tail of H2B binds to linker DNA in the nucleosome filament (rather than, for example, being bound within the nucleosome core or, alternatively, not bound at all) we compared the protection from modification of H2B in extended chromatin with that in 146 bp nucleosome core particles, which have no adjacent linker DNA. Autoradiography of two-dimensional peptide maps of individual histones, after differential radiolabelling as just described, revealed peptides in spH2B that were more highly radiolabelled in chromatin than in core particles. These were mainly from the N-terminal domain, and from this we conclude that the N-terminal tail of spH2B binds to the unusually long linker DNA in sea urchin sperm chromatin, at least in the extended state.[48]

The same experiment yielded information about the other core histones and indicates distinct differences in the behaviour of the N-terminal tails (information impossible to obtain from experiments in which all four tails are removed together). We find that the N-terminal tail of H4 is protected from modification both in core particles and in extended chromatin, indicating that its primary binding site is within the core. In contrast, the N-terminal domain of H2A is not protected in core particles or in extended chromatin, suggesting that its primary role may be in the stabilization of higher-order structure. The H3 tail, like that of H2B (as well as other regions in H3), interacts with DNA outside the 146 bp nucleosome core. This may be with the linker, but in view of the location of H3 in the nucleosome,[2,11] it is likely to be partly, if not wholly, with the two 10 bp extensions to the 146 bp core DNA that complete the two superhelical turns around the histone octamer in chromatin.[48]

5 COMMON MOTIFS IN THE DNA-BINDING DOMAINS OF spH2B AND spH1

The DNA-binding extension in the N-terminal domain of spH2B relative to somatic histones (*eg* Figure 7) contains tandem repeats of sequences that have been viewed as pentapeptides of the type –Pro–Arg–Lys–Gly–Ser– or –Pro–Thr–Lys–Arg–Ser– in different variants,[49] or as tetrapeptides of the type –Ser–Pro–Basic–Basic– or –Basic–Basic–Ser–Pro– [50] (where Basic = Lys or Arg). A major component of the binding is undoubtedly electrostatic interaction between the two adjacent basic side-chains and DNA phosphates. The motifs are predicted as β-turns;[50] the specific suggestion has been made that these are based on the tetrapeptide motif -Ser-Pro-Basic-Basic-, that they are of type 1, with serine in the first position and proline in the second, stabilized by an additional hydrogen bond from the serine OH to the peptide NH two residues away, and that they form hydrogen bonds (from the serine NH groups) with acceptors in the minor groove of DNA, much like netropsin.[51] Multiple copies of these motifs also occur in the N-terminal domain of spH1 where they are again clustered, and in the C-terminal domain where they are dispersed (Figure 6).

The structural roles *in vivo* of the N-terminal domains of spH1 and spH2B are not yet understood (whereas the C-terminal domain of spH1 probably binds mainly to linker DNA). However the N-terminal domains are multiply phosphorylated in spermatids[28] (in which the chromatin has all the structural features of sperm chromatin except for the final tight packing of fibres) and we have recently shown[52] that the phosphorylation occurs in the sequence -Ser-Pro-X-Basic- (X = Gln, Thr, Lys or Arg). If the proposed β-turn exists and is indeed normally stabilized by an additional hydrogen bond involving the serine hydroxyl group, phosphorylation would disrupt it, and would also offset the electrostatic contribution of the basic side-chains to DNA binding. Phosphorylation would thus weaken the interaction of the basic N-terminal domains of spH2B and spH1 with DNA. From the 'tightening' of the nucleus that accompanies dephosphorylation in the transition from late spermatids to mature sperm, it would seem that at least one of the roles of the N-terminal domains *in vivo* is in interactions between adjacent 30 nm filaments that are tightly packed in the final mature sperm. Interestingly, the dispersed tetrapeptide motifs in the C-terminal tail of spH1 do not appear to be phosphorylated. This suggests that the structural role of this domain is unchanged in the transition from spermatids to mature sperm (or at least that its interactions with DNA are not modulated by phosphorylation) and would be consistent with a role for the C-terminal tail *within* the 30 nm filament, in condensing linker DNA.

When a spermatozoon fertilizes an egg the sperm chromatin

decondenses. The events leading up to decondensation are rapid rephosphorylation of spH1 and spH2B and replacement of sperm-specific histones by early embryonic histones ('cleavage stage histones') present in the egg.[50] The sites phosphorylated upon fertilization have not yet been determined but it seems likely that Nature now uses the same strategy in reverse, and that they may be precisely those originally phosphorylated in spermatids and which became dephosphorylated in the final stages of sperm maturation.

6 SUMMARY AND CONCLUSION

Histones are relatively well conserved proteins that package DNA in broadly the same manner in nucleosomes in different species, tissues and cell types. Subtle variants exist in most cells, both of the core histones and more especially of H1, and these are assumed to be related in some way, not yet understood in detail, to subtle structural changes that lead to functional changes. More extreme variants occur in some cell types, notably those that are transcriptionally repressed, either permanently as in the case of mature chicken erythrocytes (in which H5 largely replaces H1), or temporarily as in the case of sea urchin sperm (in which the sperm-specific spH2B and spH1 replace their somatic counterparts); the appearance of these variants on the chromatin in these cell types correlates well with chromatin inactivation. A comparison of normal somatic H1 with H5 and spH1 has led to the identification of, first, differences in the cooperativity of binding to DNA that may be relevant to differences in the stability of higher-order structures in the chromatins from which they originated; and, secondly, of an unique structural feature (a long α-helical segment) in the C-terminal basic domain of spH1 which may be instrumental in organizing the long linker DNA in sea urchin sperm chromatin. The extended N-terminal basic domain in sperm-specific H2B also interacts with the long linker in extended sperm chromatin. This domain and the N- and C-terminal domains of spH1 contain repeated tetrapeptide DNA-binding motifs of the type -Ser-Pro-X-Basic- whose interactions with DNA in the N-terminal domains may be modulated by phosphorylation: a chemical change with structural and biological consequences.

ACKNOWLEDGEMENTS

The work carried out in this laboratory was supported by the Science and Engineering Research Council and The Wellcome Trust. I am grateful to Caroline Hill for Figure 7 and for comments on the manuscript.

REFERENCES

1 J O Thomas, *J Cell Sci*, 1984, **Suppl 1**, 1.
2 T J Richmond, J T Finch, B Rushton, D Rhodes, and A Klug, *Nature*, 1984, **311**, 532.
3 J Widom, *Ann Rev Biophys Biophys Chem*, 1989, **18**, 365.
4 J Widom and A Klug, *Cell*, 1985, **43**, 207.
5 J T Finch and A Klug, *Proc Natl Acad Sci USA*, 1976, **73**, 1897.
6 F Thoma, T Koller, and A Klug, *J Cell Biol*, 1979, **83**, 403.
7 G Felsenfeld and J D McGhee, *Cell*, 1986, **44**, 375.
8 J O Thomas and A J A Khabaza, *Eur J Biochem*, 1980, **112**, 501.
9 A C Lennard and J O Thomas, *EMBO J*, 1985, **4**, 3455.
10 S M Gasser and U K Laemmli, *Trends Genet*, 1987, **3**, 16.
11 A Klug, D Rhodes, J Smith, J T Finch, and J O Thomas, *Nature*, 1980, **287**, 509.
12 J O Thomas, in 'Eukaryotic Genes: Their Structure and Function', eds N Maclean, S P Gregory, and R A Flavell, Academic Press, London, 1983, Chapter 2, p 9.
13 R W Lennox and L H Cohen, *Biochem Cell Biol*, 1988, **66**, 636.
14 C von Holt, W N Strickland, M S Brandt, and M S Strickland, *FEBS Lett*, 1979, **100**, 201.
15 D E Wells, *Nucleic Acids Res*, 1986, **14 (suppl)**, r119.
16 P G Hartman, G E Chapman, T Moss, T and E M Bradbury, *Eur J Biochem*, 1977, **77**, 45
17 R D Cole, *Anal Biochem*, 1984, **136**, 24.
18 J Allan, P G Hartman, C Crane-Robinson, and F X Aviles, *Nature*, 1980, **288**, 675.
19 L Bohm and T C Mitchell, *FEBS Lett*, 1985, **193**, 1.
20 J Allan, T Mitchell, N Harborne, L Bohm, and C Crane-Robinson, *J Mol Biol*, 1986, **187**, 591.

21 S Ichimura, K Mita, and M Zama, *Biochemistry*, 1982, **21**, 5334.
22 E C Pearson, D L Bates, T D Prospero, and J O Thomas, *Eur J Biochem*, 1984, **144**, 353.
23 D L Bates and J O Thomas, *Nucleic Acids Res*, 1981, **9**, 5883.
24 C S Hill and J O Thomas, unpublished results.
25 J O Thomas, C Rees, and P J G Butler, *Eur J Biochem*, 1986, **154**, 343.
26 P J G Butler, *CRC Crit Rev Biochem*, 1983, **15**, 57.
27 H Weintraub, *Nucleic Acids Res*, 1978, **5**, 1179.
28 D L Poccia, M V Simpson, and G R Green, *Dev Biol*, 1987, **121**, 445.
29 J O Thomas, *Meth Enzymol*, 1989, **170**, 549.
30 R Reeves, *Biochim Biophys Acta*, 1984, **782**, 343.
31 J O Thomas and C Rees, *Eur J Biochem*, 1983, **134**, 109.
32 D J Clark and J O Thomas, *J Mol Biol*, 1986, **187**, 569.
33 M Renz and L A Day, *Biochemistry*, 1976, **15**, 3220
34 D J Clark and J O Thomas, *Eur J Biochem*, 1988, **178**, 225.
35 A Stein and P Kunzler, *Nature*, 1983, **302**, 549.
36 A Stein and M Mitchell, *J Mol Biol*, 1988, **203**, 1029.
37 D J Clark, C S Hill, S R Martin, and J O Thomas, *EMBO J*, 1988, **7**, 69.
38 J Garnier, D J Osguthorpe, and B Robson, *J Mol Biol*, 1978, **120**, 97.
39 C S Hill, S R Martin, and J O Thomas, *EMBO J*, 1989, **8**, 2591.
40 M Sundaralingham, W Drendel, and M Greaser, *Proc Natl Acad Sci USA*, 1985, **82**, 7944.
41 S Marqusee and R Baldwin, *Proc Natl Acad Sci USA*, 1987, **84**, 8898.
42 S Marqusee, V H Robbins, and R L Baldwin, *Proc Natl Acad Sci USA*, 1989, **86**, 5286.
43 S F Lambert and J O Thomas, *Eur J Biochem*, 1986, **160**, 191.
44 J O Thomas and C M Wilson, *EMBO J*, 1986, **5**, 3531.
45 N Jentoft and D G Dearborn, *J Biol Chem*, 1979, **254**, 4359.
46 T A Gerken, J E Jentoft, N Jentoft, and D G Dearborn, *J Biol Chem*, 1982, **257**, 2894.
47 J O Thomas, *Meth Enzymol*, 1989, **170**, 369.
48 C S Hill and J O Thomas, *Eur J Biochem*, 1990, in the press

49 C von Holt, P de Groot, S Schwager, and W F Brandt, in 'Histone Genes', eds G S Stein, J L Stein, and W F Marzluff, Wiley, New York, 1984, Chapter 3, p 65.
50 D L Poccia, in 'Molecular Regulation of Nuclear Events in Mitosis and Meiosis', eds R A Schlegel, M S Halleck, and P N Rao, Academic Press, New York, 1987, Chapter 6, p 149.
51 M Suzuki, *EMBO J*, 1989, **8**, 797.
52 C S Hill, L C Packman, and J O Thomas, *EMBO J*, 1990, in press.

NUCLEIC ACID RECOGNITION BY SMALL MOLECULES

Stephen Neidle

Cancer Research Campaign Biomolecular Structure Unit
The Institute of Cancer Research, Sutton, Surrey SM2 5NG, UK

1 INTRODUCTION

The expression of the genetic information residing in the bases of a DNA sequence is controlled by means of the specific recognition of appropriate regulatory sequences by DNA-binding proteins. To date, detailed three-dimensional structural information is only available on a handful of such proteins and complexes with oligonucleotides – these are exclusively bacterial in origin.[1] The dominant theory of sequence recognition has been that the hydrogen bonding potential in base pairs,[2] which differs between AT and GC ones, as well as between GC and CG, can be sensed by amino-acid side-chains, and is responsible for specific recognition. The contrary view, which has received some support from the crystal structure of the trp repressor complex,[3] is that the details of the sequence-dependent structure of DNA itself are responsible. Most likely, both sets of factors play a role. DNA is a conformationally flexible molecule, and its local structure is clearly influenced by particular sequences. A well-established example is the narrowing of the minor groove in an AT-rich region of a DNA duplex.[4]

DNA-protein interactions appear to involve the major groove specificity is required, otherwise the minor groove is the important site of contacts. This favouring of the major groove is due to its greater inherent difference in hydrogen bonding between AT, TA, GC and CG, as well as its larger diameter. For low molecular weight ligands, the minor groove is primarily implicated. Many of these ligating molecules are of considerable current interest as potential new chemotherapeutic agents and/or artificial gene regulators.[5,6] They are the principal theme of this chapter.

Minor groove binders typically have hydrophobic groups such as phenyl rings, together with one or more cationic charge centres. A crucial feature is an overall concave shape on one side or edge of the molecule which can complement the shape of the floor of the minor groove.[7] The extent of this 'isohelicity' property is an important factor in defining the level of drug–DNA binding. In general, the ligands preferentially bind to AT regions of DNA duplexes that are in the B–form, albeit with differing degrees of preference.[8] This selectivity can be attributed to various causes:
(i) steric effects with the exocyclic 2–amino group of guanine hindering effective interaction.
(ii) electrostatic effects due to the strong electrostatic potential of AT regions compared to GC ones.
(iii) displacement of water molecules, which form an especially structured network in AT regions.[9]

2 MINOR GROOVE BINDING IN SOLUTION

Interactions of ligands with DNA sequences have utilised measurements of equilibrium binding constants, increases in melting temperature and ethidium–displacement assays in order to assess effects at the polynucleotide level.[8] These results are inherently averaged over all binding sites, but nonetheless provide clear evidence of relative sequence preferences, at least for AT *versus* GC (Table 1). There are large differences between these compounds in their ability to discriminate between AT and GC base pairs, but all show the effect to at least some degree. The interactions of the pyrrolic compounds netropsin and distamycin have been especially well studied, and both require B–form duplex DNA.[8] Typically, several distinct ranges of association constant are found for these ligands, suggestive of binding to distinct sequences.

Footprinting methods enable drug binding sites on a DNA sequence to be located. They have been invaluable in defining the fine detail of sequence selectivity for these compounds. Typically, a sequence of biological DNA (possibly from a regulatory region) of 150–200 base pairs is used. The technique uses either enzymatic or

Netropsin

Berenil

Distamycin

Hoechst 33258

Table 1: Relative binding ability to alternating polynucleotides, as measured by ethidium displacement. The values of concentration required for 50% displacement have been normalised to the netropsin/poly[d(A-T)] system. Adapted from values given in reference 8.

Drug	Poly[d(G-C)]	Poly[d(A-T)]	GC/AT
Netropsin	27.3	1.0	27.3
Distamycin	65.5	0.1	655.0
Hoechst 33258	2.5	0.4	6.3
Berenil	10.4	1.6	6.4

chemical cleavage of DNA, which is then inhibited solely at the drug binding sites.[10,11] Visualisation is by means of DNA sequencing gels. Typically netropsin produces cleavage blocks at runs of at least three A or T nucleotides, with the other drugs in Table 1 giving overall similar results. There are some differences in detail, which in part are ascribable to different sizes of the various ligands. Invariably, drug binding produces enhancements of DNA cleavage at non-binding sites. It has been suggested that these represent transmitted conformational effects resulting in, for example, minor groove widening. Such effects, which are sometimes surprisingly distant from actual binding sites, may be of considerable biological significance.

Netropsin and its tripyrrolic analogue distamycin are the best-studied groove binders, with footprinting showing interaction preferentially with the sequence 5'-AATT. A number of NMR studies have concurred with this conclusion, with nOe data being used to provide detail of drug-DNA interaction geometry.[12] The recent controversial report[13] of two distamycin molecules binding to a single, highly enlarged minor groove site, is clearly at variance with these observations.

3 CRYSTALLOGRAPHIC STUDIES OF GROOVE-BINDING COMPLEXES

The crystal structure of netropsin itself shows it to be a crescent-shaped molecule, with the three amide groups pointing in and towards its concave face.[14] This shape is retained in its complex with the sequence dCGCGAATTCGCG, which forms a B-DNA double helix.[15] The drug is located in the minor groove, in the central AATT region, where it is held by three strong hydrogen

Table 2: Hydrogen-bonding interactions observed in minor-groove crystal structures. Distances are in A. Only those less than van der Waals separation have been selected for inclusion. Bases of the first strand in the oligonucleotide duplexes are numbered 1-12, and 13-24 for the second one.

		Drug donor atom	DNA acceptor atom	Distance	Ref
Netropsin		amide N4	O2 of T20	2.68	15
			N3 of A6	3.28	
dCGCGAATTCGCG		amide N8	O2 of T8	2.56	
		amidinium N10	N3 of A17	2.65	
Hoechst		imidazole N1	O2 of T19	3.21	17
+		imidazole N3	O2 of T8	2.87	
dCGCGAATTCGCG		imidazole N1	N3 of A6	3.16	16
			O2 of T20	2.78	
		imidazole N3	O2 of T19	3.00	
Netropsin	(a)	amide N4	O2 of T6	3.2	19
+		amide N6	N3 of A19	3.2	
dCGCGATATCGCG		amidinium N10	O2 of T8	2.7	
			O2 of C9	2.7	
	(b)	amide N6	N3 of A19	3.3	
		amide N8	O2 of T6	3.3	
		amidinium N1	O2 of T8	3.2	
			O2 of C9	2.6	
Distamycin		amide N1	O2 of T8	3.2	20
+			N3 of A18	3.3	
dCGCAAATTTGCG		amide N5	N3 of A6	3.2	
			O2 of T20	3.3	
		amide N7	N3 of A5	3.2	
			O2 of T21	2.9	
		amidinium N9	N3 of A4	3.1	
Berenil		amidinium N10	N3 of A18	2.4	18
+		amidinium N20	Water	2.7	
dCGCGAATTCGCG		Water	N3 of A5	3.0	
			O4' of A6	2.5	

bonds to N3 of an adenine and to O2 atoms of the thymines at each end of the tetranucleotide binding site (Table 2). It is noteworthy that there are no direct interactions between the phosphate groups of the DNA and the drug's cationic terminal amidinium groups - indeed drug-DNA complexes in general do not have this feature in spite of the cationic nature of almost all such drugs. The DNA minor groove is narrowest in the drug-bound region and there are extensive non-bonded close van der Waals contacts between netropsin and the walls and floor of the minor groove. These general features have subsequently been observed in complexes of Hoechst 33258[16,17] and berenil[18] to the same sequence. The latter drug only hydrogen-bonds the DNA directly at the 3'-end of its binding site - the 5'-end has a water molecule mediating between amidinium group and an adenine N3 atom. Further aspects of the berenil complex are discussed in Section 4. There are two reported crystal structures of the Hoechst-dodecanucleotide complex, apparently crystallised under very similar conditions. In one the drug is bound to the sequence 5'-AATT;[16] the other[17] has it displaced one base pair in the 5' direction so that it appears in the 5'-ATTC site, with weak interactions to the terminal GC base pair in the site. It is not known whether these results represent real differences in the drug's ability to discriminate between sites with subtle changes in conditions such as pH, or whether they are the consequence of differing interpretations of the (relatively weak) electron density in the minor groove, where both drug and bound water molecules may be present as a virtual continuum of electron density and thus not readily discriminated between.

Two crystal structures have been reported with non-'Dickerson-Drew' sequences. That with netropsin and dCGCGATATCGCG has the drug located as expected in the central AT region, albeit in two equally-populated orientations.[19] It has been plausibly proposed that this is due to the preference of netropsin for a non-alternating sequence, which enables geometrically more acceptable hydrogen bonds to be formed (Table 2). The alternating sequence ATAT accommodates two equivalent arrangements for hydrogen bonding involving O2 and N3 acceptor atoms. In other words, the shape of the netropsin molecule is optimally complementary to N3 .... N3, O2 ... O2 and N3 ... O2 inter/intrastrand separations in the non-alternating sequence. This subtle effect is seen more clearly in the distamycin complex,[20] with a central AAATTT sequence which enables three three-centre hydrogen bonds to be formed. There is also stabilisation from the high propellor twist and major groove bifurcated base-base hydrogen bonds in this region of the structure.

4 STUDIES ON BERENIL

Berenil[21] is an aromatic diamidinium compound with a central triazene linkage. It has marked anti-trypanocidal properties and is used extensively for the treatment of bovine trypanosomiasis. It has high neurotoxicity in man. More recently, berenil has been found to have some degree of oncoviral DNA polymerase inhibitory activity.[22]

The drug binds to double-stranded DNA and stabilises the double helix to thermal denaturation. The interaction is not an intercalative one, since unwinding has not been detected. As shown in Table 1, binding occurs preferentially to AT sequences, with behaviour very similar to that of other established groove binders.[23]

Several theoretical studies of berenil-DNA groove binding have been reported.[24-26] The initial model, based on a molecular mechanics force-field with a multipole expansion of the electrostatic terms and explicit polarisation contributions, found that the drug could fit snugly into a B-DNA minor groove, with the amidinium groups forming hydrogen bonds to successive thymine O2 atoms on opposite strands.[24] This implies preferred binding to an alternating AT sequence, in accord with footprinting data.[27] This 1,2 binding model was confirmed by two subsequent molecular mechanics studies;[25,26] however, a re-examination of the problem with systematic examination of a large number of potential binding modes has concluded that a 1,3 model is energetically more favoured, by about 5 kcal/mole. This has a berenil molecule interacting with an AT sequence by means of hydrogen bonds to an adenine or thymine, and to the next but one adenine (or thymine).[28] Strong support for the model has come from the *X*-ray analysis of a berenil-oligonucleotide complex. In this the drug hydrogen-bonds in a 1,3 manner to the DNA, although an intervening water molecule is required in order to span effectively to the 5'-end adenine at the binding site. Presumably, replacement of this by thymine would remove the necessity for the water molecule as a bridge. Thus, the 'isohelicity' concept for groove binders turns out to have an unexpected subtlety; even if complementary fitting between ligand and DNA is not quite possible, then a water molecule can fulfil this role.

5 CONCLUSIONS

X-ray crystallography and NMR studies are providing definitive data on AT-selective groove binders, and are showing that slight differences in sequence can result in significant recognition differences. Theoretical approaches of greater sophistication than have been hitherto used are required in order to explain these findings, and to predict new patterns of sequence selectivity. The problem of selective GC recognition is a yet more difficult one, and as yet no satisfactory general rules have been developed. The steric difference of the N2 amino group of guanine in the minor groove, a hydrogen bond acceptor, is an obvious one to exploit,[6] but it is clear that roles are played by the more complex factors of enlarged GC groove width, reduced electrostatic potential in a GC region and changes in displaceable water structure.

REFERENCES

1 J A McClarin, C A Frederick, B-C Wang, P Greene, H W Boyer, P Grable, and J M Rosenberg, *Science*, 1986, **234**, 1526.

2 N C Seeman, J M Rosenberg, and A Rich, *Proc Natl Acad Sci USA*, 1976, **73**, 804.

3 Z Otwinowski, R W Schevitz, R-G Zhang, C L Lawson, A Joachimiak, R Q Marmorstein, B F Luisi, and P B Sigler, *Nature*, 1988, **335**, 321.

4 H R Drew and A A Travers, *Cell*, 1984, **37**, 491.

5 P B Dervan, *Science*, 1986, **232**, 464.

6 J W Lown, K Krowicki, U G Bhat, A Skorobogaty, B Ward, and J C Dabrowiak, *Biochemistry*, 1986, **25**, 7408.
7 D Goodsell and R E Dickerson, *J Med Chem,* 1986, **29**, 727.
8 C Zimmer and U Wahnert, *Prog Biophys Mol Biol,* 1986, **47**, 31.
9 H R Drew and R E Dickerson, *J Mol Biol*, 1981, **151**, 535.
10 M W Van Dyke, R P Hertzberg, and P B Dervan, *Proc Natl Acad Sci USA*, 1982, **79**, 5470.
11 M J Lane, J C Dabrowiak, and J N Vournakis, *Proc Natl Acad Sci USA*, 1983, **80**, 3260.
12 D J Patel and L Shapiro, *Biochimie*, 1985, **67**, 887.
13 J G Pelton and D E Wemmer, *Proc Natl Acad Sci USA*, 1989, **86**, 5723.
14 H M Berman, S Neidle, C Zimmer, and H Thrum, *Biochim Biophys Acta*, 1979, **561**, 124.
15 M L Kopka, C Yoon, D Goodsell, P Pjura, and R E Dickerson, *J Mol Biol,* 1985, **183**, 553.
16 P E Pjura, K Grzeskowiak, and R E Dickerson, *J Mol Biol,* 1987, **197**, 257.
17 M Teng, N Usman, C A Frederick, and A H J Wang, *Nucleic Acids Res,* 1988, **16**, 2671.
18 D G Brown, M R Sanderson, J V Skelly, T C Jenkins, T Brown, E Garman, D I Stuart, and S Neidle, *EMBO J*, in press.
19 M Coll, J Aymami, G A van der Marel, J H van Boom, A Rich, and A H J Wang, *Biochemistry,* 1989, **28**, 310.
20 M Coll, C A Frederick, A H J Wang, and A Rich, *Proc Natl Acad Sci USA*, 1987, **84**, 8385.
21 B A Newton, in 'Antibiotics III, mechanism of action of antimicrobial and antitumour agents', ed J W Corcoran and F E Hahn, Springer-Verlag, Berlin, 1975, p 34.
22 E De Clercq and O Dann, *J Med Chem,* 1980, **23**, 787.
23 A W Braithwaite and B C Baguley, *Biochemistry,* 1980, **19**, 1101.
24 N Gresh and B Pullman, *Mol Pharmacol,* 1984, **25**, 452.
25 L H Pearl, J V Skelly, B D Hudson, and S Neidle, *Nucleic Acids Res,* 1987, **15**, 3469.
26 F Gago, C A Reynolds, and W G Richards, *Mol Pharmacol,* 1989, **35**, 232.
27 J Portugal and M J Waring, *Eur J Biochem,* 1987, **167**, 281.
28 T C Jenkins and S Neidle, to be published.

DESIGNED ENZYMES: NEW PEPTIDES THAT FOLD IN AQUEOUS SOLUTION AND CATALYSE REACTIONS

Kai Johnsson, Rudolf K Allemann, and Steven A Benner

Laboratory for Organic Chemistry, Swiss Federal Institute of Technology, CH–8092 Zurich, Switzerland

1 INTRODUCTION

The design of polypeptides that fold in solution is a central goal of modern protein chemistry, and is being pursued in many laboratories. Degrado, Eisenberg, Richardson, Erickson, and their colleagues have made substantial progress in this area.[1,2] Yet, at present, the folded structures of designed proteins have been characterized only by circular dichroic and other physical studies. No structure of a designed peptide has yet to be proven, and the question remains 'Was the design successful?'

An equally important goal of protein chemists is to design polypeptides that catalyse chemical reactions. Catalytic peptides are well known in the literature, including several that display impressive rate enhancements.[3,4,5] However, little is known about the importance of folded structure (and hence, the success of the design) to this catalytic activity. Indeed, in the case of a model for ribonuclease, impressive catalytic activity was observed somewhat unexpectedly in a dimeric form of a peptide that was designed to bind nucleic acids as a monomer.[4]

Some time ago, David Rozzell decided to apply enzymes that decarboxylate β–keto acids to a biotechnological problem, the industrial synthesis of amino acids (Figure 1). Transaminases were found that catalyse the formation of amino acids from α–keto acids and L–aspartate, both readily available (the second from a biotechnological process). However, most biochemical pathways have evolved to comprise isoenergetic reactions, and these transaminases are no exception. With an equilibrium constant close to unity, the

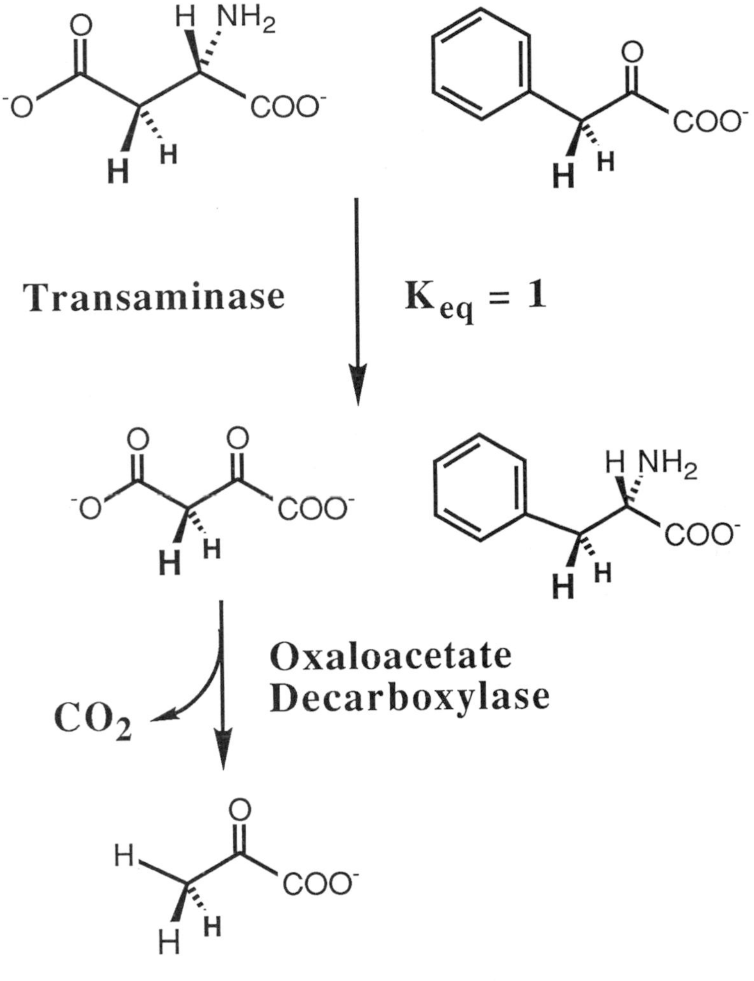

Transaminase $K_{eq} = 1$

Oxaloacetate Decarboxylase

CO_2

Figure 1: A biotechnological process[6] for making amino acids *via* transamination. Like most enzymes catalysing single steps in metabolic pathways, transaminases catalyse a reaction that is close to equilibrium, making the highest yield theoretically possible close to 50%. To drive the reaction to completion, the oxaloacetate derived from aspartate is decarboxylated by an oxaloacetate decarboxylase.

expected yields were only *ca* 50%, certainly unsatisfactory for an industrial process.

Rozzell recognized that the reaction could be driven by decarboxylating oxaloacetate, one product of the transamination, to pyruvate, using an oxaloacetate decarboxylase. He isolated an oxaloacetate decarboxylase from *Pseudomonas putida* and developed a rather successful process for the production of phenylalanine using it.[6]

The oxaloacetate decarboxylases known from Nature use manganese or magnesium as a cofactor. Presumably, the metal coordinates to the α-keto acid moiety of oxaloacetate in the active site, acting as an electron sink (Figure 2a).[7] Indeed, most β-decarboxylases that act on substrates having this α-keto acid unit have evolved to use a metal ion as a cofactor,[8] as this appears to be the way to obtain the behaviours (in particular, rapid turnover at a low cost)[9] that are desired by natural selection. Enzymes acting on substrates that do not have an α-keto acid moiety (such as acetoacetate decarboxylase) cannot so easily use a metal ion as cofactor, as there is no chelating group on the substrate. These enzymes have evolved an alternative mechanism, in which a lysine at the active site forms an imine (Schiff's base) with the substrate (Figure 2b).[10]

2 DESIGN OF AN ARTIFICIAL OXALOACETATE DECARBOXYLASE

The requirement of the natural enzyme for a metal cofactor means that the biotechnological process for making amino acids must be run in a solution containing a saturating concentration of metal ion. This is more inconvenient for the biotechnologist than for a cell, and the increased catalytic power of a metal-containing decarboxylase may not be sufficient to compensate for the inconvenience. In other words, the 'mechanistic imperative' felt by natural selection is not the same as that felt by the biotechnologist.[11] Applying the rule (which also applies to site-directed mutagenesis experiments) that protein engineering is likely to produce in the laboratory a better product than that produced by Nature only when the goals of the engineer are different from the goals of Nature,[12] this might well be a place for

(a)

(b)

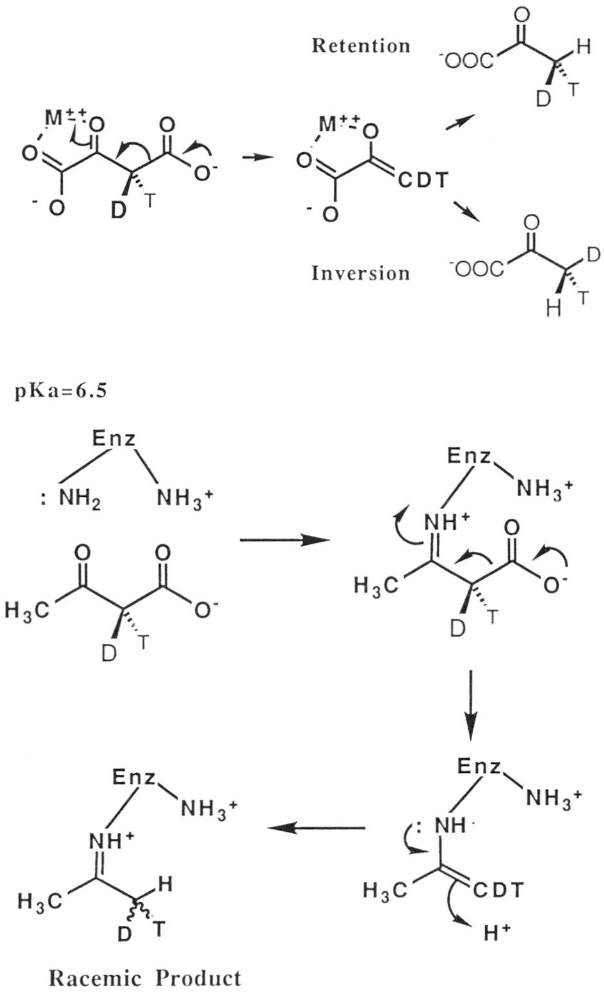

Figure 2: All oxaloacetate decarboxylases suitable for the biotechnological process require a divalent metal ion (manganese or magnesium). Presumably, the metal acts as an 'electron sink' after coordinating to the α–keto acid moiety of the substrate. The overall stereospecificity of the reaction differs for different metal–dependent oxaloacetate decarboxylases, suggesting that the stereospecificity is not a selected trait. An alternative mechanism is used by natural enzymes that catalyse a β–decarboxylation on a substrate that lacks an α–keto acid moiety (and therefore presumably cannot coordinate a metal in a manner effective to catalysis). This involves an imine (Schiff's base) intermediate between the substrate and a lysine in the active site of the enzyme.

a protein designer to have an impact. Thus, we set out to design *de novo* an enzyme to catalyse the last step in Rozzell's process *without* a metal, using instead an imine mechanism.

Several difficulties in the design were foreseeable. The ε-amino group of a normal lysine is not a good catalyst for the decarboxylation of oxaloacetate, at least at pH 7. To be an active nucleophile, the amine must be unprotonated. With a pK_a of 10.5, the ε-amino group of a normal lysine is largely protonated at physiological pH. Not surprisingly, Cbz-lysine (pK_a 10.5) catalyses the decarboxylation of oxaloacetate at pH 7 only poorly, with a second order rate constant of 0.08 $M^{-1}min^{-1}$.

Enzymes that use an active-site lysine to form an imine to decarboxylate a β-keto acid have evolved an active site specifically to circumvent this problem. In acetoacetate decarboxylase (AAD), the pK_a of the reactive lysine has been perturbed by over 4 pK_a units, to 6.0.[10] While we do not know for certain how this perturbation is effected, the reactive lysine is adjacent to a *second* lysine in the polypeptide sequence. A reasonable hypothesis is that a positive charge on the second lysine discourages the protonation of the reactive lysine, and may help bind the negatively-charged substrate as well.

Our substrate, oxaloacetate, has two negative charges. We need to design a polypeptide that, upon folding, brings together at least three lysines in the active site. Two of these should bind the substrate, whilst the pK_a of the third lysine should not be lowered, because it must be available as a nucleophile to form an imine at neutral pH. Unlike natural enzymes, our designed catalyst will be small, meaning that there will not be a large mass of protein to force the highly flexible side chains of lysine close together in space. Indeed, in polylysine, the pK_a's of the residues are quite normal (above 10), showing that the side chain amines can readily *escape* the unfavourable interactions that might lower their pK_a's. Therefore, to be safe, we placed four lysines around the intended reactive lysine to *buttress* the pK_a-perturbing groups around the central lysine (Figure 3).

One must note the *dis*advantages of this design from the point of view of forming a stable folded structure. Any interaction between lysines in the folded form of a peptide that lowers the pK_a of one,

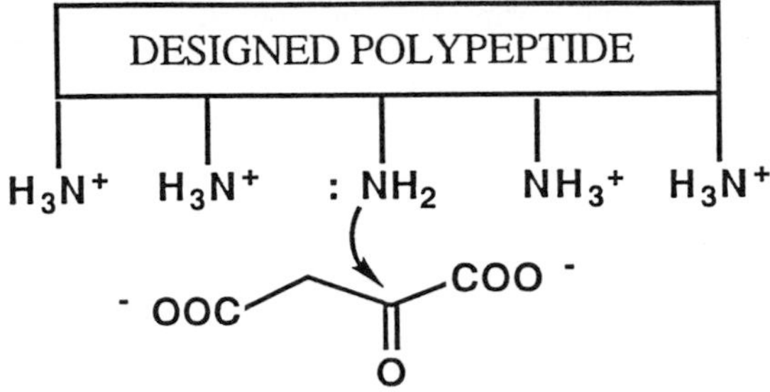

NH_2-Leu-Ala-Lys-Leu-Leu-Lys-Ala-Leu-Ala-Lys-Leu-Leu-Lys-Lys-$CONH_2$

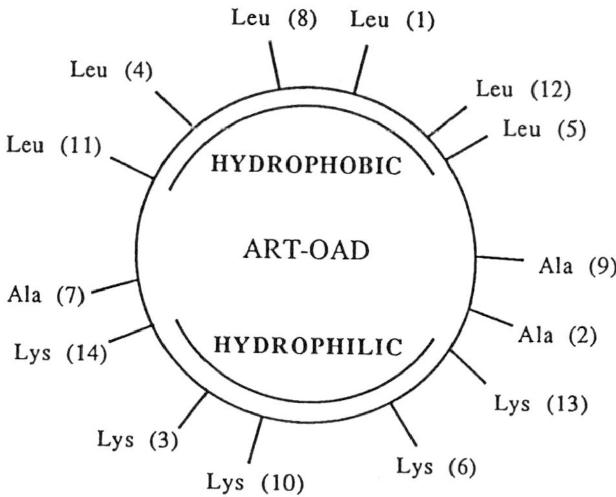

Figure 3: A schematic diagram of a peptide designed to catalyze the decarboxylation of oxaloacetate, where the folded structure brings 5 lysine side chains together in space. The central lysine is intended to have a low pK_a, and form an imine with the substrate. The lysine residues on each side lower the pK_a of the central lysine, and bind to substrate. The outside pair of lysines serves to buttress the inner lysines. Below is the sequence of the designed peptide (ART–OAD), and a representation of the peptide as a helical wheel. The helical form is amphiphilic, with lysines together on one side, and hydrophobic residues on the other. The peptide was synthesized on an Applied Biosystems 430A polypeptide synthesizer using standard chemistry (*p*–benzhydrylamine resin, using *t*–butyloxycarbonyl protected symmetric anhydrides of the amino acid building blocks, and 2–chlorobenzyloxycarbonyl protecting groups for the lysine side chains), cleaved and deprotected with trifluoromethanesulfonic acid in trifluoroacetic acid containing thioanisole and ethanedithiol, and purified to homogeneity by reverse phase HPLC. The structure was proved by automated sequencing (Applied Biosystems 470A Protein Sequencer) and FAB mass spectroscopy.

destabilizes the folded form of the peptide with respect to the unfolded form. The laws of thermodynamics require that the increased reactivity of the lysine with the low pK_a must come, kilocalorie for kilocalorie, from the conformational energy of the peptide. This fact is not appreciated by some workers in the field; it is a general principle for enzyme design.

Lowering the pK_a by 3 orders of magnitude does not necessarily increase the reactivity of an amine by 3 orders of magnitude, as the reactivity of amines as nucleophiles decreases with decreasing pK_a. This relationship is expressed in the form of a Bronsted plot. For the decarboxylation of oxaloacetate catalysed by simple amines, the slope of the Bronsted plot is *ca* 0.5.[13] This implies that for each unit the pK_a of the amine is lowered, one gains an order of magnitude in the concentration of the reactive species, but loses a half of an order of magnitude in the reactivity of that species. Thus, lowering the pK_a of a lysine from 10.5 to 7.5 is expected to increase its catalytic power (at pH 7) by a factor of *ca* 30, not by a factor of 1000.

One may wonder why we are focusing on the amine with a low pK_a required for the *formation* of the imine, while the reaction sequence has other steps (Figure 2b) that might need catalysis. There are two reasons. First, in solution, the formation of the imine is rate-limiting. Second, and more subtly, our work with natural decarboxylases suggests that the overall stereospecificity of decarboxylation is not an evolutionarily selected trait.[14] This implies (as a 'soft' hypothesis, one not intended to be proven, but rather simply to guide evolutionarily the semi-random walk involved in enzyme design) that the second step in the decarboxylation reaction (the protonation of an enamine intermediate) is not in critical need of catalysis, and can be ignored in the design of an artificial enzyme.

The first sequence that was examined is shown in Figure 3. It was synthesized by standard methods,[15] and purified to homogeneity by HPLC. The peptide is short, only 14 amino acids long. However, it can form an amphiphilic helix which brings five lysine residues together in three-dimensional space. 'Helix forming' amino acids[16] incorporated into the designed peptide encourage this secondary structure. However, in contrast to other workers in the field,[1] we have not tried to make the most stable helix possible, but rather one with a relatively low conformational stability. Low conformational stability appears to be a selected trait of natural proteins,[17] and, as discussed below, conformational instability is essential in demonstrating that the peptide catalyses decarboxylation because it adopts a folded form. Several interactions destabilize the helical conformer, including those between the lysines themselves (mentioned above), and an unfavourable interaction between the charged *N*-terminal amino group and the helix dipole.[18]

The stability of the amphiphilic helical structure should be increased in the presence of an aqueous-organic interface,[19] by quaternary interaction with other helices,[20] when covalently linked to a polypeptide fragment that is capable of forming a hydrophobic binding site,[21] or in 'structure forming solvents' such as 2,2,2-trifluoroethanol (TFE).[22] Not surprisingly, CD spectra show that the fraction of the peptide in the helical conformation increases with increasing fraction of TFE and at increasing pH (Figure 4). Further, the fraction of helix increases with increasing concentration of the peptide. The dependency of ellipticity upon the concentration of the peptide,[23] however, suggested that a *tetramer* was formed upon aggregation, rather than a dimer, with an effective disassociation constant of *ca* 1 mM.

3 CHARACTERISATION OF THE ARTIFICIAL OXALOACETATE DECARBOXYLASE BY NMR

At this point, the structure has met the same standards of structural characterization as met by previous workers.[1] However, CD data are not a proof of conformation for a peptide in solution. Therefore, we obtained two dimensional NMR spectra of the designed polypeptide. Figures 5, 6, and 7 show respectively the NH–C(α) region of a phase sensitive DQF-COSY spectrum, the NH–NH region of a NOESY spectrum, and the NH–C(α) and NH–C(β) region of a ROESY spectrum of the peptide in water–TFE (80:20) mixtures at pH 7.0.[24] The data from these (and other) spectra used to assign the secondary structure of the peptide are collected in Figure 8.

The following facts were used to assign the signals in the spectrum. The COSY spectrum (Figure 5) shows 11 cross peaks

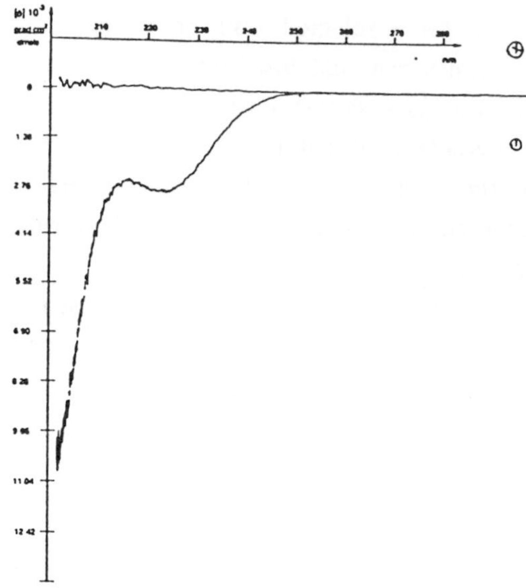

B)

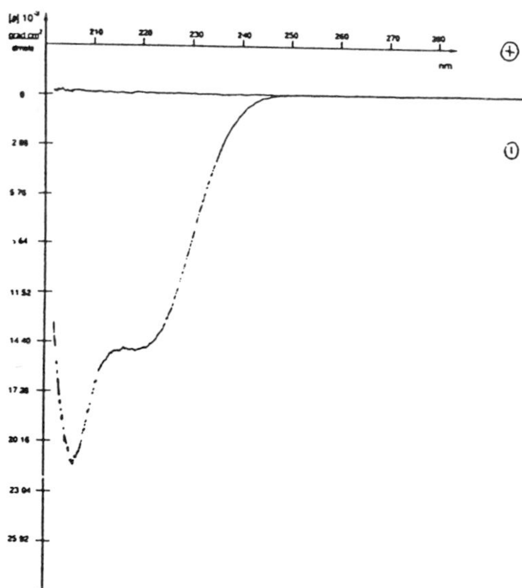

Figure 4:
A Circular dichroism spectrum (Jobin–Yvon Mark III CD spectrophotometer) of the designed peptide (35 μM) in aqueous potassium phosphate buffer (1 mM, pH 7.0 at 25 °C) containing 100 mM potassium perchlorate. The ordinate scale reflects mean residue ellipticity (deg cm ^{2}/dmol).
B Circular dichroism spectrum of the designed peptide (23.3 μM) in a 1:1 mixture of 2,2,2–trifluoroethanol and aqueous potassium phosphate buffer (1 mM, pH 7.0 at 25 °C) containing 100 mM potassium perchlorate.

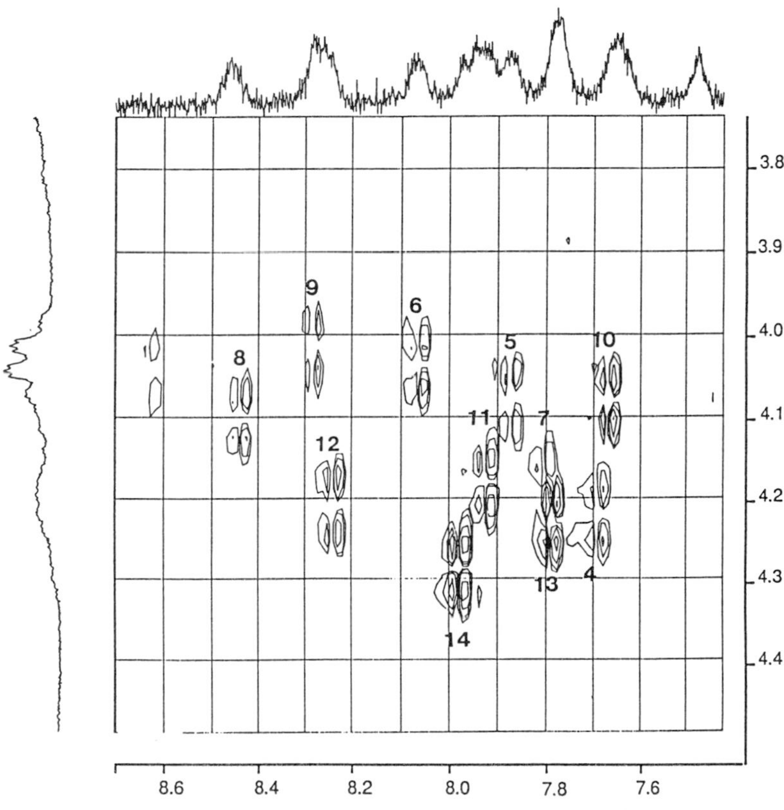

Figure 5: The spectral region of a 300 MHz phase sensitive DQF-COSY (Bruker 300W spectrometer) of the designed oxaloacetate decarboxylase in a 20:80 mixture of trifluoroethanol and aqueous potassium phosphate buffer (1 mM, pH 7.0, containing 10% D_2O) containing the NH-C(α) cross peaks. Numbers correspond to the position of the amino acid giving rise to the signal in the sequence, assigned as discussed in the text.

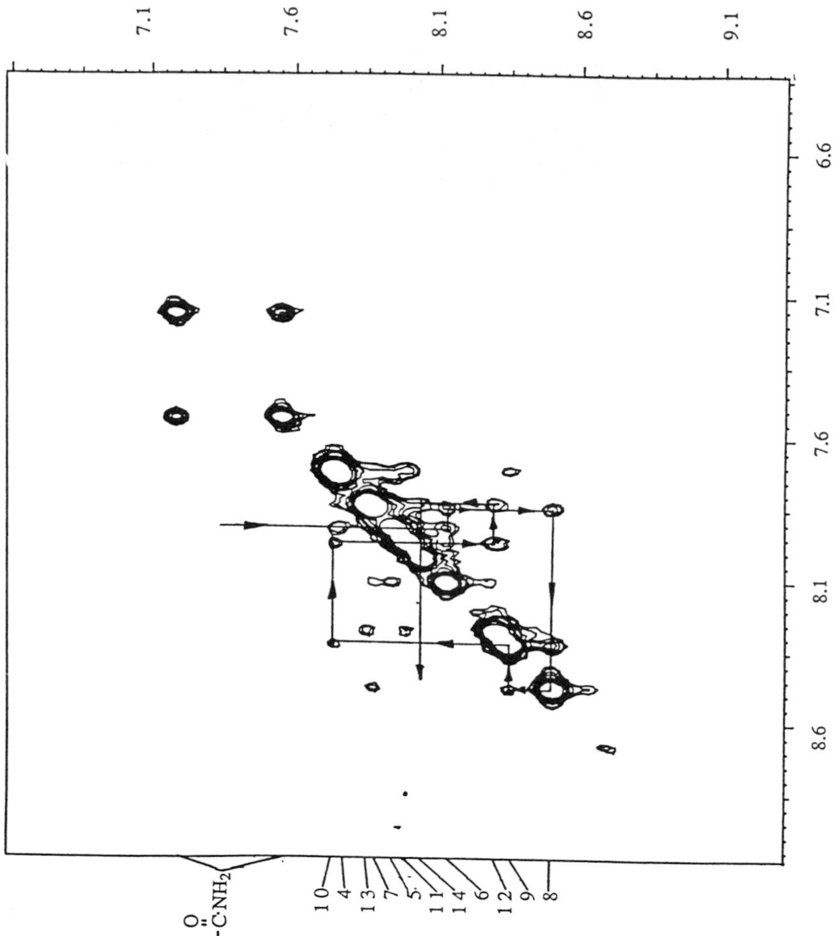

Figure 6: The spectral region of a 300 MHz NOESY (Bruker 300W spectrometer, mixing time 150 ms) of the designed oxaloacetate decarboxylase in a 20:80 mixture of trifluoroethanol and aqueous potassium phosphate buffer (1 mM, pH 7.0, containing 10% D_2O) containing the NH–NH cross peaks. d_{NN} connectivities of the sequential assignment pathway in the helix are identified with arrows that connect the corresponding cross peaks.

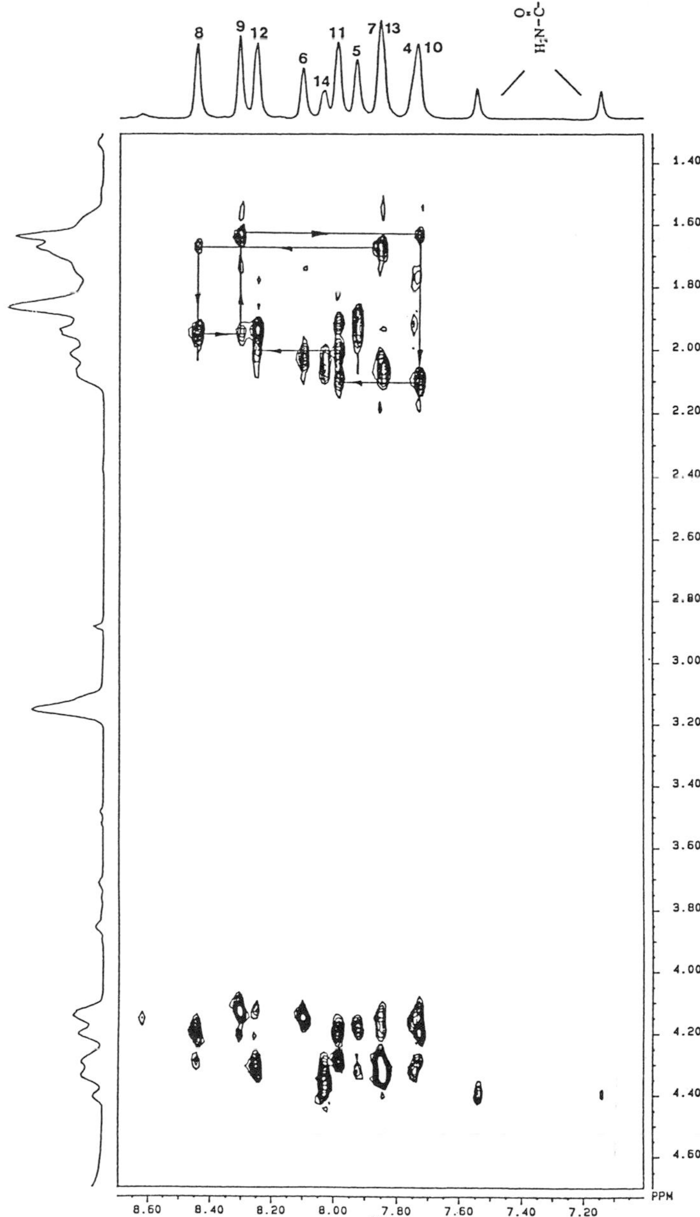

Figure 7: The spectral region of a 600 MHz ROESY (Bruker AM600 spectrometer, mixing time 200 msec) of the designed oxaloacetate decarboxylase in a 20:80 mixture of trifluoroethanol and aqueous potassium phosphate buffer (1 mM, pH 7.0, containing 10% D_2O) containing both the NH–C(α) and NH–C(β) cross peaks. $d_{\beta N}$ connectivities of the sequential assignment pathway in the helix are identified with arrows that connect the corresponding cross peaks. Inter-residual cross peaks within the NH–C(α) region are labelled.

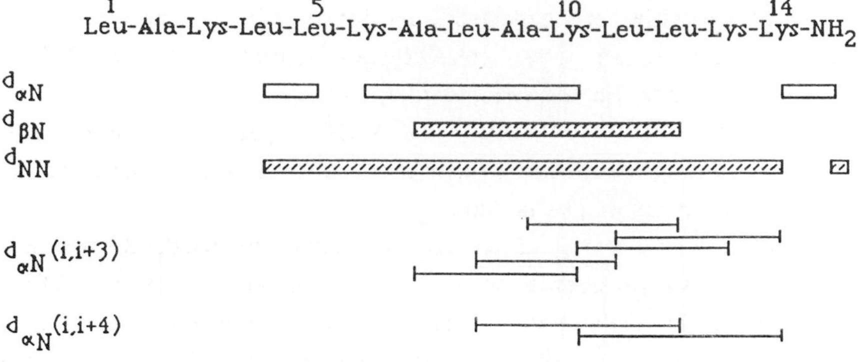

Figure 8: Amino acid sequence and survey of sequential and medium range NOE connectivities observed in the spectra of the artificial decarboxylase. Hatched and opened bars indicate sequential $d_{\alpha N}$, $d_{\beta N}$ and d_{NN} connectivities represented by strong and weak NOEs, respectively. $d_{\alpha N}$(i, i+3) and $d_{\alpha N}$(i, i+4) connectivities are represented by lines below the sequential connectivities.

between C(alpha) protons and backbone amide protons. The three amino terminal amide protons are not expected to be easily observed, as these do not form intramolecular hydrogen bonds, and therefore exchange more rapidly at pH 7.0. Connectivities between the backbone amide protons observed in the NOESY spectrum (Figure 6) permitted the assignment of these signals to 11 consecutive amino acids in the sequence. Cross peaks in the ROESY spectrum involving the carboxy-terminal amide group definitively established that these 11 residues were at the carboxyl terminus of the polypeptide. Further evidence for these assignments was gained from connectivities apparent in the ROESY and NOESY spectra between the amide protons and the C(β) protons. The TOCSY spectrum permitted assignment of the spin systems in the peptide.

The following facts were used to assign a helical structure to the peptide (Figure 8). In the NH–C(alpha) region of the NOESY spectrum, only intraresidual cross peaks were observed. However, in the ROESY spectrum, five $d_{\alpha N}$,[25] five $d_{\alpha N}$(i, i+3) and two $d_{\alpha N}$(i, i+4) connectivities were clearly resolved, again consistent with a helical structure. The helix is very likely of the α type, as $d_{\alpha N}$(i, i+4) connectivities are expected only for α helices, and there is no evidence for $d_{\alpha N}$(i, i+2) connectivities indicative of 3_{10} helices. Further evidence for helical secondary structure is gained from $d_{\alpha N}$ connectivities in the ROESY and NOESY spectra. These results establish firmly a helical structure for the 11 amino acids at the carboxyl terminus of the peptide.

With the structure of the small peptide in hand, it was next appropriate to investigate some of its physical properties. Direct titration of the polypeptide (4 mM in water) detected several amino groups with pK_a's of *ca* 10.6, one amino group with a pK_a of 9.5, and one amino group with a pK_a of 7.2. Of these last two, it was tempting to assign the former to the *N*-terminal amino group, and the latter to the amino group of lysine 10. However, the reversed assignment remains a good possibility, as the pK_a of the amino terminal amino group might be influenced by the helix dipole (*vide infra*).[18,26]

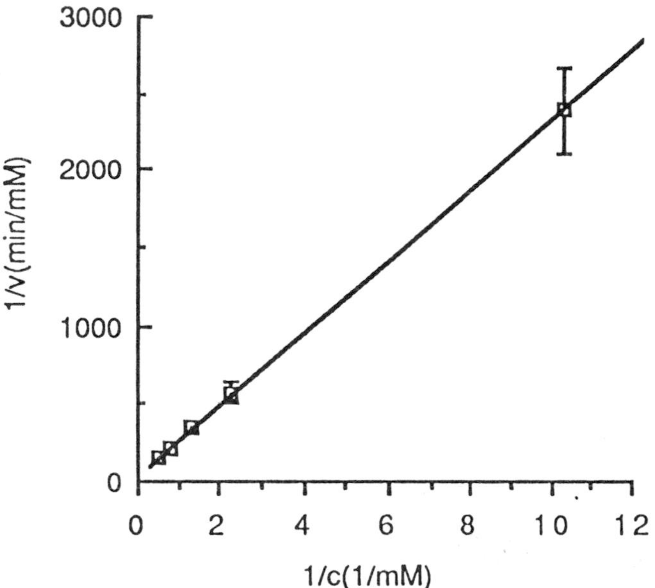

Figure 9: Lineweaver–Burk plot for the decarboxylation of oxaloacetate catalysed by the designed oxaloacetate decarboxylase (0.1 mM) in a 5:95 mixture of trifluoroethanol and aqueous BES buffer (20 mM, pH 7.0, 20°C).

4 CATALYTIC ACTIVITY OF THE ARTIFICIAL OXALOACETATE DECARBOXYLASE

The designed peptide is catalytically active as an oxaloacetate decarboxylase at 100 μM concentration in pure water (where the helix is only *ca* 15% formed). Saturation kinetics are observed (Figure 9), indicating a change in rate-determining step consistent with a Michaelis-Menten kinetic scheme with a K_M of 8 mM. This compares with a K_M of the natural enzyme (using manganese) of 4 mM.[7]

How well have we done in terms of rate enhancement? Here, there is no accepted standard in the literature that defines a 'satisfactory' rate enhancement for a designed enzyme. For catalytic antibodies, rate enhancements of 200-fold are welcomed; the best that are observed are 10,000-100,000 fold (although these often depend on rough estimates for the rate of the uncatalysed reaction).[27] Thus, we report simply that the artificial oxaloacetate decarboxylase (ART-OAD) has a k_{cat} of 0.6 min^{-1}. This is 2.8 orders of magnitude faster than the rate of spontaneous decarboxylation (0.0009 min^{-1}), but is still 5 orders of magnitude slower than that of the natural enzyme (35,000 min^{-1}). Whether or not one is impressed depends on one's prejudices. We must note however, that the natural enzyme is nearly 2 orders of magnitude larger and, as discussed above, uses a metal chelate (not an imine) as an intermediate evolved for high catalytic turnover. Therefore, we believe that we have done fairly well.

More important are the scientific questions: Does catalysis arise because the artificial enzyme adopts the folded form and has a lysine group of low pK_a? A slight curvature upwards of catalytic activity as a function of the concentration of the designed polypeptide (corresponding to an increase in the fraction of folded form as the peptide aggregates to form tetramer) suggested that it did. Thus, we investigated the impact on catalytic activity of increasing concentrations of TFE. In 5% TFE, where the fraction of helix is greater, the catalytic activity of ART-ORD is also greater. This suggests that the catalytically active species is the helical form. In contrast, the catalytic activity of simple amines (such as Cbz-lysine or Phe-OMe) is constant over the same range of concentrations of TFE, ruling out the possibility that the influence of cosolvent on catalytic activity in the designed peptide is simply a bulk solvent effect.

We next compared the catalytic power of ART–OAD with that of other amines. Here we must compare second order rate constants (or, for the enzymes, k_{cat}/K_M). Phenylalanine methyl ester (also bearing an amine with a pK_a of 7.2) was used as one model (Table 1). Consistent with a Bronsted slope of *ca* 0.5 (see above), phenylalanine methyl ester is about an order of magnitude better as a catalyst than Cbz–lysine, but still nearly 2 orders of magnitude worse than ART–OAD (which itself is 3 orders of magnitude better as a catalyst than Cbz–lysine). However, the k_{cat}/K_M of ART–OAD is 5 orders of magnitude lower than that of the natural enzyme (where the k_{cat}/K_M approaches the diffusion limit). Thus, ART–OAD is only 30% as good a catalyst (in energetic terms) as the natural enzyme. Again, however, the natural enzyme is larger, and uses manganese as a cofactor.

Table 1: Catalytic Power of ART-ORD, Natural Oxaloacetate Decarboxylase and Selected Model Compounds

	k_{cat}/K_M ($M^{-1}min^{-1}$)	log relative rate constant	structural features
Cbz-Lys pK_a 10.53	8.0×10^{-2}	0	an amine
Phe-OMe pK_a 7.2	1.1×10^{0}	1.1	an amine with a low pK_a
Oxaloacetate decarboxylase (designed) pK_a 7.2	7.5×10^{1}	3	an amine with a low pK_a and a binding site (?)
Oxaloacetate decarboxylase (natural)	7.8×10^{6}	8.0	manganese MW 40,000

One straightforward interpretation is that Cbz–Lys is simply an amine, Phe–OMe is an amine with a low pK_a, polylysine is an amine with a normal pK_a and (possibly) a binding site, and ART–OAD is (presumably) an amine with *both* a low pK_a *and* a binding site. Simply lowering the pK_a appears to be worth an order of magnitude in rate enhancement. A binding site appears to be worth 2–3 orders of magnitude.

However, the alternative possibility must be considered, that the amine with the pK_a of 7.2 is the α terminal amino group. This possibility is also consistent with the data that show that the folded form is the catalytically active form, as the helix dipole is also expected to lower the pK_a of the terminal amino group. Experiments now in progress will provide unambiguous assignments of the various pK_a's. However, should the assignment be reversed, it remains necessary to determine unambiguously which amine is the reactive amine. Problems of this type are difficult to solve, as the mechanisms involving attack by different amines are kinetically indistinguishable, and perturbing the structure of the catalyst might influence the catalytic mechanism.

Nevertheless, this is the first time a peptide has been designed to adopt a particular secondary structure, and where the folded structure has been proven by 2-dimensional NMR. It is also the first peptide designed to have catalytic activity where the formation of folded structure has been directly correlated with catalytic activity. Further, these studies offer a reasonably coherent hypothesis explaining the structural origin of catalytic activity. This hypothesis is now open to further experimental test.

5 DESIGN OF FOLDED PEPTIDES

The helix can be stabilized by making it part of a larger peptide that contains a hydrophobic binding site. This is simple in conception, but one of the central problems in peptide design centres around controlling the beginning and end of secondary structural units. While many investigators rely on statistical rules to incorporate 'helix-forming', 'sheet-forming' and 'turn-forming' residues at the desired positions in the structure,[1] these propensities are normally quite small in natural proteins, and predictions are unreliable.

We have invented onc solution to this problem involving 'non-standard structures.' For example, peptides with the sequence (Pro–Xxx–Xxx)$_n$ can neither form an α–helix nor participate in a β–sheet. Instead, they form a 'collagen helix', an extended structure with approximately 3 residues per turn.[28] The proline side-chains contribute to a hydrophobic face of the collagen helix which can interact with and stabilize an amphiphilic helix. While relatively rare in globular proteins (hence the term non-standard), a collagen helix is found in pancreatic polypeptide (PP).[29]

To examine the value of a collagen helix as a structural building block that separates other regions intended to form helices and/or β structures, we constructed in 1987 the sequence shown in Figure 10. The sequence was abstracted from those of several homologous pancreatic polypeptides, with some important differences. First, the prolines present in all of these proteins at either position 13 or position 14 (the so called 'distributed parsing unit')[30] were not present in the designed peptide. This is a significant alteration. Then, for synthetic, catalytic, and structural reasons, several residues not found in any natural PP were introduced, including the His at position 6, Ile at position 14, Glu's at positions 18, 22, and 29, and a Lys at position 25. Finally, no residues were introduced to correspond to residues 32–36 of PP. Thus, although the inspiration for the design came from a natural peptide, the final sequence was different from the consensus sequencing for PP, and quite different from the sequence of any individual PP. However, structural contacts presumably important for dimer formation were left intact, to provide a rapid way to determine whether folding occurred.

```
AJB   APSEPHYPGDDATIEELERFYEDLKEYIEVI

APP   GPSQPTYPGDDAPVEDLIRFYDDLQQYLNVVTRHRY

BPP   APLEPEYPGDNATPEQMAQYAAELRRYINMLTRPRY

PPP   APLEPVYPGDDATPEQMAQYAAELRRYINMLTRPRY
```

Figure 10: The sequence of the designed peptide (AJB), compared with the sequences of avian (APP), bovine (BPP), and porcine (PPP) pancreatic polypeptides.

The peptide was synthesized and examined. The CD spectrum of the peptide was consistent with a superimposition of that expected for a collagen helix and an α-helix, again consistent with the design. The conformation of the peptide was then investigated in aqueous solution by two-dimensional NMR in collaboration with R Baumann and K Wuethrich at Zurich. Judging simply from the appearance of cross peaks including the N-H protons at pK 7, the peptide adopts a folded structure. Further NMR studies are needed to confirm this conclusion.

However, one piece of indirect evidence strongly supports the conclusion that a structure close to the desired one has been achieved. Gel filtration studies show that the peptide has a molecular weight of *ca* 8,000. This molecular weight corresponds to the dimeric form of the peptide, as designed. The dimer is tightly bound in aqueous solution, without cosolvents, and at low concentrations of peptide. It is difficult to see how the peptide would form such a dimer were a stable secondary structure not formed. Further, the dimer contains 62 amino acids, one of the largest peptides designed to date.

6 CONCLUSIONS

Is the 'design' problem solved for enzymes? Yes, and no. We have presented here the first reasonably complete story combining design, catalysis, structure proof, and kinetic investigation. We believe, perhaps immodestly, that this represents substantial progress over the already substantial work in the literature.[1] Important for this progress has been a concern for reactivity in addition to a concern for structure, a sensible use of information provided by evolution, and a small amount of good fortune.

However, it remains difficult to find a true distinction between '*de novo* design' and 'copying Nature'. The α-helix, the reactive lysine with a pK_a perturbed by an adjacent lysine, the four helix bundle, and the collagen helix all come from Nature. Of these, only the α-helix can be truly said to have been *designed* (by Linus Pauling), in that it was an idea in the mind of a scientist before it was known in Nature.

Further, an artificial enzyme presents structural and catalytic problems that can be as complex as those presented by a natural enzyme. For example, our work has not yet described the catalytic activity of ART–OAD in a truly fundamental sense, nor has it rigorously ruled out many complex mechanistic possibilities that are different from the simple picture proposed here. While many of these possibilities imply that ART–OAD is a more active catalyst than we have described (and therefore do not endanger the standing of the protein in the 'numbers game' that concerns many in the field), they are of scientific interest, and emphasize an important point. Just as studies of a single natural enzyme can be a lifelong pursuit, so can the study of an artificial enzyme. For those of us who find that very few problems in structural or mechanistic organic chemistry are 'solved' to the level that we desire, this is a sobering thought.

ACKNOWLEDGEMENTS

We are indebted to Drs B Jaun, C Griesinger and Ruegger (Spectrospin) for obtaining and transforming the two-dimensional NMR spectra reported here. This work was supported in part by a gift from Sandoz AG.

REFERENCES

1 D L Oxender and C F Fox, eds, 'Protein Engineering', New York, A R Liss Inc, 1987.

2 P S Kim, *Prot Eng*, 1988, **2**, 249.

3 P K Chakravarty, K B Mathur, and M M Dhar, *Experientia*, 1973, **29**, 786.

4 B Gutte, M Daeumigen, and E Wittschieber, *Nature*, 1979, **281**, 650

5 B Robson and J Garnier, 'Introduction to Proteins and Protein Engineering', Amsterdam, Elsevier, 1986.

6 J D Rozzell Jr, *Meth Enzymol*, 1987, **136**, 479.

7 S Seltzer, G A Hamilton, and F H Westheimer, *J Am Chem Soc*, 1959, **81**, 4018.

8 J D Rozzell Jr and S A Benner, *J Am Chem Soc*, 1984, **106**, 4937.

9 S A Benner, *Chem Rev*, 1989, **89**, 789.

10 F H Westheimer, 'Proceedings of the Robert A Welch Foundation', 1971, **15**, 7.

11 S A Benner, K P Nambiar, and G K Chambers, *J Am Chem Soc*, 1985, **107**, 5513.

12 K P Nambiar, J Stackhouse, S R Presnell, and S A Benner, 'Enzymes as Catalysts in Organic Synthesis', ed, M Schneider, Reidel, 1986, p325.

13 Data from S P Bessman and E C Layne Jr, *Arch Biochem*, 1950, **26**, 25.

14 J A Piccirilli, J D Rozzell Jr, and S A Benner, *J Am Chem Soc*, 1987, **109**, 8084.

15 J P Tam, W F Heath, and R B Merrifield, *J Am Chem Soc*, 1986, **108**, 5242.

16 P Y Chou and G D Fasman, *Adv Enzymol*, 1978, **47**, 45.

17 S A Benner, *Adv Clin Enzymol*, 1988, **6**, 14.

18 C Mitchinson and R L Baldwin, *Proteins*, 1986, **1**, 23.

19 E T Kaiser, 'Redesigning the Molecules of Life', ed, S A Benner, Springer Verlag Berlin Heidelberg, 1988, p25.

20 D Eisenberg, W Wilcox, S M Eshita, P M Pryciak, S P Ho, and W F DeGrado, *Proteins*, 1986, **1**, 16.

21 V I Lim, *J Mol Biol*, 1974, **88**, 857.

22 J W Nelson and N P Kallenbach, *Proteins*, 1986, **1**, 211.

23 S P Ho and W F DeGrado, *J Am Chem Soc*, 1987, **109**, 6751.

24 S Macura and R R Ernst, *Mol Phys*, 1980, **41**, 95.

25 K Wuethrich, 'NMR of Proteins and Nucleic Acids', New York, Wiley Interscience, 1986.

26 E Ellenbogen, *J Am Chem Soc*, 1952, **74**, 5198.

27 D Hilvert, S H Carpenter, K D Nared, and M M Auditor, *Proc Nat Acad Sci USA*, 1988, **85**, 4953; S J Pollack, J W Jacobs, and P G Schultz, *Science*, 1986, **234**, 1570; A Tramontano, K Janda, and R Lerner, *ibid,* p1566; D Y Jackson, J W Jacobs, R Sugasawara, S H Reich, P A Bartlett, and P G Schulz, *J Am Chem Soc*, 1988, **110**, 4841; D Hilvert, S H Carpenter, K D Nared, and M T M Auditor, *Proc Nat Acad Sci USA*, 1988, **85**, 4953.

28 G E Schulz and R H Schirmer, 'Principles of Protein Structure', Springer-Verlag, Heidelberg, 1985.
29 T L Blundell, J E Pitts, I J Tickle, S P Wood, and C W Wu, *Proc Nat Acad Sci USA*, 1981, **78**, 4175.
30 S A Benner, *Adv Enzyme Reg*, 1989, **28**, in press.

ASPECTS OF THE ENZYMIC HYDROLYSIS OF GLYCOSIDES

Trevor Selwood and Michael L Sinnott

Department of Organic Chemistry, University of Bristol, Cantocks Close, Bristol BS8 1TS, UK

1 INTRODUCTION

At least in terms of simple tonnage, glycosyl transfer must be counted amongst the most important of biochemical reactions, since it is estimated that around two thirds of the carbon in the biosphere exists as carbohydrate (largely cellulose and hemicelluloses).[1] Biochemically, although glycosyl residues can be transferred between oxygen and nitrogen nucleophiles (as in the nucleoside phosphorylases), between two nitrogen nucleophiles (as in the ADP-ribosylation of protein guanidino residues),[2] or even between a nitrogen and a sulphur nucleophile,[3] the most well-studied of such reactions are between oxygen nucleophiles. Of those enzymes catalysing glycosyl transfer between oxygen nucleophiles, most attention has been devoted to the hydrolases. This is for technical reasons: the substrates are usually non-ionic and readily purified and the hydrolases are only rarely membrane bound (they can even be extracellular). Glycosyl transfer in the anabolic direction (*eg* from nucleotide diphosphosugars to a growing polysaccharide chain) is catalysed by enzymes which are membrane bound in an ordered way, and present in very low concentrations[4] in the cell.

Enzymic glycoside hydrolysis, like all nucleophilic displacement reactions, can in principle proceed with retention or inversion of the reaction centre. The initial products of the action of glycosidases do of course subsequently mutarotate to give eventually a solution containing all components of the mutarotation equilibrium (straight chain sugar and its hydrate as well as α and β furanose and pyranose forms[5]). In the case of enzymes liberating glucose or its simple derivatives, it is generally fairly easy to discover conditions in which enzymic hydrolysis is fast compared to mutarotation, with accessible concentrations of enzyme. The mutarotation of other pyranoses, and of all furanoses, may be inconveniently fast. It is commonly possible, however, to show that a glycosidase proceeds with retention of the anomeric configuration if it has transferase activity (as demonstrated, for example, by the detection of methyl glycoside when the hydrolysis of a good substrate is carried out in dilute methanol[6]). In strict logic the observation of transferase activity by itself is adequate evidence for retention, since by microscopic reversibility, an inverting glycoside hydrolase with some transferase activity would have to work on both anomers of the substrate.

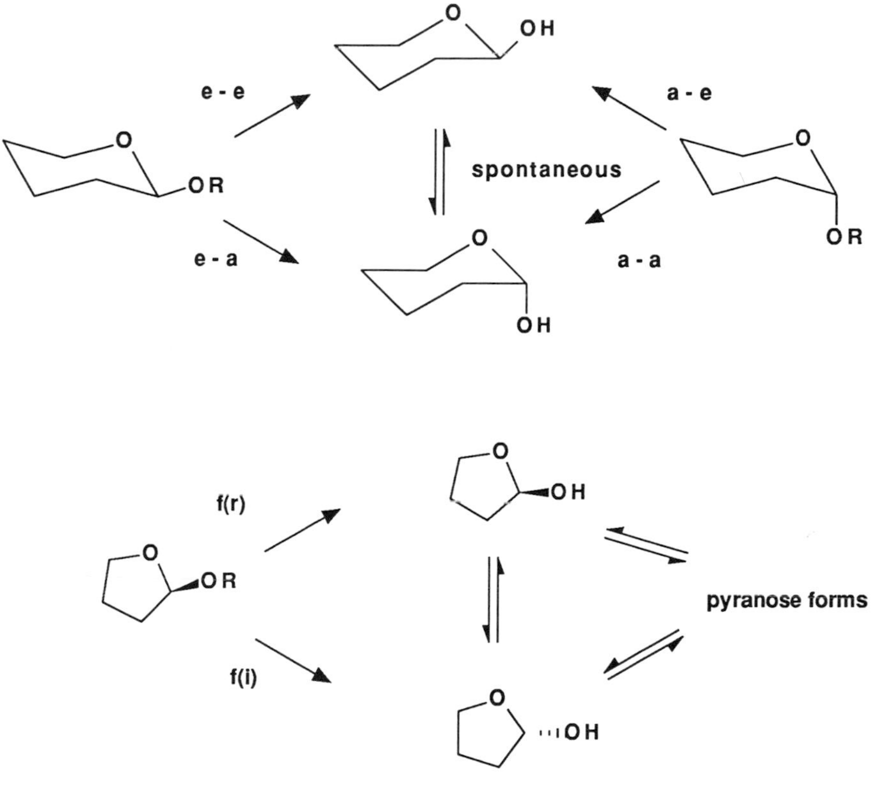

e - e
a - e
spontaneous
e - a
a - a
f(r)
f(i)
pyranose forms

Scheme 1: Types of glycosidase

One can therefore classify glycosidases as either retaining or inverting. Additionally, it is useful to classify them further according to the structure of the substrate, furanoside, or pyranoside with an axial or with an equatorial leaving group. The types of glycosidases are shown in Scheme 1. This article will summarise the evidence that all retaining pyranosidases so far examined work through the same general double-displacement mechanism, and will review in more detail published evidence from this laboratory that a representative furanosidase works by a broadly similar mechanism. Finally, we shall present unpublished work which indicates that the role of proton transfer in two inverting enzymes is not as simple as had been thought.

2 RETAINING PYRANOSIDASES

A general mechanism for retaining pyranosidases which accommodates all the data currently available is that in Scheme 2. Key features of the mechanism are an enzyme carboxylate on the other side of the sugar ring to the leaving group, and a covalent glycosyl-enzyme intermediate formed from this carboxylate and the pyranose ring of the substrate. Although the transition states leading to and from this intermediate have substantial oxocarbonium ion character, glycosyl cation intermediates are not involved. The role of acid catalysis of

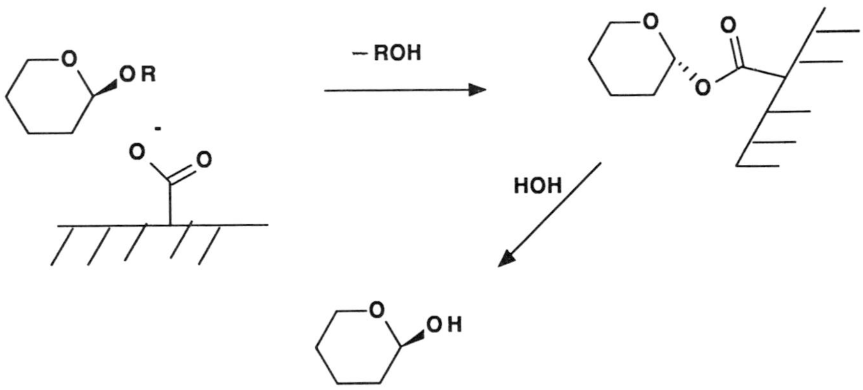

Scheme 2: A general mechanism for retaining pyranosidases

aglycone departure is not crucial, and efficient catalysis can be observed without it. In accord with the idea that electrophilic pull on the aglycone is not an essential component of the catalytic machinery, we have recently advanced evidence that the *lacZ* β-galactosidase of *Escherichia coli* uses Lewis (Mg^{2+}) rather than Brønsted acid catalysis.[7]

The evidence for this general mechanism published before mid-1985 has been reviewed previously by one of us.[8] Important subsequent evidence includes *X*-ray crystallography of barley malt α-amylase,[9] and influenza neuraminidase,[10] site-directed mutagenesis studies on human lysozyme,[11] influenza neuraminidase,[12] and the *lacZ* β-galactosidase of *E coli*,[13] and extensive structural and genetic studies on the cellulase complex of *Trichoderma reesei.*[14] An active site carboxylate has been identified in the retaining cellobioside hydrolase from *T reesei*,[15] and in an endoglucanase from *Schizophyllum commune.*[16] Additionally, Withers and his coworkers have devised suicide inactivators for retaining glycosidases which are based upon a chemical principle which relies heavily on there being glycosyl-cation-like transition states but covalent glycosyl-enzyme intermediates.[17] These compounds are the 2-deoxy-2-fluoroglycosyl fluorides and 2,4-dinitrophenolates. Replacement of the 2-hydroxyl of the pyranoside ring of the substrate with a fluorine atom results in a system less capable of sustaining a positive charge than the unmodified substrate, because of the greater inductive effect of fluorine. However, the presence of a good leaving group makes the first step of the normal turnover sequence take place at a reasonable rate. Consequently the compounds are suicide inactivators by virtue of their forming a covalent glycosyl enzyme intermediate which turns over only slowly because the glycosyl-cation-like transition state leading from the glycosyl-enzyme intermediate is destabilised. Were the glycosyl-enzyme intermediates ionic, this selective deceleration of the second stage would not be achieved.

The double displacement mechanism, acting through covalent intermediates of opposite anomeric configuration to the substrate, presents insuperable difficulties to simple stereoelectronic theory if oxocarbonium-like transition states are involved.[18] Such theory envisages a dihedral angle of 180° between the leaving group and an sp^3 lone pair on the ring oxygen. This arrangement is possible with

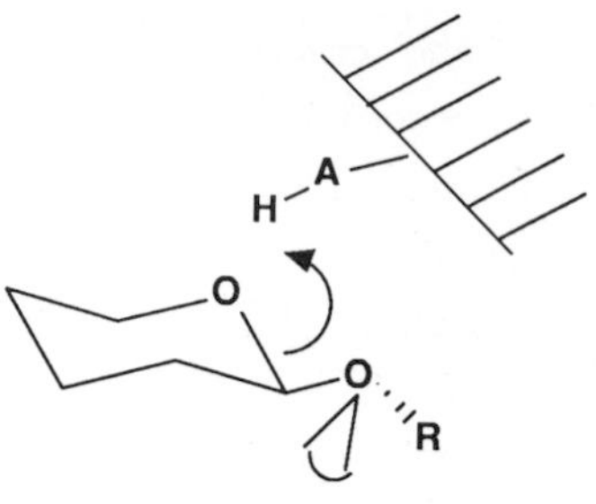

Scheme 3: Ring–opening mechanism of β–pyranosidases

a β–glucosyl derivative only in a non–chair conformer, and with an α–glucosyl derivative only in the $^{4}C_{1}$ chair. Since each of the two chemical steps of a retaining pyranosidase involve an α–β conversion, quite implausible convolutions of the pyranose ring are required. Although it has been argued that this type of stereoelectronic theory represents an over–interpretation of least motion effects,[19] it is in principle possible to rescue the theory in respect of glycopyranosidases by invoking a ring–opening mechanism for β–glycopyranosidases[20,21] (Scheme 3). Such a mechanism requires all substrates on which the enzymes act efficiently to have a structure such that the aglycone can support a positive charge. This is plainly not so, since glycosidases hydrolyse glycosyl fluorides and glycosyl pyridinium salts with high efficiency.[8] Nonetheless, the idea has its supporters, despite the lack of chemical precedent for a ring–opening mechanism during the acid–catalysed hydrolysis of pyranosides, at least in unbiassed systems in water.[22] We reasoned that, if the ring opening mechanism for a glycosidase was to be observed anywhere, it would be observed where there was chemical precedent for it. We therefore examined a typical retaining furanosidase.

3 THE α–L–ARABINOFURANOSIDASE III OF *MONILINIA FRUCTIGENA* DOES NOT OPEN THE FURANOSE RING OF ITS SUBSTRATES

Capon and Thacker[23] originally ascribed the generally faster rates of acid–catalysed hydrolysis of alkyl furanosides, compared to their pyranoside isomers, to the adoption by the furanosides of a ring-opening mechanism, in which nucleophilic participation by water was important. Lönnberg and coworkers subsequently showed that in the case of aldofuranosides there was a competition between the ring-opening mechanism and a mechanism analogous to that adopted by pyranosides, in which initial C–O cleavage takes place.[24,25] The balance between the two mechanisms was dependent on the ring substitution pattern and more importantly, on the electronegativity of the aglycone, with acidic aglycones favouring initial exocyclic cleavage. In the case of arabinofuranosides, the isopropyl glycoside hydrolysed by the ring-opening mechanism. These conclusions were based on solvent effects and entropies of activation, but at least in the case of arabinofuranosides were subsequently confirmed by the observation of 1-^{18}O kinetic isotope effects on the hydrolysis of isopropyl and *p*-nitrophenyl α-arabinofuranosides in different senses.[26] Initial ring opening during acid-catalysed hydrolysis of an arabinofuranoside is therefore firmly chemically precedented.

The fungus *Monilinia* (formerly *Sclerotinia*) *fructigena* is a plant pathogen which is responsible for the brown rot of apples: it excretes a cocktail of glycoside hydrolases which attack the plant cell wall.[27] Amongst these are two α-L-arabinofuranosidases, AFI and AFIII (AFII being intracellular), which attack the arabinogalactan component.[28] Both AFI and AFIII work with retention of the anomeric configuration,[6] and AFIII has been purified:[29] it is a monomer of M_R 40,000. It hydrolyses α-L-arabinofuranosyl pyridinium ions with high efficiency (the ratio of k_{cat} to the spontaneous hydrolysis-rate for the 4-bromoisoquinolinium ion being 2.5×10^9). Breakage of the exocyclic C–O bond limits k_{cat} for the hydrolysis of the *p*-nitrophenyl arabinofuranoside, as is shown by a 3% ^{18}O kinetic isotope effect. This isotope effect makes it likely that the absence of a detectable dependence of k_{cat} for a series of aryl arabinofuranosides on the acidity of the aglycone represents a true β_{lg} of zero, rather than the kinetic dominance of non-bond-breaking steps. The initial bond-breaking step of any ring-opening mechanism would be expected to exhibit an inverse ^{18}O leaving group kinetic isotope effect and a strongly negative β_{lg} value. Although these data with aryl arabinofuranosides could by themselves be rationalised according to the ring-opening mechanism on the assumption that decomposition of some form of acyclic hemiacetal was rate-determining, such a supposition could not also accommodate the data on the pyridinium salts. We therefore conclude that even where there is chemical precedent for the ring-opening mechanism, it is not adopted by a retaining glycosidase, and suggest that it be given no further consideration.

The near-zero β_{lg} value obtained for the hydrolysis of aryl arabinofuranosides by this enzyme, coupled with the ^{18}O kinetic isotope effect which indicates glycone-aglycone bond cleavage is rate determining for the *p*-nitrophenyl compound, suggest that proton donation to the leaving group is relatively far advanced in the first chemical transition state. This particular enzyme-substrate system was therefore considered a promising vehicle on which to try and detect general acid catalysis of aglycone departure using solvent isotope effect probes. A solvent isotope effect on an enzymic reaction can arise from perturbation of essential ionisations or from an accumulation of many small effects arising from exchange at many sites, as well as from proton transfer in a chemical step; it is therefore imperative to perform an extended series of measurements to address both the pK shift problem and the many site problem.[30] The variation of both k_{cat} and k_{cat}/K_m for hydrolysis of *p*-nitrophenyl α-L-arabinofuranoside with a pH or pD ('pL') is sigmoid, with the single pK increasing by the expected 0.5 pK unit in D_2O. There is a modest (1.45) solvent isotope effect on k_{cat} but no significant one on k_{cat}/K_m, in the plateau region. We addressed the many site problem by performing a proton inventory, a measurement of relative rate in mixtures of H_2O and D_2O. In general, the variation of rate is given by:

$$v_n/v_o = \Pi(1-n+n\varphi_T)/\Pi(1-n+n\varphi_R)$$

where n is the atom fraction of deuterium in the solution and φ_T and φ_R are the fractionation factors of exchangeable protons in the transition state and the ground state, respectively. In practice the exchangeable protons in proteins, except those in thiol groups, have fractionation factors near unity. AFIII contains no free thiols, since it is extracellular. Therefore the denominator of this expression can be taken as unity, and the experiment resolves itself into determining whether the experimental variation of the isotope effect is described by a straight line (corresponding to one-proton catalysis), a parabola (corresponding to two-proton catalysis), or a catenary (corresponding to many-proton catalysis). The proton inventory we found was strictly linear, indicating that the plateau isotope effect was indeed caused by the transfer of a single proton, which, from the other evidence outlined above, must be that transferred from the acid-catalytic group to the departing aglycone.[31]

The question as to whether the transition state(s) for this furanosidase were of the oxocarbonium-ion-like type that had long been posited for pyranosidases was addressed by a study of transition state analogue inhibitors.[32] The powerful inhibition of pyranosidases by 1,5-dideoxy-1,5-imino-alditols is well known and thoroughly investigated.[33] The original suggestion that these compounds were powerful inhibitors because of the resemblance of their conjugate acids in charge, though not precise geometry, to aldopyranosyl cations[34] fell into disfavour because of the pH variation of their inhibitory potency, which generally increased with increasing pH.

A more detailed analysis of the whole catalytic apparatus, however, suggests that this pattern of behaviour is what would be expected if these compounds were, as originally suggested, transition state analogues. At the transition state, not only does the ring oxygen of the substrate carry a positive charge, but the acid catalytic group of the enzyme will be partially deprotonated. The transition state will be mimicked, therefore, by the protonated inhibitor binding to a deprotonated enzyme. This situation is indistinguishable by simple binding experiments from deprotonated inhibitor binding to protonated enzyme. Since by definition enzymes with an acid-catalytic group deprotonated are well away from their pH optimum, studies of the pH variation of inhibitory power of iminoalditol inhibitors lend themselves to interpretation in terms of the binding of deprotonated inhibitor to protonated enzyme.

In the case of AFIII, the studies outlined above had indicated a transition state in which the catalytic proton was largely transferred to the leaving group, and had also given indication of the pK of this acid catalytic group. Therefore, if at the transition state for this enzyme the arabinofuranose ring resembles an arabinofuranosyl cation, we would expect the conjugate acids of amine (I) and (II) to bind tightly to deprotonated enzyme. Indeed, the K_i values for both inhibitors show a bell-shaped dependence on pH, with the acid limb of the bell being governed by the pK value obtained from the catalysis studies, and the alkaline limb being governed by the pK of the inhibitors, as measured by titration. The minimal K_i for (II) is 1μM, well below K_m values for substrates; the minimal K_i for (I) is 1.2mM, likewise well below the values for *trans*-1,2-dihydroxy-cyclopentane derivatives.[6] The tighter binding of (I) and (II) is therefore plausibly attributed to their resemblance to the arabinofuranosyl cation (III).

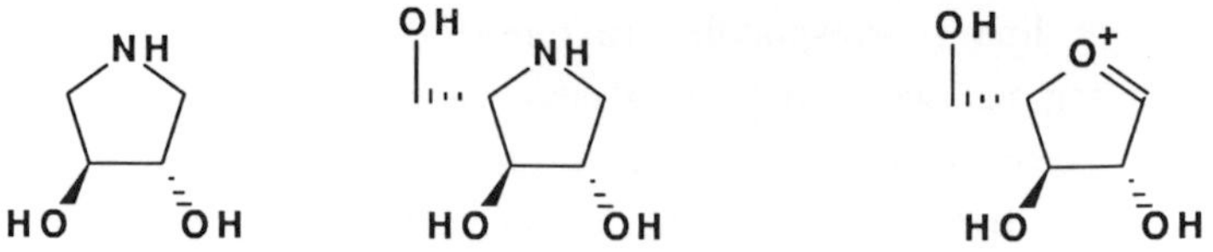

We conclude that this furanosidase reacts through a mechanism closely similar to that of Scheme 1.

4 PROTON-TRANSFER IN THE ACTION OF TWO INVERTING GLYCOSIDASES

The mechanism of those glycosidases acting with inversion of the anomeric configuration has for many years been discussed as the single nucleophilic displacement by a water molecule first suggested by Koshland.[35] The departure of the aglycone is envisaged as being assisted by a general acid catalytic group on the enzyme and the nucleophilicity of the water molecule as being enhanced by an enzymic general base (Scheme 4).

Quite the best evidence for the operation for this mechanism with at least some inverting enzymes comes from Hehre's discovery that inverting glycosidases hydrolyse the appropriate glycosyl fluorides of *both* anomeric configurations.[36] The hydrolysis of the fluoride of the same anomeric configuration as the O-glycoside substrate exhibits normal Michaelis-Menten kinetics, whereas the hydrolysis of the 'wrong' fluoride shows at low substrate concentration a greater than linear increase in the rate of release of fluoride ion relative to substrate concentration. Furthermore, fluoride ion release is

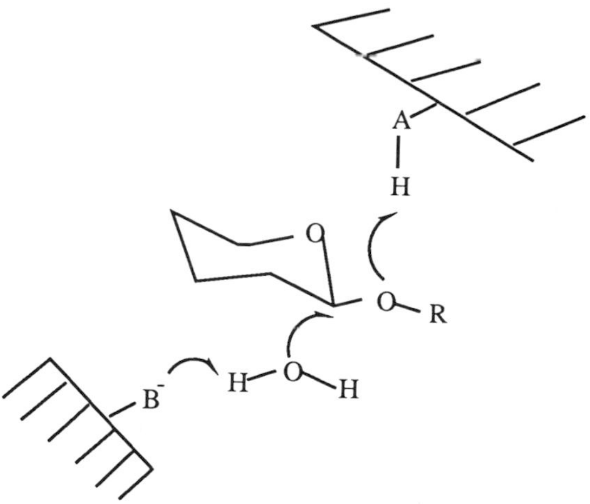

Scheme 4: Koshland's mechanism of action of inverting glycosidases[35]

stimulated by the addition of glycosyl derivatives which would ordinarily be considered competitive inhibitors. In these cases, and also in some of the cases where the enzyme acted upon the wrong fluoride alone, it was possible to isolate transfer products. Thus, in the case of the glucoamylase of *Rhizopus niveus*, addition of methyl α–D–glucopyranoside to a solution of enzyme and β–glucopyranosyl fluoride led to the production of methyl α–maltoside.[37] These data are elegantly rationalised by the suggestion that the 'wrong' fluorides are hydrolysed by a resynthesis–hydrolysis mechanism as shown in Scheme 5, in the first step of which the active form of the enzyme is

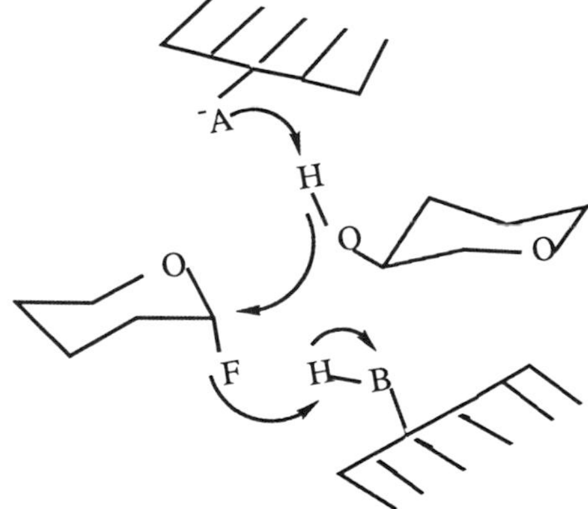

Scheme 5: Resynthesis–hydrolysis mechanism for hydrolysis of 'wrong' glucopyranosyl fluorides

the tautomer of the form which is active in the normal hydrolytic reaction. The special features of the hydrolysis of 'wrong' fluorides provide telling evidence for the presence of both an acid and a basic group in the single displacement mechanism with natural substrates.

The success of this mechanism in rationalising what is known about inverting glycosidases led us to make three predictions:

(i) Two-proton catalysis should be observed during the action of a glucoamylase on a substrate for which bond-breaking is rate-determining.

(ii) The inverting cellobiohydrolase from the cellulolytic fungus *Trichoderma reesei* should hydrolyse α-cellobiosyl fluoride by the resynthesis-hydrolysis mechanism.

(iii) 1,1-Glucopyranosyl difluoride (1-fluoro-D-glucopyranosyl fluoride) may be a powerful inhibitor of glucoamylase, since the β-fluoro group may be able to hydrogen bond to the group AH, whilst at the same time the α-fluoro group may be able to hydrogen bond to BH^+.

In the event none of these predictions was fulfilled. We first examined the question of possible two-proton catalysis by glucoamylase. The enzyme we used is the separated isoenzyme II from the mixture of amyloglucosidases (glucoamylases) produced by *Aspergillus niger*, which is a well-characterised enzyme of known primary structure.[38] Although this enzyme exhibits maximal activity towards α-glucopyranosides with a second glucopyranose residue as the leaving group, it will act slowly on nitrophenyl glucosides. We therefore chose *p*-nitrophenyl α-D-glucopyranoside as the substrate, not only to have a chromophoric change, but also to make it likely that bond-breaking would be rate-determining, rather than the non-covalent events which commonly obscure the chemistry with good substrates.

Figures 1 and 2 show the variation with pL of k_{cat} and k_{cat}/K_m for hydrolysis of the nitrophenyl glucoside by amyloglucosidase II; protein concentrations were converted into concentrations of active sites assuming completely active protein, which is reasonable given the industrial use of the enzyme in bioreactors operating at quite elevated temperatures. Both profiles are sigmoid, with the pK governing k_{cat} shifting from 6.7±0.1 in H_2O to 7.4±0.1 in D_2O, and that governing k_{cat}/K_m shifting analogously from 6.5±0.1 in H_2O to 6.7±0.1 in D_2O.

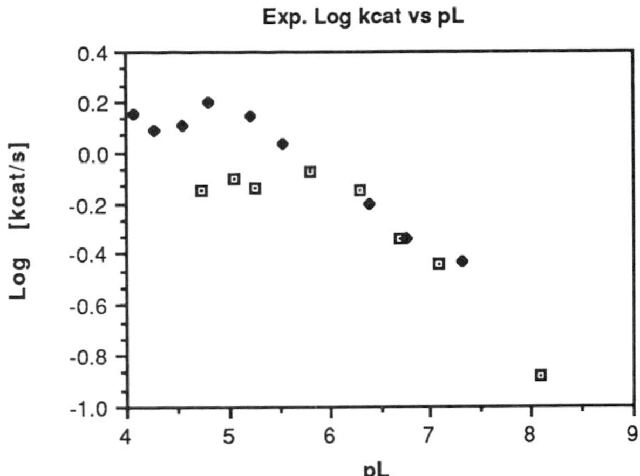

Figure 1: First-order rate constant for the hydrolysis of *p*-nitrophenyl α-D-glucopyranoside by amyloglucosidase II of *Aspergillus niger* (the full symbols represent points for H_2O and the open symbols points for D_2O).

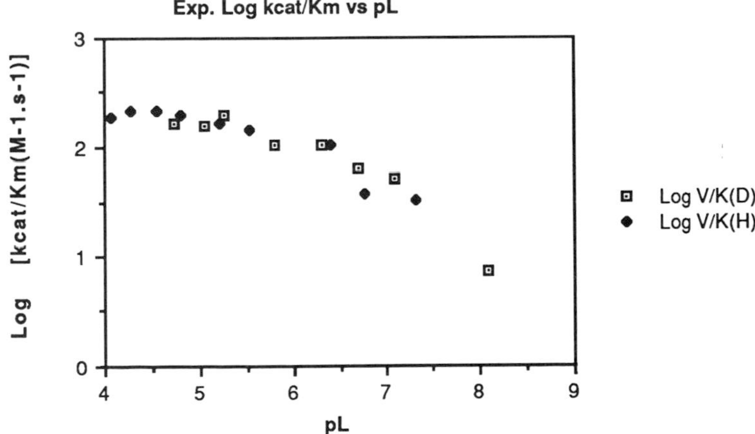

Figure 2: Second-order rate constant for hydrolysis of *p*-nitrophenyl α-D-glucopyranoside by the amyloglucosidase II of *Aspergillus niger*.

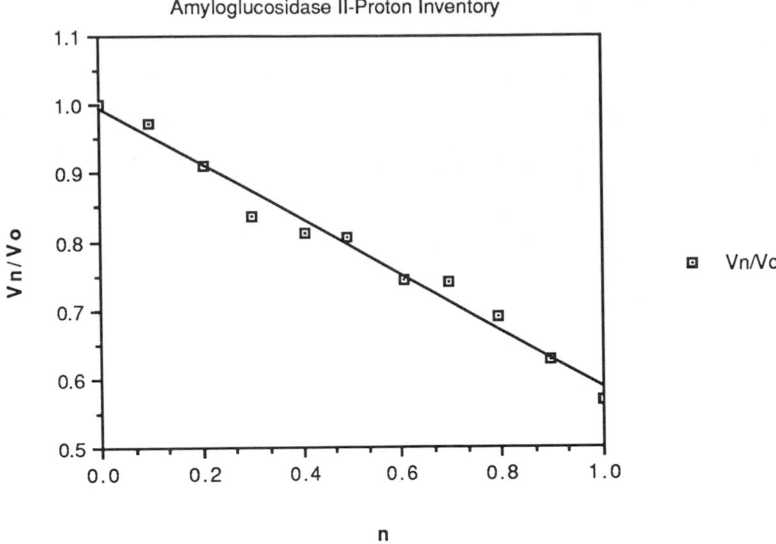

Figure 3: Proton inventory for k_{cat} for hydrolysis of *p*–nitrophenyl α–D–glucopyranoside by the amyloglucosidase II of *Aspergillus niger*

In the plateau region there is only a small isotope effect on k_{cat}/K_m, but a reasonably significant one (1.8) on k_{cat}. The proton inventory on k_{cat} (Figure 3) indicates that this isotope effect arises from a single proton: fitting to a straight line gives φ_1 = 0.58±0.01, with the fit 99.9999% significant by the *F*–test; fitting to a parabola gives φ_1 = 0.64±0.05, φ_2 = 0.92±0.05, with the second term only 21% significant by the *F*–test. One–proton catalysis alone is therefore involved.

It could be argued that nitrophenolate will be a sufficiently good leaving group that any recruitment of the catalytic machinery of the enzyme that transfers the second proton will be redundant. This objection has some force, but could be made with equal force with respect to the proton inventory for the arabinofuranosidase described above, where again a *p*–nitrophenolate leaving group was employed, but in this case partial proton transfer to it was detected. Both proton inventories were performed at buffer ratios which gave a pH below 5, *ie* at least 2 pK units below the pK_a of *p*–nitrophenol, so there is no *a priori* reason why general acid catalysts should not be observed.

The tetraacetate of 1-fluoro-D-glucopyranosyl fluoride has recently become available,[39] and Zemplén deacetylation of this compound yields a non-crystalline material of essentially identical ^{19}F NMR spectrum to that of the tetraacetate, which however fails to show any significant inhibitory potency against amyloglucosidase II (K_i≈36mM).[40]

As part of its machinery for degrading crystalline cellulose, the fungus *Trichoderma reesei* (formerly *viride*) excretes at least five enzymes which hydrolyse β-glucosyl linkages:[41] two endoglucanases, a β-glucosidase, and two enzymes with an exo cellobiohydrolase action on soluble substrates, designated CBHI and CBHII. We recently showed that, acting on β-cellobiosyl fluoride, CBHI gave β-cellobiose as first product and CBHII gave α-cellobiose.[42]

In accord with expectation, CBHII does indeed hydrolyse α-cellobiosyl fluoride as well as β-cellobiosyl fluoride. At the time of writing, however, we have been unable to obtain any evidence whatsoever that this reaction proceeds through the resynthesis-hydrolysis mechanism.[43] The kinetics of hydrolysis of *both* anomers of cellobiosyl fluoride by CBHII are Michaelian, with K_m=8.2±0.1 mM for the α-anomer and 0.282±0.004 mM for the β-anomer (at 25°C and pH 5.2). The V_{max} value for the α-anomer is, surprisingly, 2.0 times greater than that for the β-anomer. If the hydrolysis of the α-fluoride in D_2O is followed essentially to completion by NMR spectroscopy, the production of α-cellobiose parallels the disappearance of α-cellobiosyl fluoride, and there is no indication of any accumulation of cellotetraosyl fluoride. Glucose inhibits, rather than stimulates, the production of fluoride ion from the α-fluoride.

With this data to hand, it is tempting to speculate that the mechanism of hydrolysis of α-cellobiosyl fluoride may be related to the hydrolysis of normal substrates only insofar as the active site of CBHII is conducive to cellobiosyl-cation-like transition states. It is possible to envisage an 'internal return'-like mechanism for the hydrolysis of α-cellobiosyl fluoride by CBHII, in which the water molecule which normally acts as a nucleophile with this substrate first acts as an acid (possibly assisted by partial proton donation by BH^+), and then some ion-pair-like species (be it discrete ion-pair, or ill-defined transition state on a plateau on an energy surface), involving the cellobiosyl cation and H-O-H······F^- on its α face, collapses directly to α–cellobiose and HF. This type of mechanism has close chemical precedent in the solvolysis of α–glucopyranosyl fluoride in mixtures of ethanol, trifluoroethanol, and phenol: most of the product in such solvolyses is the α–glucoside obtained from the most acidic component of the solvent mixture.[44]

5 MATERIALS AND METHODS

Amyloglucosidase II was supplied by NOVO A/S as a lyophilised solid and was used without further purification. *p*–Nitrophenyl α–D–glucopyranoside was obtained from the Aldrich Chemical Co and used without further purification.

H_2O was distilled, deionised and degassed. D_2O (99.8%) was obtained from Aldrich and used without further purification. Buffer solutions were made from 0.1 M sodium acetate/acetic acid (pH 4.07 to 5.52 and pD 4.74 to 5.81) and 0.1 M NaH_2PO_4/Na_2HPO_4 (pH 6.76 to 7.32 and pD 6.38 to 8.08) with analytical grade reagents.

Michaelis–Menten parameters were determined by continuous monitoring of absorbance at 370nm in a Pye–Unicam PU 8800 uv/visible region spectrophotometer with a thermostatically controlled cell–block maintained at 50.0±0.05°C; 1mm path length cells were used for all kinetic runs. Only the first 5–10% of the reaction was followed to obtain the initial rates. The slopes were analysed using the manufacturer's software. All rates were corrected for any change in extinction coefficient of the product on changing pL and solvent.

Michaelis–Menten parameters were calculated by using the program HYPER.[45] The program HBBELL[45] was used to calculate the pKa values governing the pL–dependencies of V_{max} and V_{max}/K_m in H_2O and D_2O. Both programs were translated into BASIC from FORTRAN to run on a BBC master microcomputer.

The pH values of the buffer solutions were measured with a radiometer PM62 standard pH–meter equipped with a glass combination electrode; pD was taken as the pH–meter reading +0.4. Measurements in mixtures of H_2O and D_2O were carried out in 0.1 M sodium acetate/acetic acid at a constant buffer ratio that gave a pH of 4.86 in pure H_2O; isotopic composition was calculated gravimetrically. Initial rate measurements were made at a substrate concentration of 6.77 x 10^{-2} M, by injecting 20 μl of a 0.914 M solution of the substrate in dimethyl sulphoxide into 250 μl of the appropriate buffer solution. These runs were followed at λ = 400nm (to allow a high substrate concentration) at 49.8±0.05°C. There is no solvent isotope effect on the extinction coefficient of the product at 400 nm, or K_m.

ACKNOWLEDGEMENTS

We thank the Science and Engineering, and Agricultural and Food Research Councils for financial support of this work, Professor Jonathan Knowles and Dr Päivi Lehtovaara, VTT, for providing CBHI and CBHII, and Dr Martin Schulein, NOVO A/S, for amyloglucosidase II.

REFERENCES

1 E Gruber, *Papier*, 1976, **30**, 533.
2 G Soman, J Narayan, B L Martin, and D J Graves, *Biochemistry*, 1986, **25**, 4113.
3 R E West, J Moss, M Vaughan, T Liu, and T–Y Liu, *J Biol Chem*, 1985, **260**, 14428.
4 See *eg* (a) V N Shibaev, *Adv Carbohydr Chem Biochem*, 1986, **44**, 277; (b) I W Sutherland, *Ann Rev Microbiol*, 1985, **39**, 243.
5 For a modern study of mutarotation see P W Wertz, J C Garver, and L Anderson, *J Am Chem Soc*, 1981, **103**, 3916.
6 See *eg* A H Fielding, M L Sinnott, M A Kelly, and D Widdows, *J Chem Soc Perkin Trans 1*, 1981, 1013.
7 T Selwood and M L Sinnott, *Biochem J*, in press.
8 M L Sinnott in 'Enzyme Mechanisms', M I Page and A Williams, eds, Royal Society of Chemistry, London, 1987, p 259.
9 B Svensson, R M Gibson, R Haser, and J P Astier, *J Biol Chem*, 1987, **262**, 13682.
10 P J Bossart, Y Sudhakar Babu, W J Cook, G M Air, and

W G Laver, *J Biol Chem*, 1988, **263**, 6421.
11 M Muraki, Y Jigami, M Morikawa, and H Tanaka, *Biochim Biophys Acta*, 1987, **911**, 376.
12 M R Lentz, R G Webster, and G M Air, *Biochemistry*, 1987, **26**, 5351.
13 (a) M Ring, D E Bader, and R E Huber, *Biochem Biophys Res Commun*, 1988, **152**, 1050; (b) D E Bader, M Ring, and R E Huber, *ibid*, 1988, **153**, 301.
14 (a) J Knowles, P Lehtovaara, T Teeri, M Penttilä, I Salovuori, and L André, *Philos Trans Roy Soc London*, 1987, **A321**, 449; (b) P J Kraulis, G M Clore, M Nilges, T A Jones, G Pettersson, J Knowles, and A M Gronenborn, *Biochemistry*, 1989, **28**, 7241.
15 P Tomme and M Claeyssens, *FEBS Lett*, 1989, **243**, 239.
16 A J Clarke and M Yaguchi, *Eur J Biochem*, 1985, **149**, 233.
17 (a) S G Withers, I P Street, P Bird, and D H Dolphin, *J Am Chem Soc*, 1987, **109**, 7530; (b) S G Withers, K Rupitz, and I P Street, *J Biol Chem*, 1988, **263**, 7929; (c) S G Withers and I P Street, *J Am Chem Soc*, 1988, **110**, 8551.
18 M L Sinnott, *Biochem J*, 1988, **224**, 817.
19 M L Sinnott, *Adv Phys Org Chem*, 1988, **24**, 111.
20 G W J Fleet, *Tetrahedron Lett*, 1985, **26**, 5073.
21 C B Post and M Karplus, *J Am Chem Soc*, 1986, **108**, 1317.
22 The literature on non-enzymic glycosyl transfer prior to 1982 is reviewed by M L Sinnott, in M I Page, ed, 'The Chemistry of Enzyme Action', Elsevier, Amsterdam, 1984, p389. An example of endocyclic ring-opening in a biassed pyranosyl system is given by R B Gupta and R W Franck, *J Am Chem Soc,* 1987, **109**, 6554.
23 B Capon and D Thacker, *J Chem Soc B*, 1967, 185.
24 (a) H Lönnberg, A Kankanperä, and K Haapakka, *Carbohydr Res*, 1977, **56**, 277; (b) H Lönnberg and L Valtonen, *Finn Chem Lett* 1978, 209
25 H Lönnberg and A Kulonpää, *Acta Chem Scand Ser A*, 1977, **31**, 306.
26 A J Bennet, M L Sinnott, and W S S Wijesundera, *J Chem Soc Perkin Trans 2*, 1985, 1233.
27 R J W Byrde, A H Fielding, S A Archer, and E Davies, in

'Fungal Pathogenicity and the Plant's Response', R J W Byrde and C V Cutting, eds, Academic Press, London, 1973, p 39.

28 (a) F Laborda, A H Fielding, and R J W Byrde, *J Gen Microbiol*, 1973, **79**, 321; (b) F Laborda, S A Archer, A H Fielding, and R J W Byrde, *ibid*, 1974, **81**, 151.

29 M A Kelly, M L Sinnott, and M Herrchen, *Biochem J*, 1987, **245**, 843.

30 K S Venkatasubban and R L Schowen, *CRC Crit Rev Biochem*, 1984, **11**, 1.

31 T Selwood and M L Sinnott, *Biochem J*, 1988, **254**, 899.

32 M T H Axawamaty, G W J Fleet, K A Hannah, S K Namgoong, and M L Sinnott, *Biochem J*, in press.

33 Reviews: (a) G Legler, *Pure Appl Chem*, 1987, **59**, 1457; (b) L E Fellows, *Chem Brit*, 1987, **23**, 842.

34 P Lalégerie, G Legler, and J M Yon, *Biochimie*, 1982, **64**, 977.

35 D Koshland, *Biol Rev*, 1953, **28**, 416.

36 (a) E J Hehre, C F Brewer, and D S Genghof, *J Biol Chem*, 1979, **254**, 5942; (b) E J Hehre, T Sawai, C F Brewer, M Nakano, and T Kanda, *Biochemistry*, 1982, **21**, 3090; (c) T Kasumi, C F Brewer, E T Reese, and E J Hehre, *Carbohydr Res*, 1986, **146**, 39; (d) T Kasumi, Y Tsumuraya, C F Brewer, H Kersters-Hilderson, M Claeyssens, and E J Hehre, *Biochemistry*, 1987, **26**, 3010.

37 S Kitahata, C F Brewer, D S Genghof, T Sawai, and E J Hehre, *J Biol Chem*, 1981, **256**, 6017.

38 E Boel, M T Hansen, I Hjort, I Høgh, and N P Fiil, *EMBO J*, 1984, **3**, 1581.

39 J-P Praly and G Descotes, *Tetrahedron Lett*, 1987, **28**, 1405.

40 A Konstaninidis and M L Sinnott, unpublished.

41 T-M Enari and M-L Niku-Paavola, *CRC Crit Rev Biotechnol*, 1987, **5**, 67.

42 J K C Knowles, P Lehtovaara, M Murray, and M L Sinnott, *J Chem Soc Chem Commun*, 1988, 1401.

43 C A Jefferson, J Knowles, P Lehtovaara, M Murray, T Selwood, and M L Sinnott, unpublished; C A Jefferson, Stage III Thesis, University of Bristol, 1989.

44 M L Sinnott and W P Jencks, *J Am Chem Soc*, 1980, **102**, 2026.

45 W W Cleland, *Methods Enzymol*, 1979, **63**, 103.

THE EFFECT OF BINDING UPON THE REACTIVITY OF BILIRUBIN

Lesley M Anderson and Anthony R Butler

Chemistry Department, The Purdie Building, The University, St Andrews KY16 9ST, UK

1 INTRODUCTION

Bilirubin (1) is produced in mammals by the breakdown of haem and its removal from the body is necessary because it is toxic. In the liver it is converted mainly into a water-soluble diglucuronide by reaction with two molecules of glucuronic acid, in reactions catalysed by the enzyme glucuronyl transferase. In contrast to the diglucuronide, bilirubin is almost completely insoluble in water and its transport around the body in serum poses a problem. In the main, this is solved by complexing the bilirubin with albumin, which renders it soluble. Binding agents other than albumin may also be present. Thus, serum contains bilirubin in two forms: the water soluble glucuronide and that bound to albumin. For a review of medical aspects of bilirubin see ref 1.

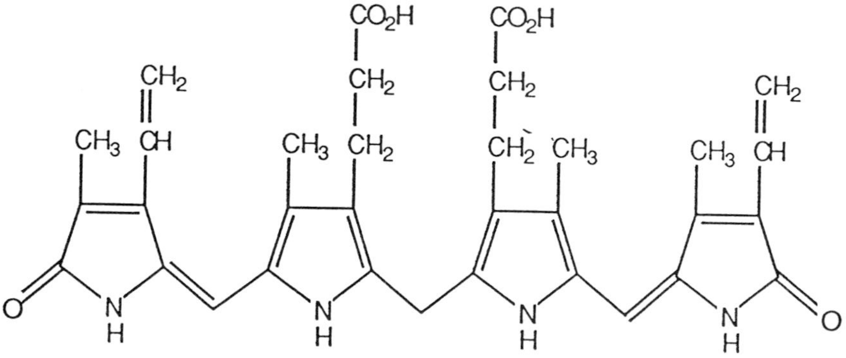

CO_2H CO_2H CH_2 CH_2 CH_2 CH_2 CH_3 CH CH_3 CH_2 CH_2 CH_3 CH_3 CH O N H N H N H N H O

(1)

Table 1: Observed ^{13}C NMR chemical shifts for pyrrole-3-propanoic acid in the absence and presence of α-cyclodextrin

Free pyrrole	Bound pyrrole	Difference
177.99	178.54	+0.55
123.48	123.56	+0.08
110.10	110.07	-0.03
46.84	47.12	+0.28
38.42	38.85	+0.43

The measurement of the bilirubin level in serum is an important tool for the diagnosis of liver dysfunction. It is done colorimetrically by reaction with a benzenediazonium ion, which cleaves the molecule about the central methylene bridge to give two highly coloured azo fragments.[2] The two forms of bilirubin in serum react with a benzenediazonium ion differently. Reaction with the diglucuronide is rapid and complete but albumin-bound bilirubin, it is claimed, is completely unreactive. It does react, however, on the addition of an 'accelerator'.[3] This is substance from a group which includes caffeine, urea and alcohol. Our aim in the study described below was to examine the nature of the binding of bilirubin to albumin and the effect of this binding on reactivity.

2 BINDING OF PYRROLE-3-PROPANOIC ACID AND BILIRUBIN TO α-CYCLODEXTRIN

The structural complexity of both bilirubin and albumin led us to examine some simpler, model compounds. As a model for bilirubin we chose pyrrole-3-propanoic acid. It is known[4] that α-cyclodextrin complexes with bilirubin. The possibility that complexation with pyrrole-3-propanoic acid will also occur was examined by a study of the ^{13}C NMR spectrum in the absence and presence of α-cyclodextrin. The results are shown in Table 1. The changes in

Table 2: Effect of addition of human albumin on the rate of reaction of 0.0005M pyrrole-3-propanoic acid with diazotised sulphanilic acid.

albumin/M	k_{obs}/s^{-1}
none	1.59
0.0001	1.36
0.0003	1.15
0.0005	0.87
0.0007	0.79
0.001	0.68

chemical shift are similar to those reported for complexation of *p*-nitrophenol with 6-*O*-α-D-glucopyranosyl-α-cyclodextrin.[5] That the biggest changes occur with the atoms in the aliphatic side-chain suggests that only part of the molecule fits into the cavity of α-cyclodextrin, and this was supported by molecular modelling with Chem-X. Although complexation occurs it was found that the presence of α-cyclodextrin had very little or no effect on the rate of reaction of pyrrole-3-propanoic acid with diazotised sulphanilic acid, as studied by stopped-flow spectrophotometry. This was in contrast to the effect of human albumin on the same reaction (Table 2). There is clearly some part of the albumin molecule, probably $-NH_3^+$ groups of lysines,[6] which is efficacious in binding the aliphatic carboxylic acid side-chain.

For a similar study involving bilirubin it was necessary to assign unambiguously all the signals in its ^{13}C NMR spectrum. Previous studies[7] have left only a few signals in doubt and these we assigned by means of a two-dimensional experiment, the COLOC (COrrelation *via* LOng range Coupling) pulse sequence optimised to *J* approximately 10Hz. The results are shown in Figure 1. When α-cyclodextrin was added to a solution of bilirubin in d_6-DMSO small changes in the chemical shifts of the atoms in the side-chains were observed (*eg* +0.13 ppm for the carbonyl group), suggesting that these again are the site of complexation. Rather surprisingly addition of α–cyclodextrin caused a small increase in the rate of reaction of bilirubin with diazotised sulphanilic acid.

It was impossible to examine profitably changes in the NMR spectrum of bilirubin on addition of albumin on account of its low solubility, and labelling of certain atoms of bilirubin is required. The key observation of clinical significance is that the reaction between bilirubin and diazotised sulphanilic acid is not retarded significantly by addition of albumin. Why bilirubin and the bilirubin/albumin complex both react much more slowly with diazotised sulphanilic acid than the diglucuronide remains a mystery.

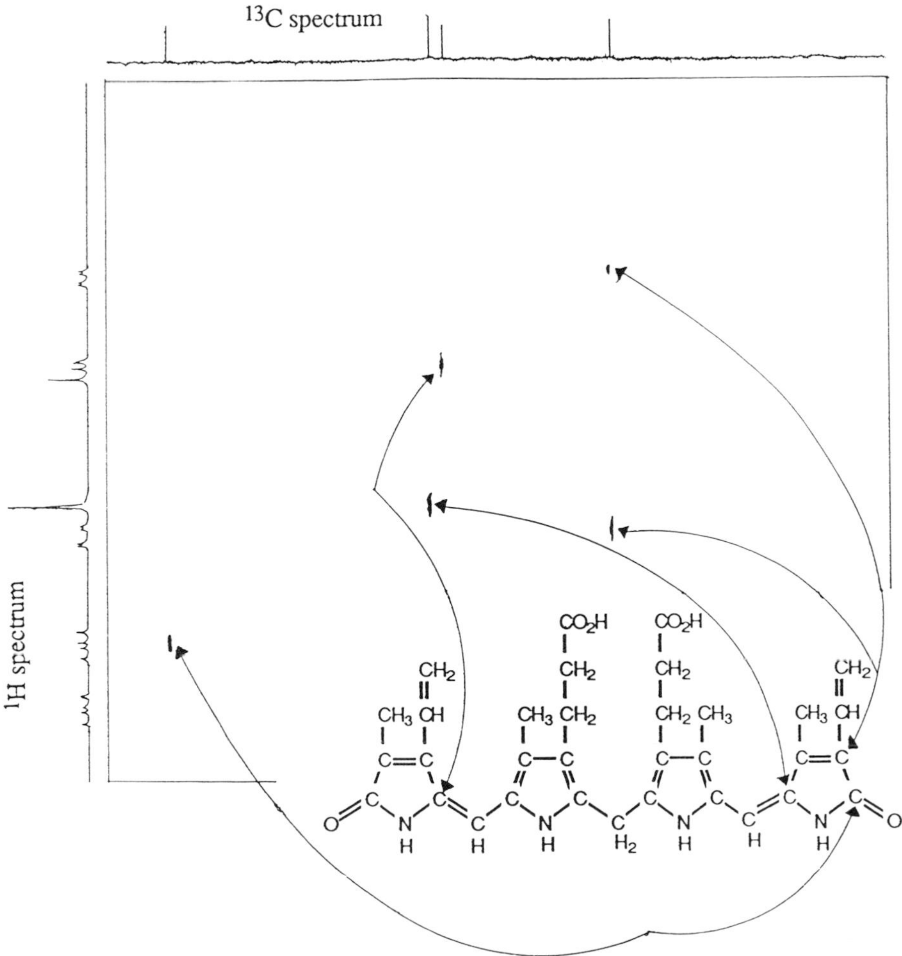

Figure 1: COLOC NMR spectrum of bilirubin showing assignments of specific carbon atoms

REFERENCES

1 L M Gartner and I M Arias, *New England J Med*, 1969, **280**, 1339.
2 H T Malloy and K A Evelyn, *J Biol Chem*, 1937, **119**, 481.
3 L Jendrassik and L Grof, *Biochem Z*, 1938, **297**, 1938.
4 D A Lightner, J K Gawrónski, and K Garónska, *J Am Chem Soc*, 1985, **107**, 2456.
5 Y Kanda, Y Yamamoto, Y Inoue, R Chûjô, and S Kobayashi, *Bull Chem Soc Japan*, 1989, **62**, 2002.
6 M Bouvier, G R Brown, and L E St-Pierre, *Can J Chem*, 1989, **67**, 596.
7 *eg* D Kaplan and G Navon, *Isr J Chem*, 1983, **23**, 177.

TRANSFER OF THE GENERAL ACYL GROUP IN SOLUTION:

Variation of Transition–state Structure Exemplified by Diphenylphosphoryl Group Transfer between Phenolate Ion Nucleophiles

Salem A Ba–Saif, Mark A Waring, and Andrew Williams

University Chemical Laboratory, Canterbury, Kent CT2 7NH, UK

1 INTRODUCTION

The transfer of the acyl group between nucleophilic donors and acceptors in aqueous solution (equation 1) is very important in many biological and chemical systems:

$$\text{A–Lg} + \text{Nu}^- \longrightarrow \text{A–Nu} + \text{Lg}^- \qquad \text{(eqn 1)}$$

Since carboxylic, phosphoric, sulphonic and other acids are under consideration it is useful to refer to 'acyl' group as a general term to denote the electrophilic moiety (A) constituting part of the acid (A–OH). The transfer of the acyl function is a special case of group transfer between nucleophiles, where the range of possible mechanisms is very similar.

Acyl group transfer is involved in processes ranging from protein biosynthesis to carbon–carbon bond formation in synthetic organic chemistry. Transfer of the sulphenyl group (R–S–) is important, for example, in disulphide interchange and in the reactions of lipoic acid. Sulphuryl (^-O_3S–) group transfer is the basis of sulphur fixation and of some detoxification reactions. Phosphoryl group transfer has manifest significance in biological systems.

Many acyl group transfer reactions require 'enabling' steps (equation 2) such as protonation or deprotonation, without which the overall reaction will not proceed:[1]

$$\text{RCONHR} \xrightarrow{+\text{ROH}} \text{R}\!-\!\overset{\text{O}^-}{\underset{\text{ROH}^+}{\text{C}}}\!-\!\text{NHR} \longrightarrow \text{R}\!-\!\overset{\text{O}^-}{\underset{\text{RO}}{\text{C}}}\!-\!\text{N}^+\text{H}_2\text{R} \xrightarrow{-\text{RNH}_2} \text{RCOOR} \qquad \text{(eqn 2)}$$

In this chapter the fundamental acyl transfer process involving simply the bond formation and bond fission steps will be discussed. A possible complication in the reactions considered is diffusion of the reactants in the solvent. However, this is not significant for the reactions described here, because for all of them diffusion is unlikely to be a rate limiting step.

The mechanistic possibilities for the fundamental reactions are summarised in the reaction map (Figure 1). Reactions with intermediate species $Nu{-}\bar{A}{-}Lg$ and A^+ have been thoroughly

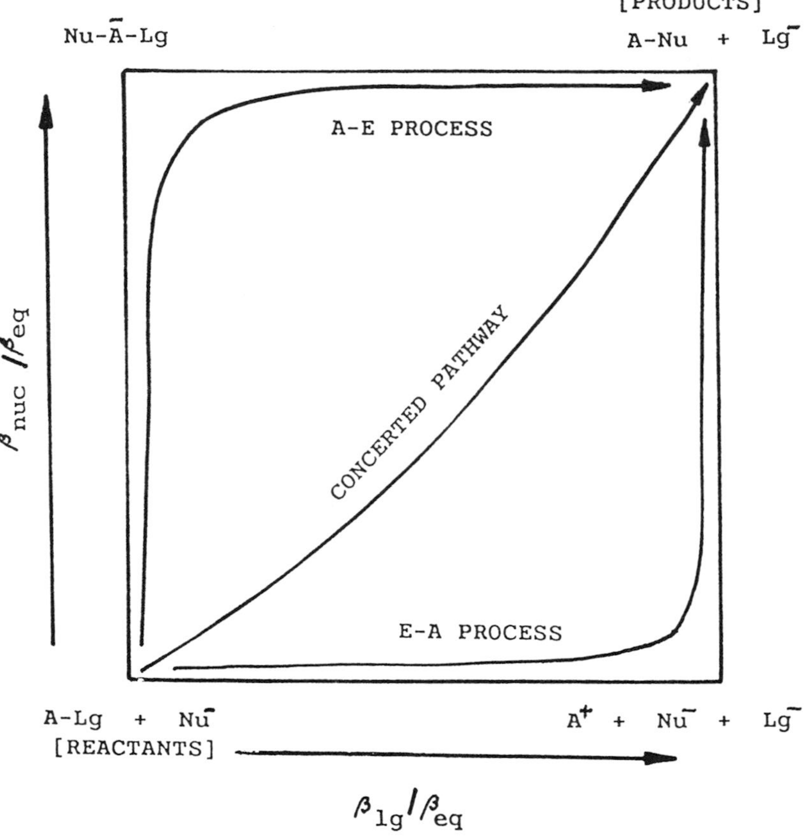

Figure 1: Reaction map for the transfer of the general acyl function (A–) between general nucleophilic donors (Lg^-) and acceptors (Nu^-). The transition state structure is documented by two 'α' parameters (see text) corresponding to the change in bond formation ($\alpha{=}\beta_{nuc}/\beta_{eq}$) and bond fission ($\alpha{=}\beta_{lg}/\beta_{eq}$). A–E and E–A refer to the stepwise addition–elimination routes, respectively.

studied;[2,3] they are represented by the shorthand terms $B_{Ac}2$, S_N2, $S_N2(P)$, $S_N2(S)$ *etc*, for the addition–elimination path,[3] and E1cB, S_N1 *etc*, for the elimination–addition route.[3] As both stepwise processes have been demonstrated for acyl group transfer in solution, an 'in–between' mechanism with a single transition state, *ie* a concerted mechanism, can be readily inferred from the map by consideration of the effect of energy variation of the four discrete states represented by the four corners (Figure 1).

2 APPLICATION OF POLAR SUBSTITUENT EFFECTS

The polar substituent effect in the form of a Bronsted β, Hammett ρ or other parameter measures *change in charge* from reactant to product or transition state when, respectively, the equilibrium or rate constant is measured.[4] The change in equilibrium or rate constant arises from an energy change caused by interaction of the substituent(s) with the charge in the various states.

The change in charge is measured by substituent effects derived from both bonding *and* solvation changes in a solution reaction. Chemists usually delineate reactions in solution as if the bonds were forming and breaking *in vacuo*, thus distorting the true picture. Solvation is a fundamental part of a reaction in solution and it is assumed in this review that the various states (reactant, transition, product or intermediate), written as *in vacuo*, carry their appropriate solvation. It is our contention that trying to attribute polar effects to bonding changes alone is a worthless exercise for reactions in solution as solvation and bonding are inextricably connected.

3 EFFECTIVE CHARGE

Charge change must be estimated relative to that of a standard reaction, which is usually the ionisation of an acid. This is because ionisation constants of acids are readily measured over a wide range of structures and there already exists a large body of data. The standard equilibrium should have a close relationship with the reaction in question. For example, the transfer of the acetyl group between phenolate ions should be compared with the ionisation of phenols (equations 3 and 4):

$$\underset{(+0.7)}{CH_3CO{-}O{-}Ar} + Nu^- \underset{}{\overset{K_{eq}}{\rightleftharpoons}} \underset{(-1.0)}{CH_3CO{-}Nu} + {}^-O{-}Ar \qquad \text{(eqn 3)}$$

$$\beta_{eq} = -1.7$$

$$\underset{(0)}{H{-}O{-}Ar} \overset{K_a}{\rightleftharpoons} H^+ + \underset{(-1.0)}{{}^-O{-}Ar} \qquad \text{(eqn 4)}$$

$$\beta_{eq} = -1.0$$

The Bronsted plot of log K_{eq} *versus* log K_a is linear and has a slope of β_{eq} (= −1.7). If the charge change on the oxygen in the standard equilibrium is *defined* as −1.0 the relative change in charge on the same oxygen in the equilibrium of equation 3 is −1.7. The absolute charge change on the oxygen in equation 4 is *not* −1.0 and the measured charge is therefore denoted as 'effective charge'[5] to distinguish it. The β value for rate constants refers to effective charge change for formation of the transition state.

A transition state with a constant structure through a series of substituents will necessarily possess a constant charge on its constituent atoms and will therefore give rise to a linear polar substituent correlation such as a Bronsted or Hammett plot. Non-linear plots or plots with two linear portions with different slope indicate change in transition state structure be it by change in rate limiting step or in mechanism.

The change in effective charge obtained from β, ρ or other parameter may be used to position the transition state in the reaction map (Figure 1) relative to the four discrete states. This can be done by comparing the effective charge change to the transition state with that to the product (β/β_{eq}); this ratio is called the Leffler index (α).[5b] It is a common (but well understood)[6] misconception that overall transition state structure is measured by Leffler's index for a single bond change. A better description of transition state structures is obtained when the Leffler index is measured for all the major bond changes in a reaction; this is indeed one of our aims. The index, α, refers to the totality of solvation and bonding changes from ground to transition state for a particular bond change compared with the overall change in these qualities. Leffler's index is loosely referred to as an index of bonding, but this is shorthand for the totality expressed above.

4 CONCERTED AND STEPWISE MECHANISMS

A non-linear Bronsted or Hammett plot with its convex side upwards is a standard diagnosis of the existence of an intermediate on a reaction path.[7] It results because the change in rate limiting step causes a different transition state to be 'read' by the substituents. On the contrary, a linear Bronsted or Hammett plot does *not* diagnose a single step mechanism. It merely indicates constancy of transition state 'read' by the substituents over the range of substituent species investigated.

(eqn 5)

A linear Bronsted or Hammett plot *is* diagnostic of a single step mechanism *if* it can be predicted that the putative two step process should change its rate limiting step in the substituent range studied.[8,9] The transfer of the phosphoryl group between pyridine nucleophiles (equation 5) was shown to involve a linear Bronsted correlation independently by our laboratory and that of Jencks[8a] over the range of pK_a values for the pyridine nucleophile, where a change in rate limiting step should have occurred for the stepwise mechanism; a concerted displacement was proposed. An absolutely reliable prediction of the breakpoint for a putative stepwise mechanism requires the reaction under investigation to be essentially symmetrical.

The nucleophile structure must be of the same type as the leaving group so that the criterion for change in rate limiting step, namely equal forward and return rate constants for the putative intermediate, occurs when the pK_a's of the nucleophile and leaving group are equal. The Bronsted plot for reaction of phenolate ions with 4-nitrophenyl diphenylphosphate (equation 6)[9] is illustrated in Figure 2 and shows no trace of non-linearity over a wide range of pK_a values for the attacking phenolate ions above and below the pK_a of the leaving 4-nitrophenolate ion:

$$ArO^- + (PhO)_2POO4NP \xrightarrow{k_{ArO}} ArOPO(OPh)_2 + {}^-O4NP \qquad \text{(eqn 6)}$$

4NP = 4-nitrophenyl

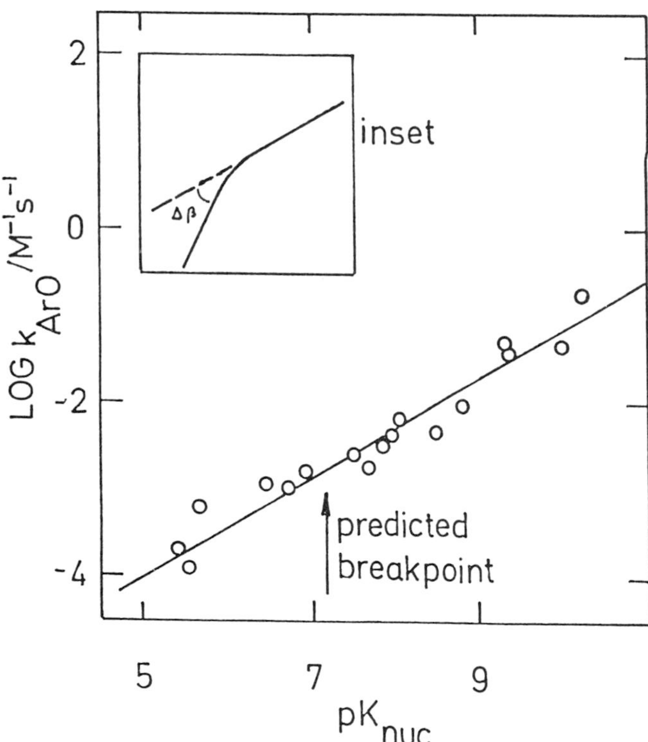

Figure 2: Bronsted dependence on the pK_a of the attacking phenolate ion for reaction with 4-nitrophenyl diphenylphosphate. The inset indicates a Bronsted line over a pK_a region involving change in rate limiting step for a *putative stepwise* process.

The mechanism involving the putative intermediate (equation 7) predicts a change in rate limiting step when the pK_a of the nucleophile equals that of 4-nitrophenol; this mechanism is not consistent with the constant transition state indicated by the linear Bronsted plot over the range:

$$(PhO)_2POO4NP \underset{k_{-1}\ (\beta_{-1})}{\overset{(\beta_1)\ k_1[Ar\bar{O}]}{\rightleftharpoons}} ArO{-}P(PhO)(O^-)(OPh){-}O4NP \xrightarrow[-\bar{O}4NP]{(\beta_2)\ k_2} ArOPO(OPh)_2 \quad \text{(eqn 7)}$$

There is *no* technique which provides absolute certainty! Equation 7 leads to a theoretical Bronsted plot[10] consisting of two intersecting linear correlations, the slope of each corresponding to the effective charge change on its respective transition state. The difference in slopes ($\Delta\beta$) of the lines (see Figure 2 inset) is the difference in effective charge between each transition state and the data cannot distinguish between a stepwise process where $\Delta\beta \simeq 0.1$ and a concerted mechanism (where, naturally, $\Delta\beta = 0$). The uncertainty of 0.1 in $\Delta\beta$ is consistent with only a small structural difference for the two transition states in an overall change of 1.4.[9] Conservatively the mechanism would be at least in the borderline region between concerted and stepwise. Dietze and Jencks[11] are of the opinion that such a small difference in effective charge is consistent with an energy difference between intermediate and transition states which makes it difficult to account for the stability of the intermediate.

We can employ a *reductio ad absurdum* argument to indicate the existence of a single transition state and this is exemplified with the diphenylphosphoryl group transfer. In the region of $pK_a < pK_{lg}$ in Figure 2 the value of β_{nuc} is 0.6 (k_2 is rate limiting in this region and only those values of k_{ArO} from this region are used in computing β_{nuc}). Thus, β_{eq} (= $\beta_1 - \beta_{-1}$) for formation of the intermediate is less than 0.6 because $\beta_{nuc} = \beta_1 + \beta_2 - \beta_{-1}$. The value of β_{eq} for formation of the product from the intermediate is >0.8 because the overall $\beta_{eq} = 1.4$.[9] Formation of the ArO–P bond (step k_1) should possess a *larger* change in effective charge than in the k_2 step (not involving ArO–P bond change), whereas the observed β_{eq} values indicate otherwise. In other words the large bond change is observed to have a small change in effective charge, whereas the small bond change has a large change in effective charge which is absurd.

5 VARIATION OF TRANSITION STATE

Linearity of Bronsted and other free energy correlations has been the subject of much robust, but friendly, debate. A Bronsted plot for the attack of phenolate anion on diphenylphosphate esters with different aryloxide leaving groups gives rise to a selectivity β_{lg} which varies as the pK_a of the nucleophilic aryloxide is changed (Figure 3).[9] The value $p_{xy} = d\beta_{lg}/dpK_{nuc}$ is not zero implying that the transition state structure *does* vary as the attacking or leaving group is altered, although the individual Bronsted plots are linear! Both Kirby[12] and Gorenstein[13a] have found similar results for attack of oxyanions on arylphosphate esters, but the use of a range of structurally unrelated nucleophiles make the results difficult to interpret. We have found transition state variation in carbonyl group transfer reactions between aryloxide nucleophiles.[13b]

The explanation of linear plots from reactions with varying transition states derives from a consideration of the energy surface of the reaction map and a knowledge of how altering the energies at the four corners will move the transition state.[14] Consideration of Figure 4 indicates that as the pK_a of the attacking phenolate ion is increased the bottom left and bottom right energies will rise. This distorts the surface such that the transition state moves *along* the reaction coordinate *towards* the bottom left corner as shown. Increase in energy of the bottom right corner will cause a change in the surface to move the transition state structure *perpendicular* to the reaction coordinate *away* from the increase in energy. The resultant movement of the structure of the transition state is to the left, making β_{lg} less negative; there could be little movement in the vertical direction causing little change in β_{nuc} (and hence no curvature in the plot as illustrated in Figure 2).

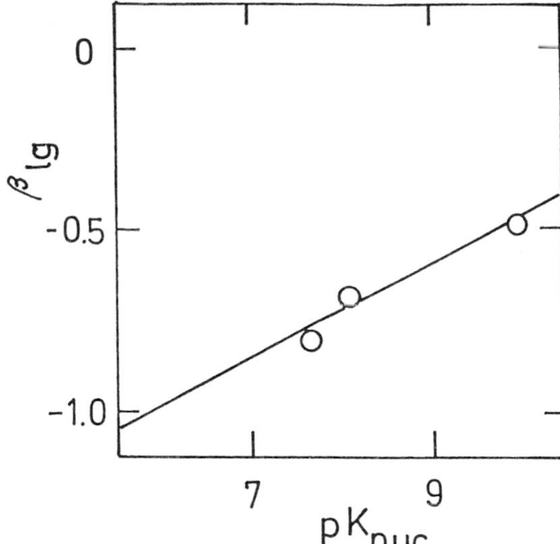

Figure 3: Dependence on pK_{nuc} of β_{lg} for attack of phenolate ions on aryl diphenyl phosphate esters.

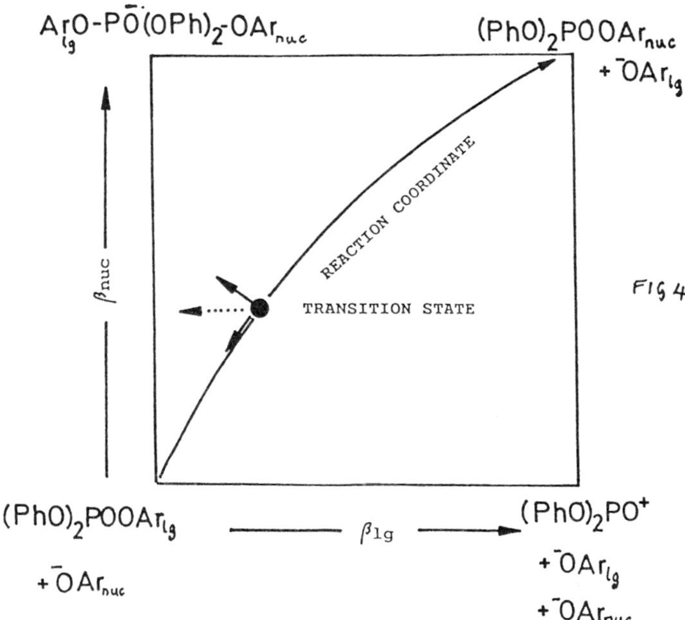

Figure 4: Variation of transition state charge structure as a function of increasing the pK_a of the phenolate ion nucleophile in its attack on aryl diphenylphosphate esters. The dotted arrow represents the resultant movement of the transition state.

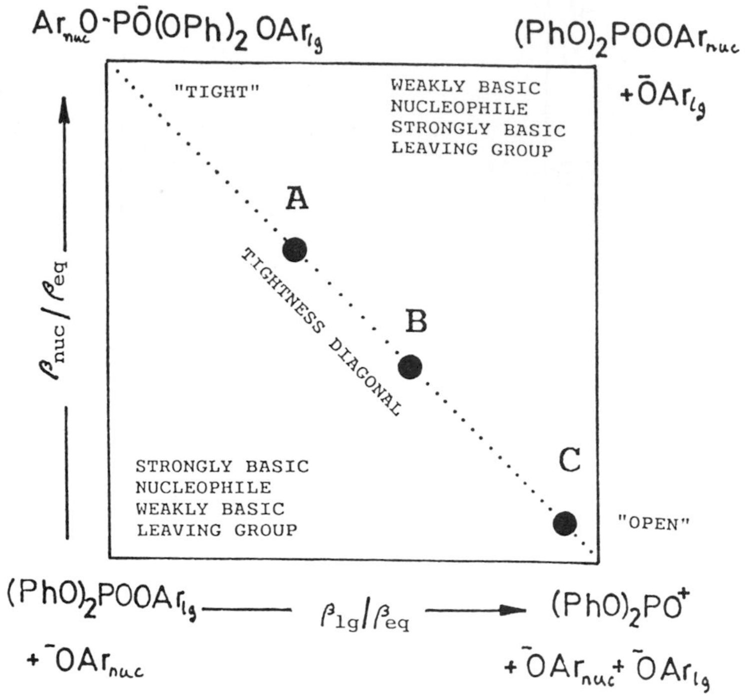

Figure 5: Transition state structures for the symmetrical transfer of the diphenylphosphoryl group between identical phenolate ion nucleophiles: A (phenolate ion),[9] B (4-nitrophenolate ion)[9] and C (2,4-dinitrophenolate ion).[17] The triangular sectors are labelled to indicate the transition state structures for transfer between non-identical donor and acceptor phenolate ion nucleophiles.

Variation of the basicity of either leaving group or nucleophile will alter the structure of the transition state. When nucleophile and leaving group are *identical* the transition state for the concerted acyl group transfer lies on the 'tightness' diagonal which is called the 'disparity mode' by Grunwald,[16] and is illustrated in Figure 5. Variation in β_{lg} for the reaction of phenolate ions with diphenylphosphate esters possessing different aryl oxide leaving groups (Figure 3) indicates that strong nucleophiles will have transition states at the upper left corner of the tightness diagonal and weak nucleophiles at the lower right. The positions of the transition states for the symmetrical transfer of the diphenylphosphoryl group are determined from the value of β_{lg} for the parent phenolate anion[9] and from β_{nuc} for 4-nitrophenolate ion[9] and 2,4-dinitrophenolate ion[17] and the value for β_{eq}.[9] Strengthening the nucleophile and weakening the leaving group will shift the transition state structure into the bottom left triangle (see Figure 5). Weakening the nucleophile and strengthening the leaving group puts the transition state structure into the top right triangle.

Concertedness and variation of transition state structure for acyl group transfer reactions in solution is not confined to the example presented here; concerted mechanisms are seen with phosphoryldianion transfer,[8a,8b,18] as well as carbonyl,[19,20,21] sulphuryl monoanion,[22] sulphenyl[23] and diphenylphosphinyl[24] group transfer reactions. Variation of transition state structure with variation of donor and acceptor nucleophiles has been demonstrated with the carbonyl, diphenylphosphinyl and phosphoryl group transfer reactions.

6 CONCLUSIONS

Acyl group transfer reactions between nucleophiles can be concerted. Strongly basic donor and acceptor nucleophiles favour either a $B_{Ac}2$ stepwise process or a concerted mechanism with a tight transition state. Weakly basic nucleophiles favour an S_N1 type of stepwise mechanism or an open concerted pathway.

Acyl group transfer reactions should not be assumed to involve addition intermediates such as those originally discovered by Bender[25] or eliminative intermediates such as the metaphosphate ion, the subject of a frustrating hunt during the past decade.[26] Discussions of the mechanism of acyl group transfer should seriously consider all the mechanistic types documented here. The stereochemistry of the transition state is straightforward when it is a pentacoordinate species. Carbonyl transfer should have tetrahedral stereochemistry for the transition state at the tight end of the tightness diagonal and planar stereochemistry at the open end. However, there does not seem at present any experimental way of delineating the stereochemistry for a particular carbonyl transfer.

SUMMARY

Acyl group transfer is reviewed in this chapter with particular reference to the reaction:

$$ArO^- + (PhO)_2PO{-}OAr' \longrightarrow ArO{-}PO(OPh)_2 + {}^-OAr'$$

The transfer of a general acyl group ($RCO-$, RSO_2-, R_2PO- *etc*) between weak nucleophiles is likely to be a concerted process in solution. The concerted mechanism lies between the classical stepwise process involving either an intermediate adduct ($Nu{-}\bar{A}{-}Lg$) or an eliminative intermediate (A^+).

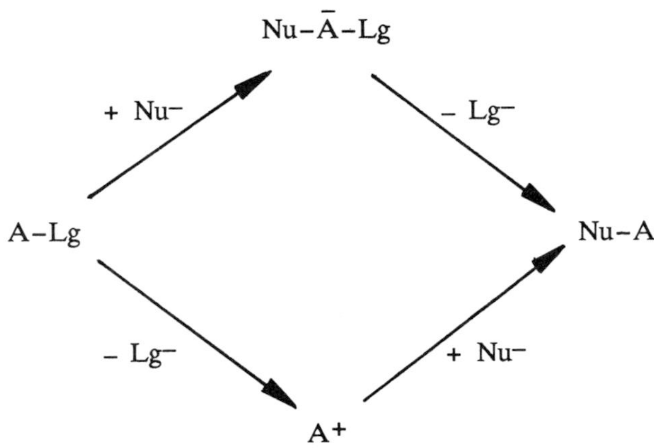

Strong nucleophiles favour an adduct-like transition state, whereas weak nucleophiles favour one like the intermediate (A^+). Variation of transition state structure occurs even in the structural range of the phenolate ion series of nucleophiles.

ACKNOWLEDGEMENTS

We are grateful to the SERC for the provision of a research grant and S A Ba-Saif thanks the Government of Saudi Arabia for a Scholarship.

REFERENCES

1 W P Jencks, *Chem Rev*, 1972, **72**, 705.
2 a) B Capon, M I Dosunnu, and M de N Matos Sanchez, *Adv Phys Org Chem*, 1985, **21**, 37; b) R A McClelland and L J Santry, *Accounts Chem Res*, 1983, **16**, 394.
3 A Williams and K T Douglas, *Chem Rev*, 1975, **75**, 627.
4 J Hine, *J Am Chem Soc*, 1960, **82**, 4877.
5 a) W P Jencks, *Brookhaven Symp Quant Biol*, 1971, **36**, 1; b) A Williams, *Accounts Chem Res*, 1984, **17**, 425; c) S Thea and A Williams, *Chem Soc Rev*, 1986, **15**, 125.
6 A Pross and S S Shaik, *Accounts Chem Res*, 1983, **16**, 363.
7 H Maskill, 'The Physical Basis of Organic Chemistry', Oxford University Press, Oxford, 1985.
8 a) M T Skoog and W P Jencks, *J Am Chem Soc*, 1984, **106**, 7597; b) N Bourne and A Williams, *ibid*, 1984, **106**, 7591; c) N Bourne, A R Hopkins, and A Williams, *ibid*, 1985, **107**, 4327.
9 S A Ba–Saif, unpublished data, University of Kent, 1988.
10 S A Ba–Saif; A K Luthra, and A Williams, *J Am Chem Soc*, 1987, **109**, 6362.
11 P E Dietze and W P Jencks, *ibid*, 1986, **108**, 4549.
12 S A Khan and A J Kirby, *J Chem Soc B*, 1970, 1172.
13 a) R Rowell and D G Gorenstein, *J Am Chem Soc*, 1981, **103**, 5894; b) S A Ba–Saif, A K Luthra, and A Williams, *ibid*, 1989, **111**, 2647.
14 W P Jencks, *Bull Soc Chem France*, 1988, 218.
15 M M Kreevoy and I–S H Lee, *J Am Chem Soc*, 1984, **106**, 2550.
16 E Grunwald, *ibid*, 1985, **107**, 4710.
17 M A Waring, unpublished observations, University of Kent, 1989.
18 a) G Lowe, *Accounts Chem Res*, 1983, **16**, 244; b) S L Buchwald, J M Friedman, and J R Knowles, *J Am Chem Soc*, 1984, **106**, 4911; c) P M Cullis and I Iagrossi, *ibid*, 1986, **108**, 7870.
19 E Chrystiuk and A Williams, *ibid*, 1987, **109**, 3040.
20 T W Bentley and H C Harris, *J Chem Soc Perkin Trans 2*, 1986, 619.
21 T C Curran, C R Farrar, O Niazy, and A Williams, *J Am Chem Soc*, 1980, **102**, 6828.

22 P D'Rozario, R L Smyth and A Williams, *ibid*, 1984, **106**, 5027.
23 D J Hupe and E R Pohl, *Israel J Chem*, 1985, **26**, 395.
24 N Bourne, E Chrystiuk, A M Davis, and A Williams, *J Am Chem Soc*, 1988, **110**, 1890.
25 M L Bender, *ibid*, 1951, **73**, 1626.
26 D Herschlag and W P Jencks, *ibid*, 1986, **107**, 7938.

2nd CLEMO MEMORIAL LECTURE:
C-NUCLEOSIDES: SYNTHESIS AND BIOSYNTHESIS

J Grant Buchanan

Heriot–Watt University, Chemistry Department, Riccarton, Edinburgh EH14 4AS, UK

1 INTRODUCTION

Professor George Roger Clemo* was born one hundred years ago, on 2 August 1889. I saw him only once, in December 1954, when he delivered his Hugo Müller lecture, 'Some Newer Aspects of the Organic Chemistry of Nitrogen'[1] in the old Chemical Society lecture theatre in Burlington House. He had just retired from the Chair of Organic Chemistry at King's College, Newcastle upon Tyne, and I moved there with his successor James (later Sir James) Baddiley a few weeks later. There were many stories of his experimental prowess. I worked in his former laboratory where, among the bottles of alkaloids, snowdrop leaves, and other potential sources of new natural products, was a large bottle of the pyridine–SO_3 complex. At that time Jim Baddiley was interested in the synthesis of the coenzyme 'active sulphate' (1) and we discovered that Professor Clemo's SO_3 complex was the reagent of choice for the formation of the sulphatophosphate linkage.[2]

---

* G R Clemo (1889–1983) was professor of organic chemistry at King's College, Newcastle from 1925 until 1954. He was a versatile organic chemist who made many contributions in natural product chemistry (terpenes and alkaloids), and was a pioneer in the 1930's with the application of deuterium labelling in organic chemistry.

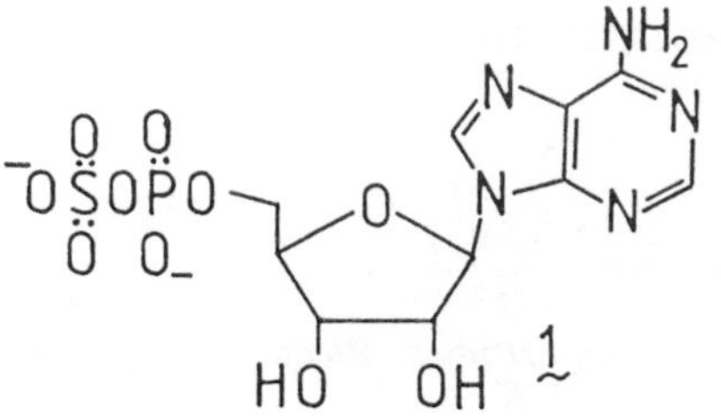

C–Nucleosides are ribonucleosides in which the sugar β–D–ribofuranose is linked to a carbon atom in a nitrogen–containing heterocycle.[3] The first to be discovered, pseudo–uridine (**2**), is a component of transfer ribonucleic acids (^{t}RNA) and is an isomer of the more common *N*–linked nucleoside uridine (**3**). Other *C*–nucleosides,[3] produced by microorganisms, show antitumour and antiviral properties probably due to their similarity in structure to the common *N*–nucleosides.[4,5] These include showdomycin (**4**), formycin (**5**) [an isomer of adenosine (**6**)], formycin B (**7**), pyrazofurin (**8**) and oxazinomycin (minimycin) (**9**). Such compounds would, I am sure, have appealed to Professor Clemo, both in their synthesis and biosynthesis.

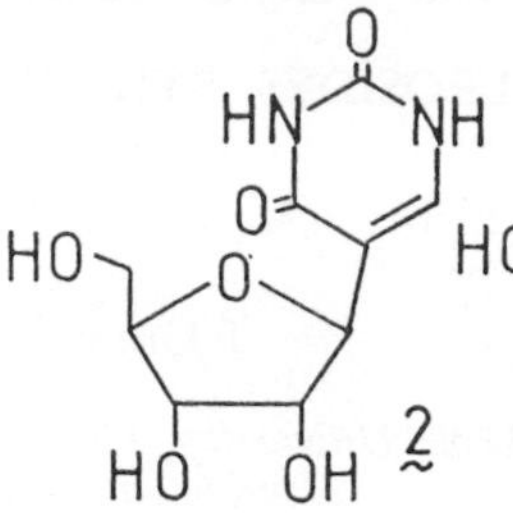

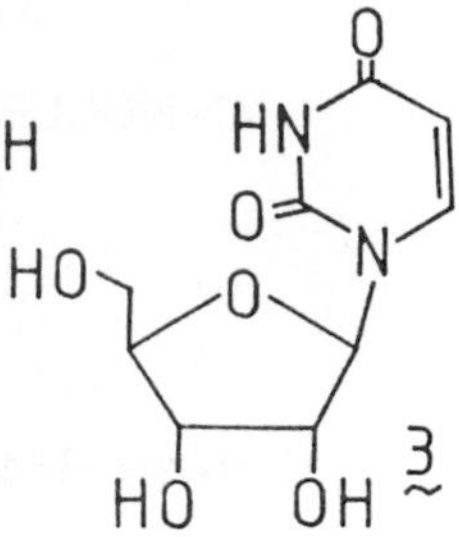

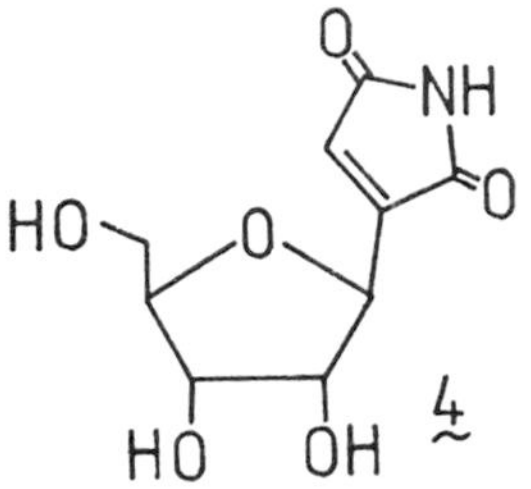

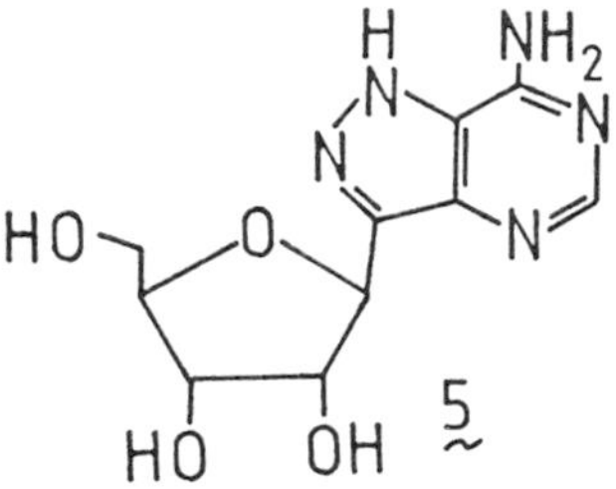

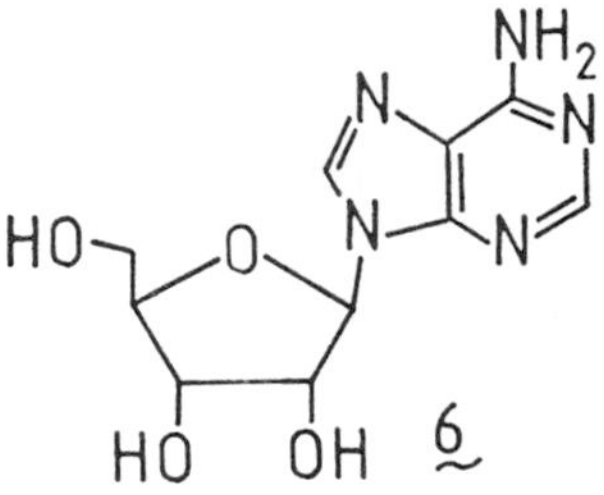

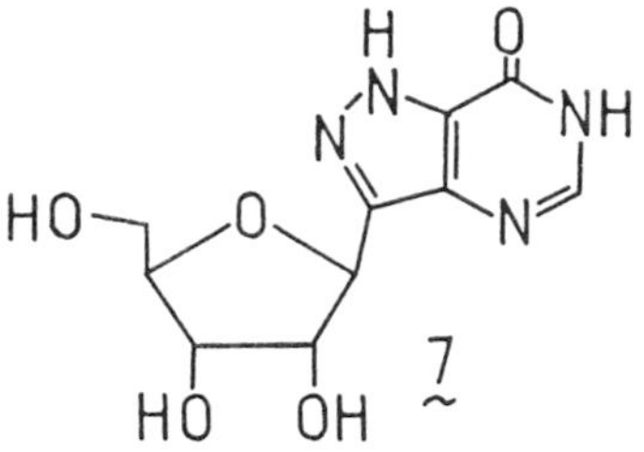

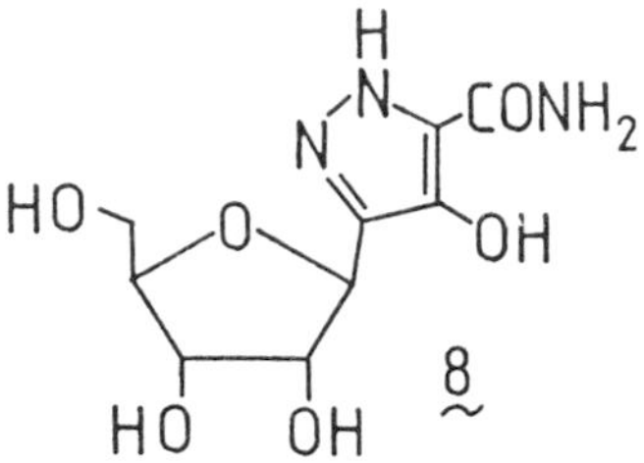

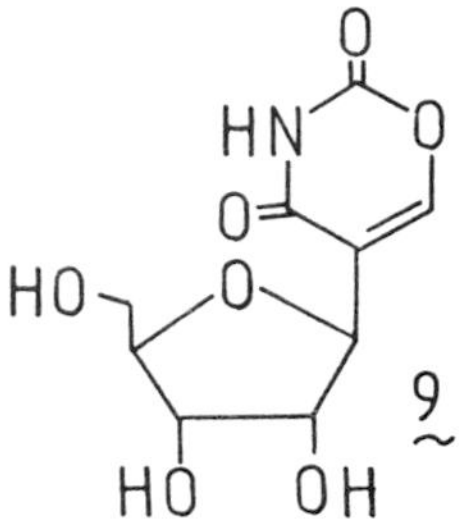

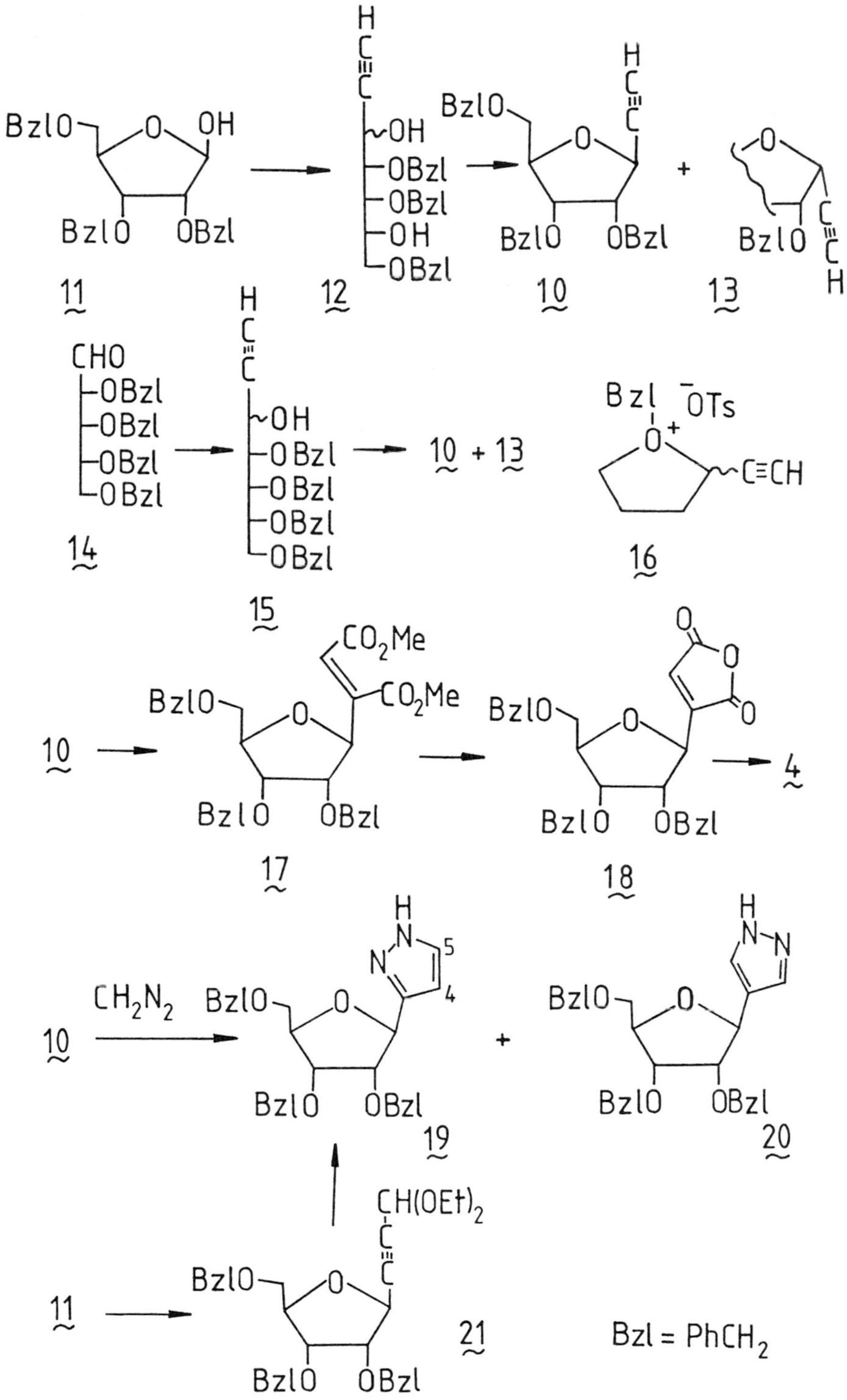
BzlO
OH
BzlO
OBzl
11
OH
OBzl
OBzl
OH
OBzl
12
10
13
+
CHO
OBzl
OBzl
OBzl
OBzl
14
15
10 + 13
Bzl
OTs
C≡CH
16
CO2Me
CO2Me
17
18
4
CH2N2
5
4
19
20
CH(OEt)2
21
Bzl = PhCH2

2 SYNTHESIS OF *C*-NUCLEOSIDES

As an overall target we wished to develop a simple and reliable synthetic route to the naturally occurring *C*-nucleosides which could easily be extended to the synthesis of closely related compounds. Our initial target was the protected β-D-ribofuranosylethyne (10) whose synthesis from the crystalline tri-*O*-benzyl-D-ribofuranose (11)[6] was achieved.[7] Reaction of the lactol (11) with ethynylmagnesium bromide afforded a mixture of diastereomeric diols (12), consisting mainly of the D-*altro* isomer (7:3). Treatment with toluene-*p*-sulphonyl chloride in pyridine gave the β-ethyne (10) (51%) through 3-*O*-sulphonylation of the *altro* isomer, followed by ring closure with inversion of configuration at C-3, together with the α-D-ribofuranosylethyne (13) derived similarly from the D-*allo* diol. Regio-control of the ring closure depends on the greater availability of HO-3 to the reagent. We therefore explored[8] a route from the *aldehydo*-ribose (14).[9] The mixture of *altro* and *allo* alcohols (15), on sulphonylation as before, gave the β- and α-ethynes, *via* the expected RO-5 benzyloxy participation[10] as in (16), in 83% yield, but in a 1:1 ratio. We have not achieved greater stereoselectivity in the Grignard reaction of (14) and the route from the more accessible lactol (11) is preferred.

The β-ethyne (10) was converted into showdomycin[11] by bis-methoxycarbonylation[12] to give the maleate (17) (80%), followed by conversion into the anhydride (18), imide ring formation and deprotection. The α-isomer, which is biologically inactive, was also prepared from the α-ethyne (13).[8]

In the synthesis of formycin (5)[13] and pyrazofurin (8)[14] we envisaged the preparation of the key pyrazole (19) followed by introduction of the necessary functionality at C-4 and C-5. Cycloaddition of diazomethane to the ethyne (10) proceeded in good yield but afforded equal amounts of the regioisomers (19) and (20).[15] The problem was solved[16] by the synthesis of the acetal (21) [51% yield from (11)], using reactions analogous to those for (10), and its conversion (90%) into the pyrazole (19) by treatment with acid and hydrazine.

Substitution in the pyrazole ring was achieved by extending the work of Habraken and Poels[17] who discovered that the

1,4-dinitropyrazole (22) reacted with a secondary amine by *cine*-substitution to give the tertiary amine (23). We showed[13] that a number of nucleophiles will react in this way and that the 5-substituted derivatives (24)–(27) can be prepared in high yield. A generalised form of the reaction of (22) with the nucleophile X^- is also shown. Of particular value to us was the formation of the nitrile (25). In the model series the dinitropyrazole (22) is prepared by

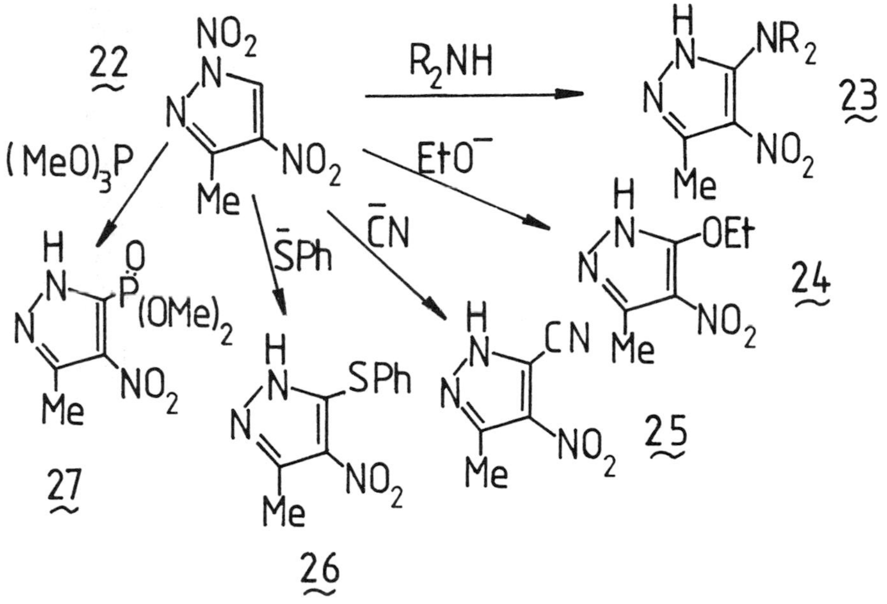

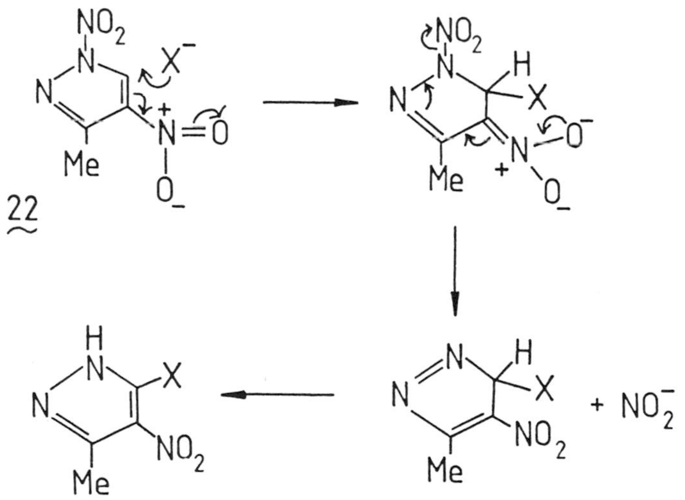

19

AcO O N N Ac AcO OAc

29

AcO O N N NO2 NO2 AcO OAc

28

AcO O N N H CN NO2 AcO OAc

30

AcO O N N H CN NH2 AcO OAc

31

AcO O N N N CN N_2^+ AcO OAc

33

RO O N N H CN OH RO OR

34 R = Ac ; 36 R = H

RO O N N H CONH2 OH RO OR

35 R = Ac

8 R = H

RO O N N H NH2 N N RO OR

32 R = Ac

5 R = H

HO O N N H CONH2 R HO OH

37 R = NH_2

38 R = H

reaction of 3-methyl-4-nitropyrazole with acetyl nitrate.[17,18] The conversion of the ribosylpyrazole (**19**) into the required dinitropyrazole (**28**) was not straightforward and our original solution to the problem required temporary *N*-protection of the pyrazole ring as the 2,4-dinitrophenyl derivative.[13,16] We have recently discovered that 3-methylpyrazole can be converted directly into the 1,4-dinitropyrazole (**22**) by treatment with trifluoroacetyl nitrate in trifluoroacetic acid,[20] developed from a reagent described by Crivello.[21] The tetra-acetyl compound (**29**), prepared by debenzylation and acetylation of pyrazole (**19**), was thus nitrated to give the dinitropyrazole (**28**) in 90% overall yield from (**19**),[22] compared with 50% over 6 steps.[13,16]

The dinitropyrazole (**28**), with an excess of potassium cyanide in aqueous ethanol, gave the nitrile (**30**) (89%) which was converted into the amine (**31**). Reaction with formamidine acetate gave formycin triacetate (**32**) which yielded formycin (**5**) on deacetylation [58% overall from (**30**)].

Pyrazofurin (**8**) was prepared[14] from the amine (**31**) by diazotisation, followed by photolysis of the diazo compound (**33**)[23] and conversion of the nitrile group in (**34**) to the amide (**35**). Deacetylation gave pyrazofurin (**8**) in 35% overall yield from the nitro-nitrile (**30**). From these intermediates we have prepared the nitrile (**36**) and amides[24] (**37**) and (**38**), none of which showed antiviral activity.

A number of nucleosides in which D-ribose is replaced by D-arabinose or D-xylose show antitumour and antiviral activity.[5] A route into *ara*-formycin (**39**) using all six carbons of the hexose D-mannose rather than a repetitive synthesis from D-arabinose was suggested when the ketose (**40**)[25] was converted by hydrazine into the pyrazole (**41**) in high yield.[26] Exploration of polyol protection and ring closure reactions,[27] followed by detailed pyrazole and pyrimidine chemistry led to *ara*-formycin (**39**).[20] Similar chemistry was used to synthesise *xylo*-formycin (**42**) from di-*O*-isopropylidene-D-gulonolactone (**43**).[28,29] Neither (**39**) nor (**42**) were biologically active.

It is well known that certain analogues of the normal nucleosides in which the ribose unit is replaced by an acyclic residue show potent antiviral activity.[5] We have prepared several compounds related to

(*S*)–9–(2,3–dihydroxypropyl)–adenine (**44**).[30] The pyrazole (**45**) was easily prepared from the 3–toluene–*p*–sulphonate of D–glucose[31] or, better, D–allose.[32] Reaction of the desired dimesylate (**46**) with sodium borohydride effected the deoxygenation at C–1'.[33] The formycin analogue (**47**) was prepared, using the methodology already described, together with the pyrazofurin analogue and several pyrazolo[4,3–*d*]pyrimidines;[19,33] none showed antiviral activity.

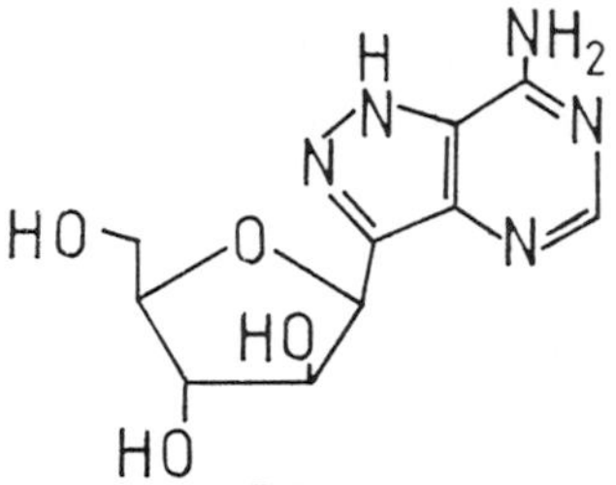

39 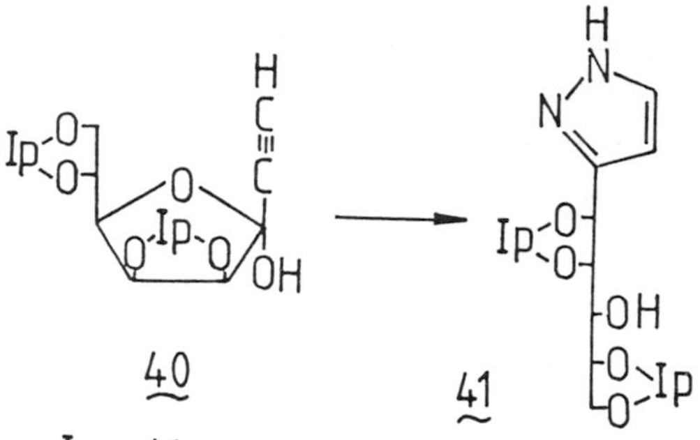 40 41

Ip = $Me_2C<$

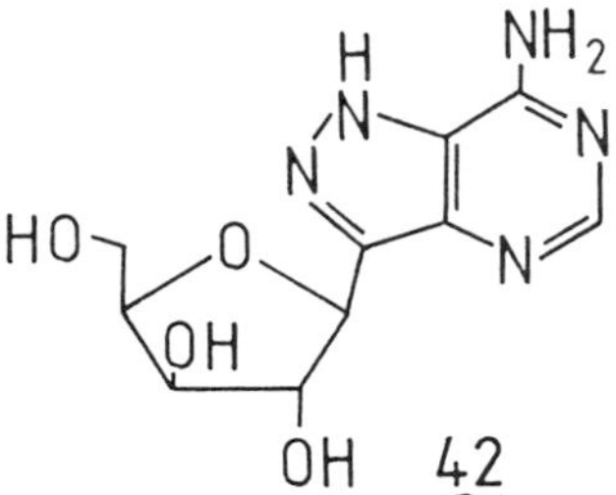

42 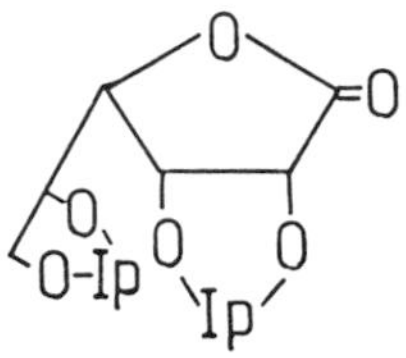 43

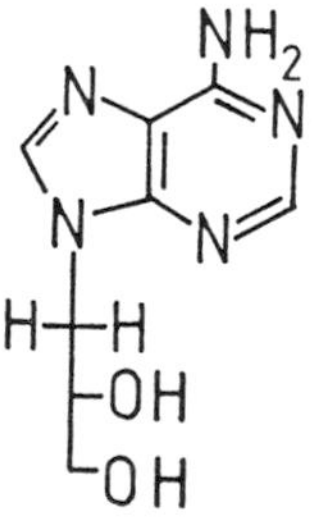

44 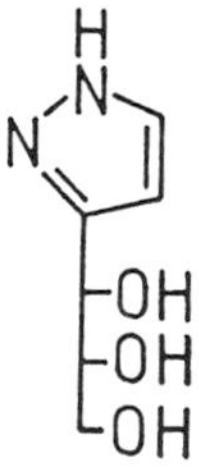 45 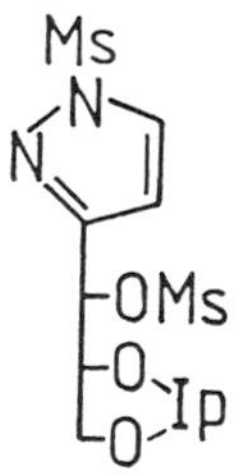 46 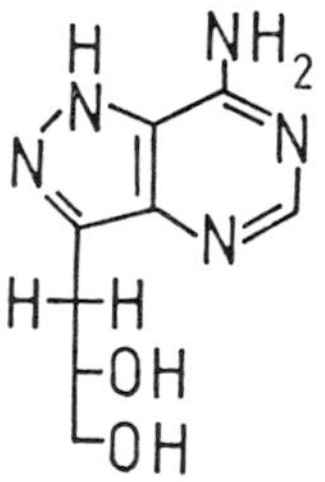 47

Ms = $MeSO_2$

3 BIOSYNTHESIS OF *C*-NUCLEOSIDES

In parallel with the synthetic work Richard Wightman and I have studied the biosynthesis of the *C*-nucleosides showdomycin (4), pyrazofurin (8) and, by implication, formycin (5). Elstner and Suhadolnik, in pioneering work using radiolabelled precursors,[34,35] demonstrated that in *Streptomyces showdoensis* showdomycin (**48**) was derived from D-ribose [presumably as phosphoribosyl pyrophosphate (**49**)] and α-oxoglutarate (**50**) or glutamate (**51**) or an unsymmetrical metabolite derived from these. Attachment involved the carbon atom corresponding to C-4 of α-oxoglutarate, which becomes C-2 of showdomycin;[34] C-1 of α-oxoglutarate is lost.[34] Later work using singly labelled ^{13}C-acetates confirmed these results and demonstrated the importance of the tricarboxylic acid (TCA) cycle in this organism.[35]

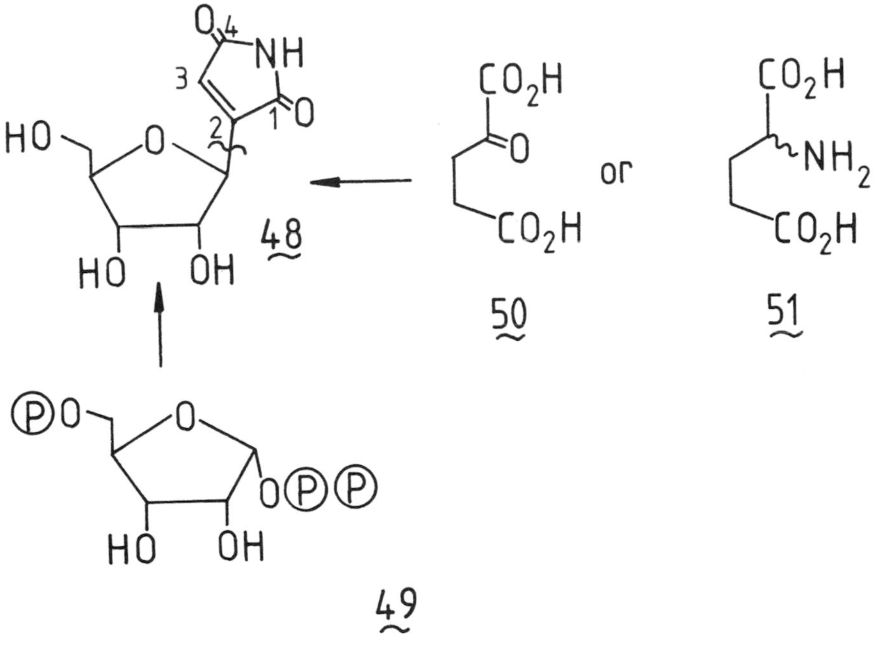

We have corroborated and extended these results, mainly using stable isotopes.[36,37] When [5-$^2H_1$]-D-ribose, shown in the furanose form (**52**), was administered to growing cultures of *S showdoensis,* it was incorporated specifically into the ribose portion of showdomycin (**53**), as shown by the ^{2}H NMR spectrum.[36] The origin of the maleimide ring was then examined in detail. The assignment of the two carbonyl signals in the ^{13}C NMR spectrum of showdomycin[35] was confirmed by their coupling constants in the proton–coupled spectrum. When [5–^{13}C]–DL–glutamic acid (**54**) was added to the medium, the isolated showdomycin (**55**) showed enhancement in the higher–field carbonyl signal, due to C–1. It was shown, in addition, that [5–^{14}C]–L–glutamic acid (**56**) was utilised 6.4 times more efficiently than its D–isomer (**57**), and, as expected, there was even less radioactivity incorporated from [1–^{14}C]–L–glutamic acid (**58**).[36]

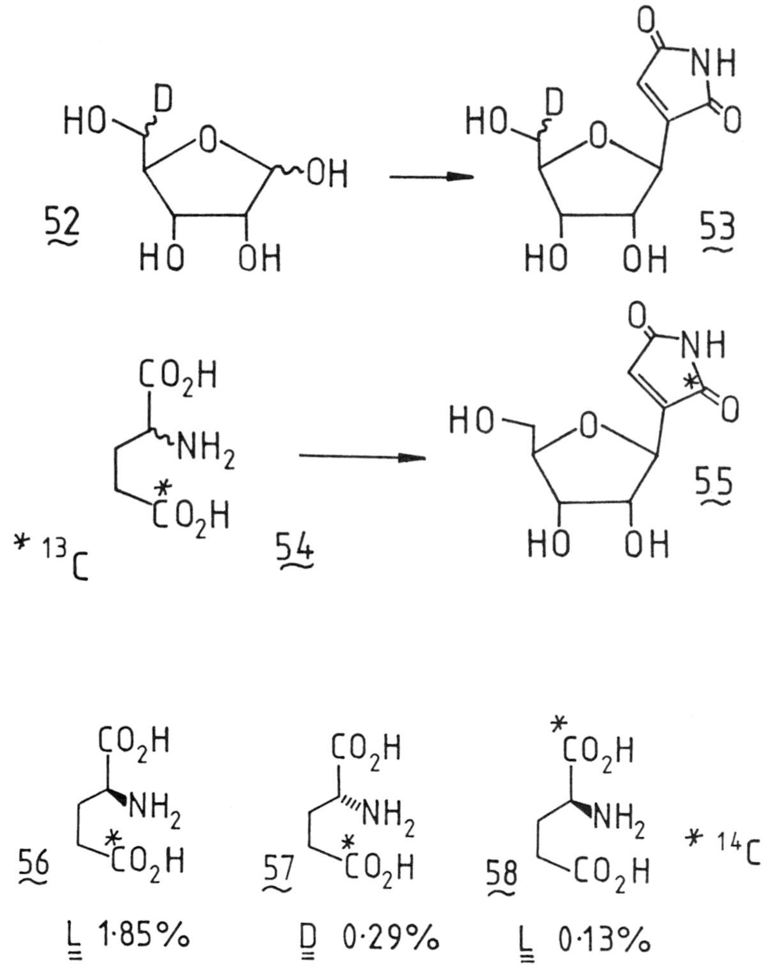

52 → 53

* ^{13}C 54 → 55

56 L 1·85%   57 D 0·29%   58 L 0·13%   * ^{14}C

Specific incorporation of ^{14}C-labelled glutamic acids into showdomycin

CO_2H 15 NH_2 13 CO_2H

59

HO O 15 NH 13 O O HO OH

60

double label —

It was then shown that the amino group in L-glutamic acid is the source of the maleimide nitrogen atom. [5-^{13}C,^{15}N]-Glutamic acid (59) was fed to *S showdoensis*. The ^{13}C spectrum of the resulting showdomycin (60) showed satellites surrounding the enriched C-1 singlet, due to ^{13}C-^{15}N couplings. From the relative intensities it was estimated that *ca* 35% of the doubly labelled glutamic acid that had been incorporated into showdomycin had retained both ^{15}N and ^{13}C. Despite this appreciable loss of nitrogen, probably due to the action of transaminases, the experiment established clearly that the nitrogen atom of L-glutamate is retained in showdomycin.[36]

Glutamic acid has five hydrogen atoms attached to carbon of which at least four must be lost in forming showdomycin. We found that [3-$^{2}H_2$]-DL-glutamic acid (61) gave showdomycin (62) labelled with deuterium.[37] Experiments in collaboration with Dr Douglas Young of the University of Sussex showed that the deuterium atom in (2*S*,3*S*)-[3-^{2}H]-glutamic acid (63)[38] is retained as the vinyl hydrogen in showdomycin (62), whereas the deuterium atom is lost from the (2*S*,3*R*)-isomer (64).[37] The stereospecific retention of the 3-*pro-S* hydrogen has implications for the mechanism of linking the ribose and glutamate units. The 3-oxo intermediate (65), a possible nucleophilic species derived from L-glutamate, has been suggested as an intermediate in the biosynthesis of muscarine, ibotenic acid and related metabolites, but must be ruled out in the case of showdomycin.

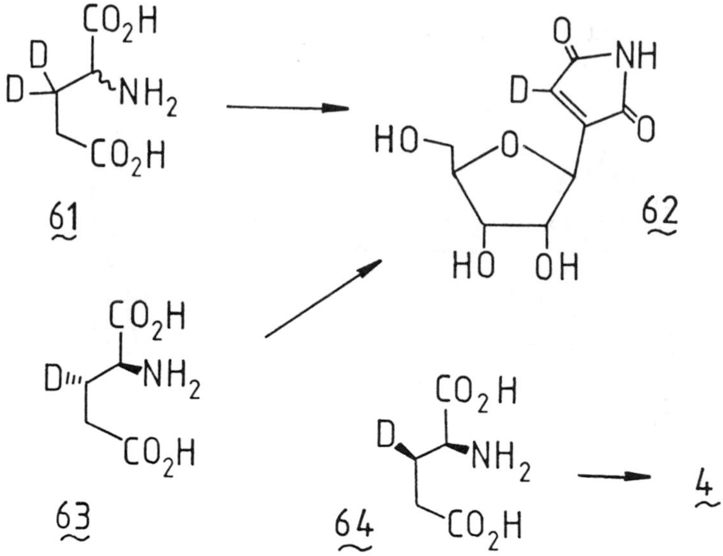

61 → 62

63 → 62

64 → 4

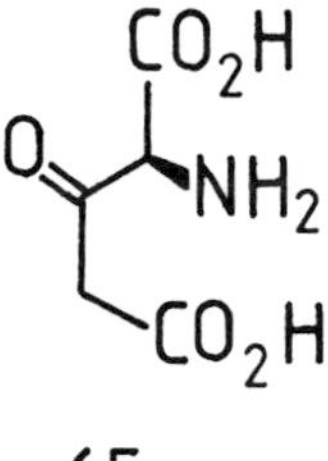

65

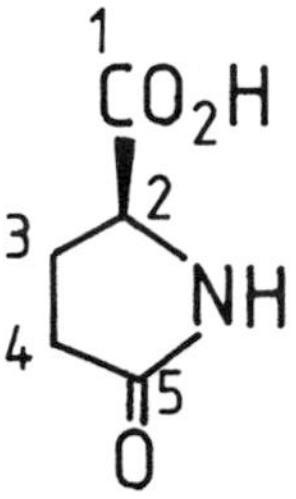

66

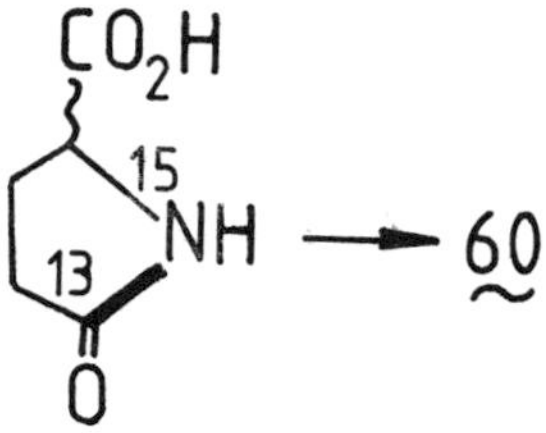

67 → 60

68

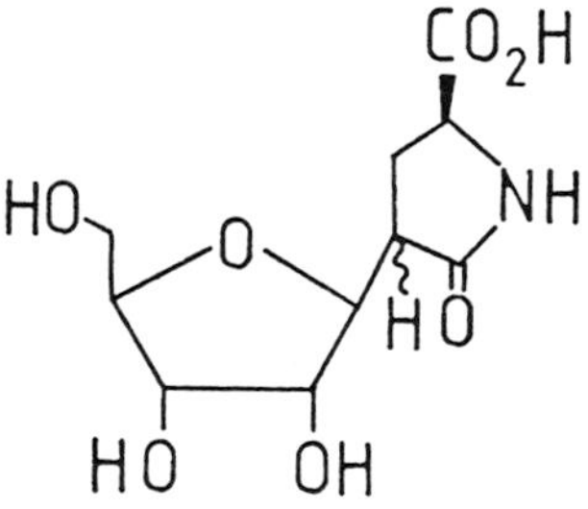

69a 4-R

69b 4-S

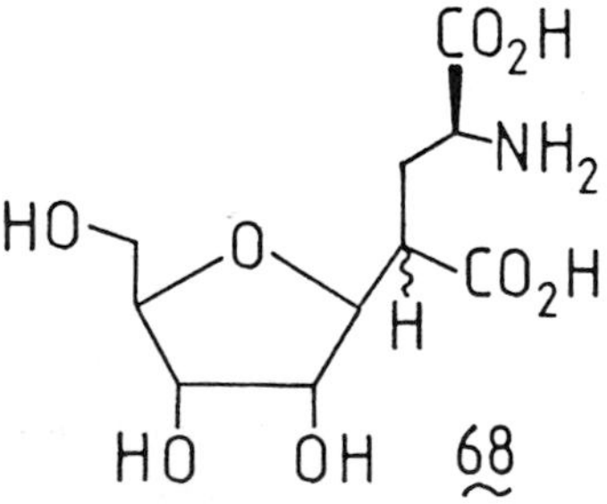

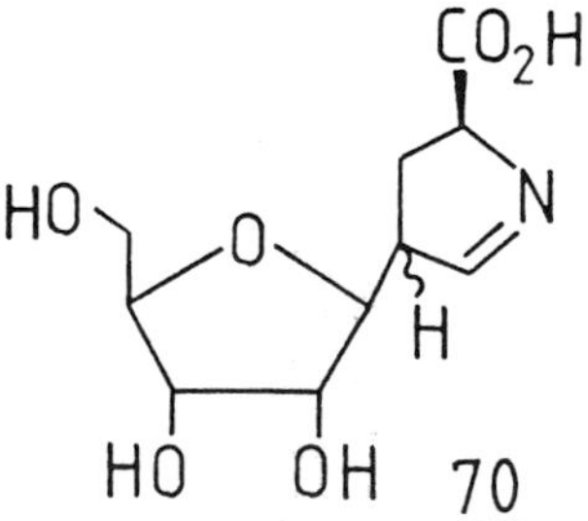

70

L-Pyroglutamic acid (**66**) appeared to be a possible precursor and, in [U-^{14}C]-form, was shown to be incorporated into showdomycin at about the same level as L-glutamate. When the doubly labelled DL-pyroglutamate (**67**) was fed, the resulting showdomycin contained some of the doubly labelled form (**60**), but to about the same extent as from the glutamate (**59**). Again there was considerable loss of nitrogen and it appears that conversion of pyroglutamate into glutamate occurs, followed by transamination. Although L-pyroglutamate is a specific precursor of showdomycin it has not been shown to be an obligatory intermediate. Our current view is that ribosyl derivatives (**68**)[40], (**69**)[41] or (**70**), presumably as 5'-phosphates, are possible intermediates. The two synthetic lactams (**69**) (unlabelled) have been given to us by Professor J E Baldwin, but we have not had the opportunity to test them.

The biosynthesis of formycin (**5**) and pyrazofurin (**8**) attracted us because of the structural similarities to adenosine (**6**) and the cytotoxic nucleoside bredinin (**71**).[42] The 5'-phosphate of the aminocarboxamide (**72**) ('AICAR') is a well known precursor of adenosine-5' phosphate[43] and we were tempted to think that formycin biosynthesis might follow a similar pathway. A further point of interest is the nitrogen-nitrogen bond in the pyrazole ring, comparatively rare in natural products. We studied pyrazofurin biosynthesis because of the accessibility of *Streptomyces candidus* and soon found[40] that [1-^{14}C]-D-ribose and [5-$^{2}H_1$]-D-ribose (**52**) were incorporated. It became apparent that glycine and bicarbonate, fundamental units in purine biosynthesis, were not precursors. On the other hand both [U-^{14}C]- and [1-^{14}C]-L-glutamic acid (**73**), but not the D-isomer, were incorporated into pyrazofurin (**74**). The higher specific incorporation found for the [U-^{14}C] material was to be expected from the loss of C-1 *via* the TCA cycle. When [1-^{13}C]-DL-glutamic acid (**75**) was fed the resulting pyrazofurin (**74**) was labelled in the amide group. We argued that an intermediate of type (**76**) [*cf* (**68**)] would account for these results, C-5 of the original glutamate being lost. In a study[44] of formycin biosynthesis it was found that four carbons of [U-^{13}C]-L-glutamic acid were incorporated into the pyrazolopyrimidine unit. Reinterpretation of the NMR data[40] resulted in the labelling pattern shown in (**77**), indicating a close analogy with pyrazofurin biosynthesis.

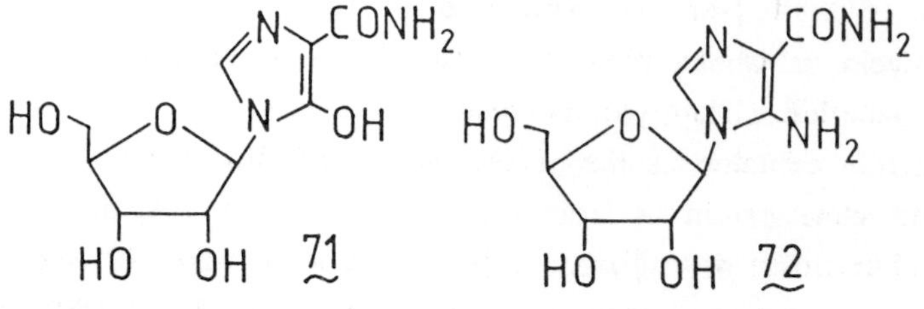

71

72

73

74

$* \; ^{14}C$ or ^{13}C

Transamination

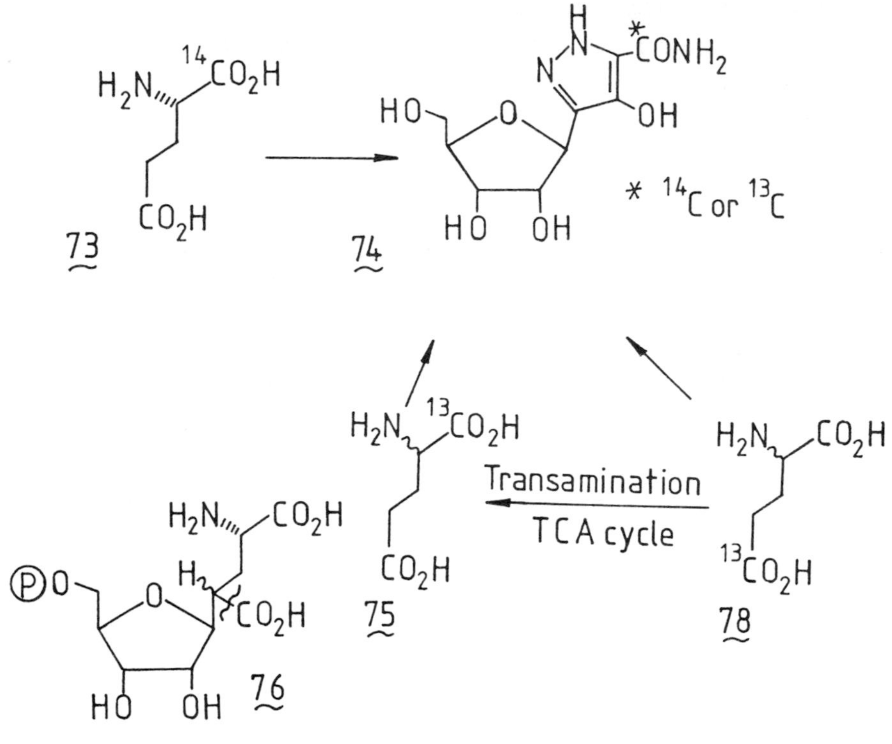

TCA cycle

75

76

78

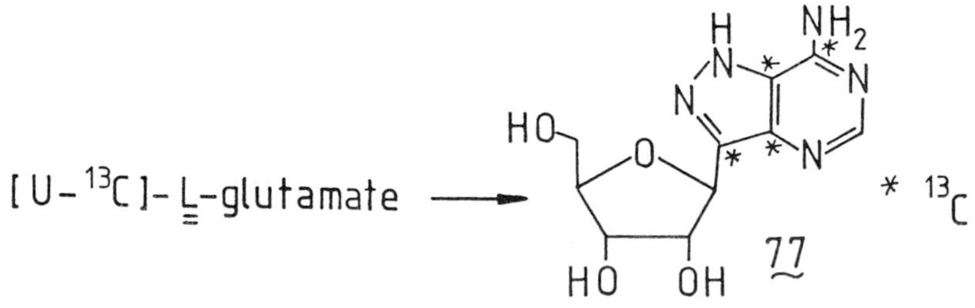

[U-^{13}C]-L-glutamate

77

$* \; ^{13}C$

We therefore fed [5-^{13}C]-DL-glutamic acid (78) and were surprised to find that ^{13}C was incorporated into the amide group slightly more efficiently than from the [1-^{13}C]-material (75). It is known that (78) may be converted into (75) by transamination followed by one turn of the TCA cycle, but this explanation is not entirely satisfactory.

An experiment with [1,2-$^{13}C_2$]-DL-glutamic acid (79) has led us to modify our earlier theory. The resulting pyrazofurin (80) carries, as expected, the double label involving the amide group, which is also labelled singly. The single label found at C-3 cannot be explained by the operation of the TCA cycle (*cf* Figure 1), but may be accommodated by the formation of a symmetrical intermediate, *eg* (81) or (82).

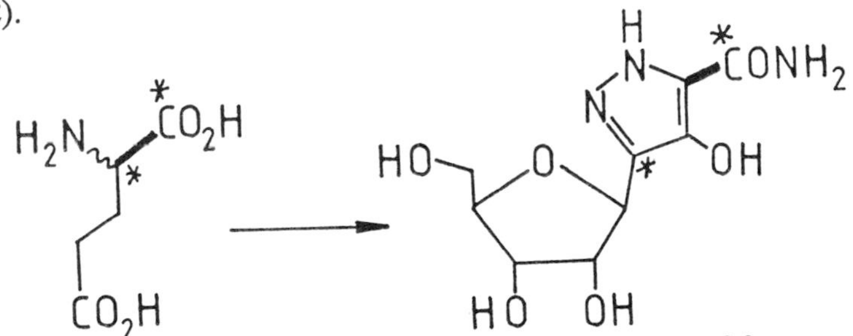

79 → 80

▮ double ^{13}C; * single ^{13}C

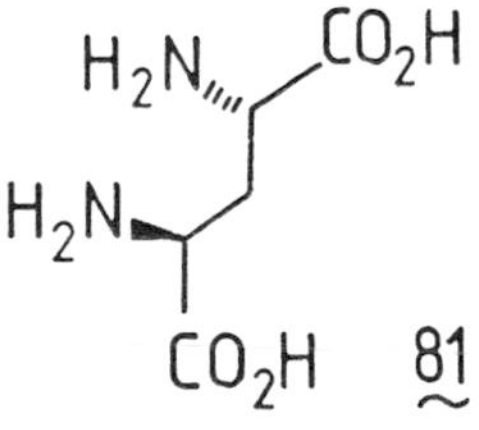

81

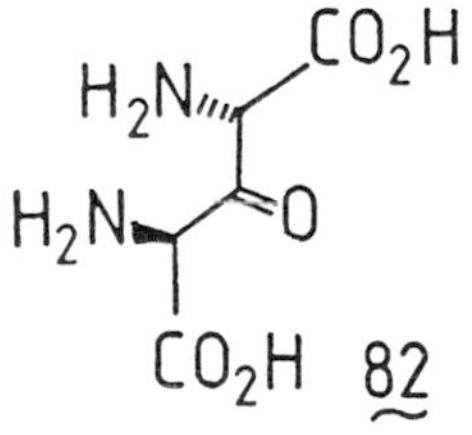

82

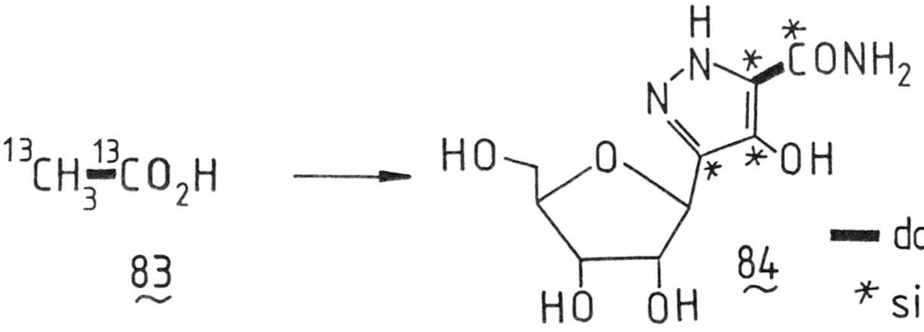

$^{13}CH_3$–$^{13}CO_2H$ (83) → 84

▬ double

* single

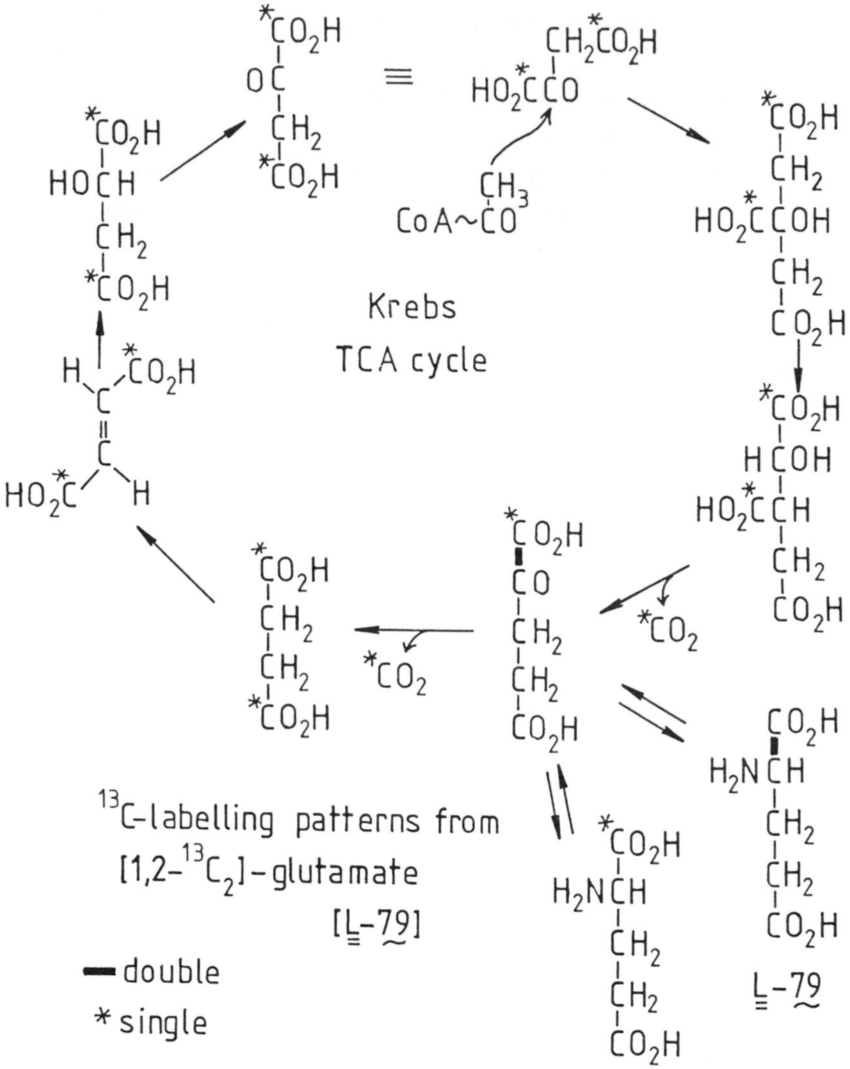

Krebs
TCA cycle

^{13}C-labelling patterns from
[1,2-$^{13}C_2$]-glutamate
[L-79]

— double
* single

L-79

Figure 1: ^{13}C–Labelling pattern from [1,2 –$^{13}C_2$]glutamate as a result of the tricarboxylic acid (TCA, Krebs) cycle.

[2-^{13}C]-Acetate has been shown by Suhadolnik[45] to be incorporated into C-3,4,5 of pyrazofurin. Using [1,2-$^{13}C_2$]-acetate (**83**) we found the labelling as in (**84**) which is consistent with the operation of the TCA cycle. Quantitative analysis of the relative incorporations is, however, in favour of the symmetrical intermediate hypothesis. It should be noted that the origin of the pyrazole ring nitrogen atoms has not yet been determined. The non-ribose portion of the nucleoside antibiotics showdomycin, formycin, pyrazofurin and oxazinomycin (minimycin)[46,47] has now been shown to be derived from L-glutamic acid or α-oxoglutarate. We[40] and others[45,48] have suggested a common ribosyl intermediate related to (**68**) [(**76**)], but this now appears to be an over-simplification.

ACKNOWLEDGEMENTS

I wish to thank my senior colleagues Dr Alan Edgar and Dr Richard Wightman in particular for their sustained contributions, and my skilled and enthusiastic collaborators whose names are recorded in the Bibliography. I thank Professor Gordon Kirby for useful discussions. The work was mainly supported by SERC in the form of studentships, and grants in the case of the biosynthetic work.

REFERENCES

1 G R Clemo, *J Chem Soc,* 1955, 2057.

2 J Baddiley, J G Buchanan, R Letters, and A R Sanderson, *J Chem Soc*, 1959, 1731.

3 J G Buchanan, *Fortschr Chem Org Naturst*, 1983, **44**, 243, and references cited therein.

4 R J Suhadolnik, *Prog Nucleic Acid Res Mol Biol*, 1979, **22**, 193, and references therein.

5 J Goodchild in 'Topics in Antibiotic Chemistry', vol 6, ed P G Sammes, Ellis Horwood, Chichester, 1982, p105.

6 R Barker and H G Fletcher Jr, *J Org Chem*, 1961, **26**, 4605; N A Hughes and P R H Speakman, *J Chem Soc (C)*, 1967, 1182.

7 J G Buchanan, A R Edgar, and M J Power, *J Chem Soc Perkin Trans 1*, 1974, 1943.

8 G Aslani-Shotorbani, J G Buchanan, A R Edgar, C T Shanks, and G C Williams, *J Chem Soc Perkin Trans 1*, 1981, 2267.

9 P J Garegg, B Lindberg, K Nilsson, and C-G Swahn, *Acta Chem Scand Ser B*, 1973, **27**, 1595.

10 S Winstein, E Allred, R Heck, and R Glick, *Tetrahedron*, 1958, **3**, 1; E Allred and S Winstein, *J Am Chem Soc*, 1967, **89**, 3991, 3998, 4008, 4012.

11 J G Buchanan, A R Edgar, M J Power, and C T Shanks, *J Chem Soc Perkin Trans 1*, 1979, 225.

12 R F Heck, *J Am Chem Soc*, 1972, **94**, 2712.

13 J G Buchanan, A Stobie, and R H Wightman, *Can J Chem*, 198, **58**, 2624.

14 J G Buchanan, A Stobie, and R H Wightman, *J Chem Soc Perkin Trans 1*, 1981, 2374.

15 J G Buchanan, A R Edgar, M J Power, and G C Williams, *Carbohydr Res*, 1977, **55**, 225.

16 J G Buchanan, A R Edgar, R J Hutchinson, A Stobie, and R H Wightman, *J Chem Soc Perkin Trans 1*, 1980, 2567.

17 C L Habraken and E K Poels, *J Org Chem*, 1977, **42**, 2893.

18 J W A M Janssen, H J Koeners, C J Kruse, and C L Habraken, *J Org Chem*, 1973, **38**, 1777.

19 J G Buchanan, M Harrison, R H Wightman, and M R Harnden, *J Chem Soc Perkin Trans 1*, 1989, 925.

20 J G Buchanan, D Smith, and R H Wightman, *Tetrahedron*, 1984, **40**, 119.

21 J V Crivello, *J Org Chem*, 1981, **46**, 3056.

22 R R Talekar, J G Buchanan, and R H Wightman, unpublished results.

23 D G Farnum and P Yates, *J Am Chem Soc*, 1962, **84**, 1399.

24 J G Buchanan, N K Saxena, and R H Wightman, *J Chem Soc Perkin Trans 1*, 1984, 2367.

25 J G Buchanan, A D Dunn, and A R Edgar, *J Chem Soc Perkin Trans 1*, 1976, 68.

26 J G Buchanan, A D Dunn, A R Edgar, R J Hutchison, M J Power, and G C Williams, *J Chem Soc Perkin Trans 1*, 1977, 1786.

27 J G Buchanan, M E Chacon-Fuertes, and R H Wightman, *J Chem Soc Perkin Trans 1*, 1979, 244.
28 J G Buchanan, S J Moorhouse, and R H Wightman, *J Chem Soc Perkin Trans 1*, 1981, 2258.
29 J G Buchanan, D Smith, and R H Wightman, *J Chem Soc Perkin Trans 1*, 1986, 1267.
30 E de Clercq, J Descamps, P De Somer, and A Holy, *Science*, 1978, **200**, 563.
31 K Freudenberg and A Dozer, *Ber*, 1923, **56**, 1246.
32 P Smit, G A Stork, and H C van der Plas, *J Heterocycl Chem*, 1975, **12**, 75.
33 J G Buchanan, A Millar, R H Wightman, and M R Harnden, *J Chem Soc Perkin Trans 1*, 1985, 1425.
34 E F Elstner and R J Suhadolnik, *Biochemistry*, 1971, **10**, 3608; *idem, ibid*, 1972, **11**, 2578.
35 E F Elstner, R J Suhadolnik, and A Allerhand, *J Biol Chem*, 1973, **218**, 5385.
36 J G Buchanan, M R Hamblin, A Kumar, and R H Wightman, *J Chem Soc Chem Commun*, 1984, 1515.
37 J G Buchanan, A Kumar, R H Wightman, S J Field, and D W Young, *J Chem Soc Chem Commun*, 1984, 1517.
38 S J Field and D W Young, *J Chem Soc Perkin Trans 1*, 1983, 2387.
39 K Nitta, R J Stadelmann, and C H Eugster, *Helv Chim Acta*, 1977, **60**, 1747.
40 J G Buchanan, M R Hamblin, G R Sood, and R H Wightman, *J Chem Soc Chem Commun*, 1980, 917.
41 J E Baldwin, R M Adlington, and N G Robinson, *Tetrahedron Lett*, 1988, **29**, 375.
42 K Mizuno, M Tsujino, M Takada, M Hayashi, K Atsumi, K Asano, and T Matsuda, *J Antibiotics*, 1974, **27**, 775,
43 J M Buchanan and S C Hartman, *Adv Enzymol*, 1959, **21**, 199.
44 K Ochi, S Yashima, Y Eguchi, and K Matsushita, *J Biol Chem*, 1979, **254**, 8819.
45 J Suhadolnik and N L Reichenbach, *Biochemistry*, 1981, **20**, 7042.
46 K Isono and R J Suhadolnik, *J Antibiotics*, 1977, **30**, 272.
47 K Isono and J Uzawa, *FEBS Lett*, 1977, **80**, 53.
48 K Isono, *J Antibiotics*, 1988, **41**, 1711.

COENZYME B_{12}-DEPENDENT REARRANGEMENT REACTIONS: CHEMICAL PRECEDENT FOR KEY ELEMENTARY STEPS AND INSIGHTS INTO ENZYME-ASSISTED COBALT-CARBON BOND CLEAVAGE

Richard G Finke

Department of Chemistry, University of Oregon, Eugene, Oregon 97403, USA

1 INTRODUCTION AND BACKGROUND

Coenzyme B_{12}, also known as adenosylcobalamin ($AdoB_{12}$; Figure 1), is a cofactor for 12 rearrangement reactions, Figure 2.[1] These remarkable reactions have long captured the interest of both chemists and biochemists, in part because they have no precedent in the non-enzymic collection of reactions known to organic chemistry.

A study of the reactions in Figure 2 reveals the well-known result that each of the rearrangements is the result of an *apparent* intramolecular 1,2-shift of a hydrogen and an electronegative group, X = OH, NH_2, or a carbon framework (equation 1):

$$\overset{X}{|}{}^*C - \overset{H}{|}C \;\underset{\text{Enzyme + } B_{12} \text{ cofactor}}{\overset{\text{Net 1,2 shift}}{\rightleftharpoons}}\; \overset{H}{|}{}^*C - \overset{X}{|}C \qquad \text{eqn 1}$$

X = OH (3 examples); NH_2 (5 examples); C framework (4 examples)

These enzymic reactions involve a C–H bond activation step, an interesting rearrangement step, and then a step in which a C–H bond is restored in the rearranged substrate. Consideration of these steps provides another reason why scientists have remained fascinated by the reactions which require $AdoB_{12}$ as a cofactor.

The structure of the B_{12} cofactor has been known since its historic crystal structure in 1961.[2] The finding of a naturally occurring Co–C bond was a surprising result, but one which

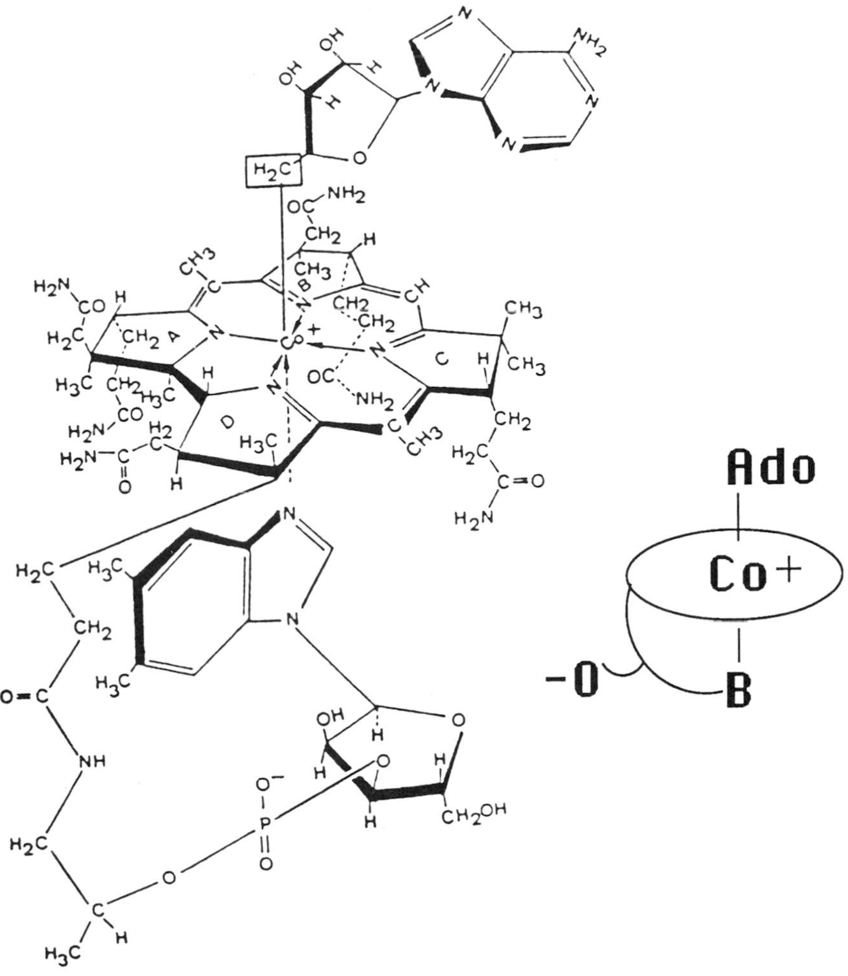

Figure 1: The structure of adenosylcobalamin (AdoB_{12}; also known as Cocnzyme B_{12}) and the abbreviated representation of AdoB_{12} that will be used throughout the text.

allows the (correct) deduction that the crucial role for the B_{12} cofactor centres around the Co–C bond.*

---

* The following prophetic 1972 quote from Robert H Abeles is noteworthy in this regard: 'You will see that Vitamin B_{12} Coenzyme has a cobalt–carbon bond. This was the first known compound that was discovered that had a water–stable cobalt–carbon bond. In this bond probably lies the secret of its reactivity.'[3]

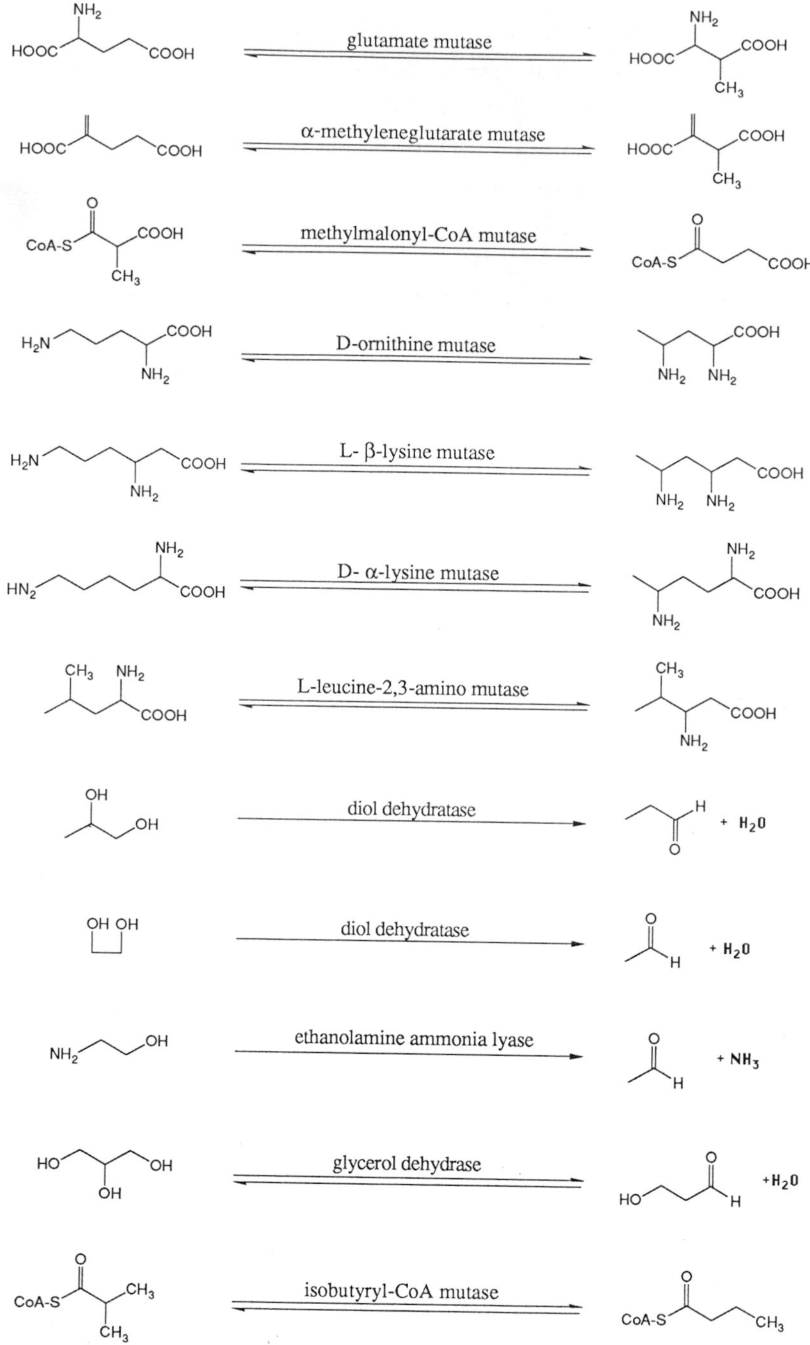

Figure 2: Twelve AdoB$_{12}$-dependent rearrangement reactions that are presently known. The isobutyryl-CoA mutase reaction is a recent discovery.[42]

On the other hand, *the* major impediment to further mechanistic understanding of this class of enzymes is the lack of an exact 3-dimensional structure for *any* of the B_{12}-dependent proteins.* Even the precise protein compositions are in doubt for many of the B_{12}-dependent enzymes. A notable exception here is the collection of very recent studies on methylmalonyl-CoA mutase,[4] work which includes: cloning and structural characterization of the genes coding for this enzyme from *Propionibacterium shermanii*,[4a] structural information (the association behaviour) of the protein's α,β subunits,[4b] studies of the 6 cysteine thiols present in the α,β oligomer,[4c] and the exciting report of crystallization of bacterial methylmalonyl-CoA mutase in a form that is suitable for *X*-ray diffraction structural analysis (crystals diffracting to 3.2Å).[4d]

The B_{12}-dependent enzyme diol dehydratase, which catalyses the 1,2-interchange of a H and a OH group (Figure 2), is of special interest to the work done at Oregon probing the mechanism of the rearrangement step. The diol dehydratase reaction was chosen because it is the B_{12}-dependent enzyme where the most extensive enzymic mechanistic studies are available for comparative purposes. Diol dehydratase is a fairly large protein, consisting of 4 subunits of approximate molecular weights of $(60K)_2$, $(51K)_1$, $(29K)_2$, $(15K)_2$, for a total molecular weight of *ca* 259,000 daltons. However, there is some disagreement over issues such as the best isolation procedure, the exact composition, the subunit structure, and the degree of membrane association of this protein.[5] A complete polypeptide sequence is not available for this enzyme, or for most of the other B_{12}-dependent enzymes.

---

* The magnitude of this impediment can be seen from the following simple but perhaps illustrative example. Imagine, for example, trying to write the arrow-pushing mechanism and transition state structure for even a one-step reaction like the Diels-Alder reaction without knowing the exact structures (or even compositions) of the reactants or products. This simple example reminds us that exact composition plus three dimensional structure of the reactants and products are crucial underpinnings of mechanistic work. Only when these are in hand, can one begin the next stage of mechanistic studies. In that stage, the usual goals are to establish the composition, the structures, and the dynamics interconnecting the intermediates and transition states of the reaction or energy surface.

The B_{12} Mechanism Most Often Cited Prior to About 1980

The mechanism most often found in the literature prior to *ca* 1980, a mechanism widely cited even today, is shown in Figure 3. This scheme is based largely on a considerable body of excellent, painstaking, and often elegant mechanistic studies of the B_{12}-dependent enzymes[6] – although there is a need to repeat some of the key experiments as larger amounts of highly purified enzyme become available (*eg* from cloning and overproduction). There is, for example, good biochemical evidence for homolytic cleavage[6a] of the Co–C bond of $AdoB_{12}$ to produce Co(II) B_{12r}* and for participation of the C5' carbon (the carbon bonded to cobalt) as a H-transfer site.[6b,c] However, there is no *direct* enzymic evidence for an Ado· radical. Furthermore, the unusually large k_H/k_D and k_H/k_T isotope effects seen for the H (D,T) transfer steps are only slowly becoming better understood; they suggest the participation of a second H-transfer site in the enzymes, an Enzyme–XH site labelled '–XH' in the literature, since the identity of 'X' is unknown.[6]

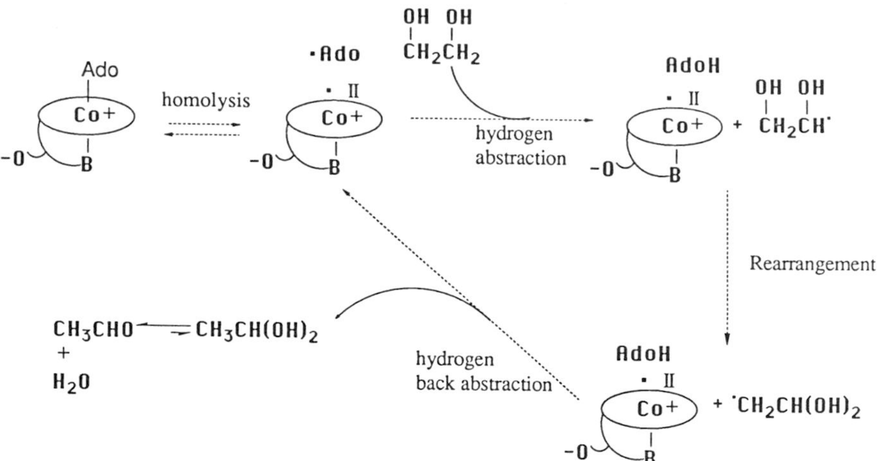

Figure 3: The mechanism most often cited prior to 1980 for the B_{12}-dependent rearrangements. The dotted lines (····) indicate steps that were without rigorous chemical precedent as of *ca* 1980.

---

* The following are standard abbreviations for the Co(III), Co(II), and Co(I) oxidation states of B_{12}: B_{12a}, B_{12r}, and B_{12s}, respectively.

Other mechanistic information has been provided through extensive enzyme stereochemical studies. For example, Arigoni, Rétey and co-workers have examined both enantiomers of propane-1,2-diol as well as the 14 different chiral and achiral ethylene glycol molecules one can make *via* D, ^{18}O, and H labels.[7] Their key finding is that (partial)[7c] stereospecificity in the H· abstraction or ^{18}O labelling experiments is seen *only* for propane-1,2-diol; with labelled and optically active ethylene glycol substrates, *racemization is observed.* This result demands planar, sp^2-hybridized substrate intermediates. Ultimately, the stereochemical results provide very strong evidence for a radical mechanism once they are combined with the spectroscopic evidence for kinetically competent radical processes in the enzymes.

However, despite this body of mechanistic data for the B_{12}-dependent enzymes, the mechanism shown in Figure 3 remains controversial, is completely accepted by few, and is likely to undergo substantial changes in the future, especially in regard to the role of the protein '-XH' H-transfer site. In fact, the B_{12}-dependent enzymic reactions have been called 'probably less well understood than almost any other group of enzyme catalysed reactions.'[8]

Why are the B_{12}-dependent enzymes so poorly understood? A major reason is the lack of protein composition and structural information as noted above; a second reason has been the lack of applicable chemical precedent for most of the proposed elementary steps in Figure 3 - despite a large body of literature in the 'B_{12} model chemistry' area. Specifically, when we began our studies good chemical precedent* was (and in some cases still is) lacking for the following proposed elementary steps in Figure 3:

(i) the $AdoB_{12}$ *thermal* homolysis step;

(ii) the H· abstraction from substrates such as ethylene glycol by a *thermally generated Ado·* (and the absolute rate constant at physiological temperatures plus the k_H/k_D for this step);

---

* By *good* chemical precedent we mean rigorous, quantitative studies of the *exact* reaction step proposed, insofar as that is possible in the absence of the enzyme. For example, the well-known photochemical homolysis of $AdoB_{12}$ and other $R-B_{12}$ compounds does not constitute acceptable chemical precedent for the first step in Figure 3.

(iii) the rearrangement step [does cobalt participate in this step or not, and is the observed stereochemistry the province of the cobalt cofactor or the protein (or both)?];
(iv) the last step, the H· abstraction from an unactivated $AdoCH_3$ group by the product radical (an enthalpically thermoneutral step at best, at least in the absence of the enzyme). This last step is probably the least understood and thus most bothersome step in this 'most often cited mechanism'.

In other words, there is a need for a chemical paradigm for the elementary mechanistic steps proposed to operate in the enzymic rearrangement reactions requiring Coenzyme B_{12} as a cofactor.

The above goal has been the focal point of our efforts in bio–organic/bio–inorganic chemistry at Oregon over the last 12 years. This goal is also the focus of the present contribution. In keeping with the design of these Symposium Proceedings, we will focus primarily on our own work, but will include lead references to related studies.*

2 THE R–Co BOND HOMOLYSIS AND BOND DISSOCIATION ENERGY (BDE) PROBLEM

Demonstration of the reversible thermal homolysis of the Co–C bond of $AdoB_{12}$ became possible only after the required methods for the study of transition metal–alkyl homolyses were developed as part of the natural evolution of organometallic chemistry. Seminal contributions by Professor J Halpern's research group are noted in this regard.[9]

---

* B_{12} chemistry is a large area where many groups continue to make important contributions. Some lead references to work in the area of chemical models for the 12 different B_{12}–dependent rearrangements are listed in references 21 and 31.

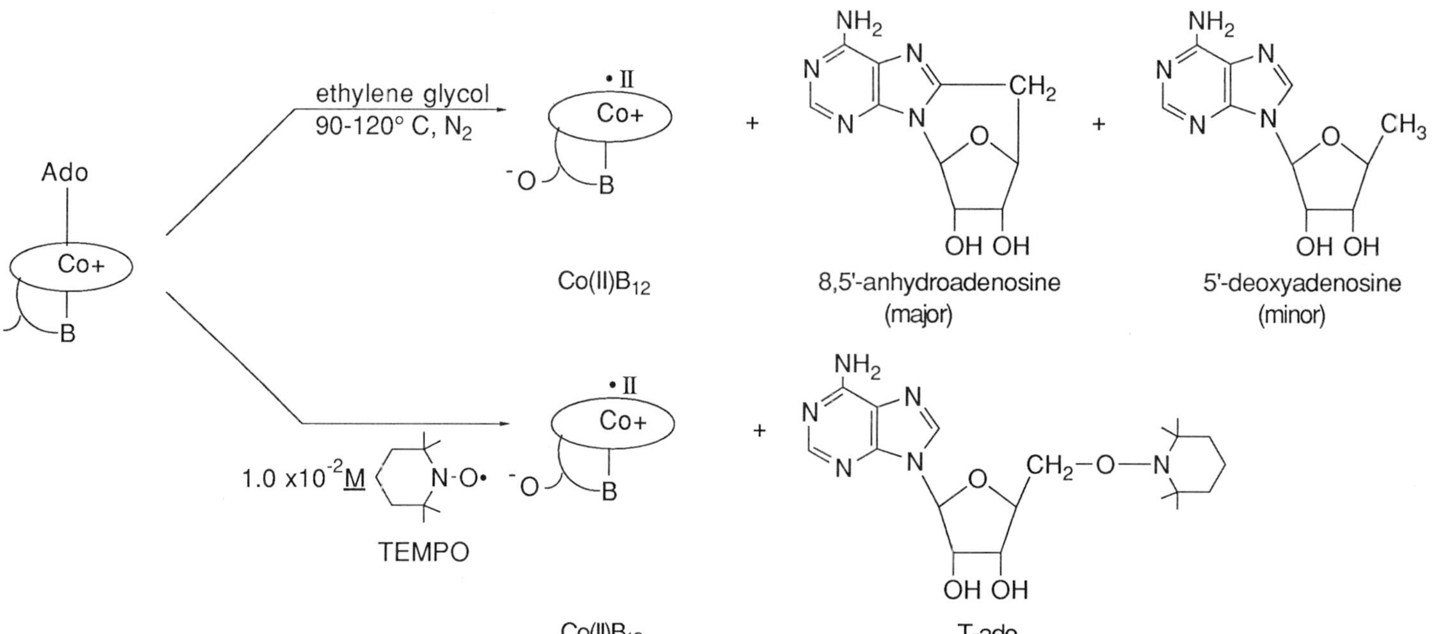

Figure 4: The observed products for $AdoB_{12}$ thermal homolysis in ethylene glycol without (top) and with (bottom) added nitroxide TEMPO. Complete mass balance is observed.[12]

Our own work has focused on the development of a radical trapping method applicable to B_{12}–alkyls, specifically one with a radical trap that would have *thermodynamic stability* towards possible redox side reactions with the thermolysis product Co(II) B_{12r}. The Ph D work of Brad Smith yielded the nitroxide radical trapping kinetic method,[10] which has subsequently proved to be the method of choice for RB_{12} thermolysis studies.[11]

The products formed in the anaerobic thermolysis of $AdoB_{12}$ in the enzymic substrate ethylene glycol as solvent, both with and without the nitroxide TEMPO, are shown in Figure 4.[12] In the presence of TEMPO, quantitative yields of Co(II) B_{12r} and TEMPO-trapped Ado· are observed. Interestingly, even in the absence of TEMPO the reaction proceeds smoothly and with 4 clean isosbestic points as documented in Figure 5. This is due to

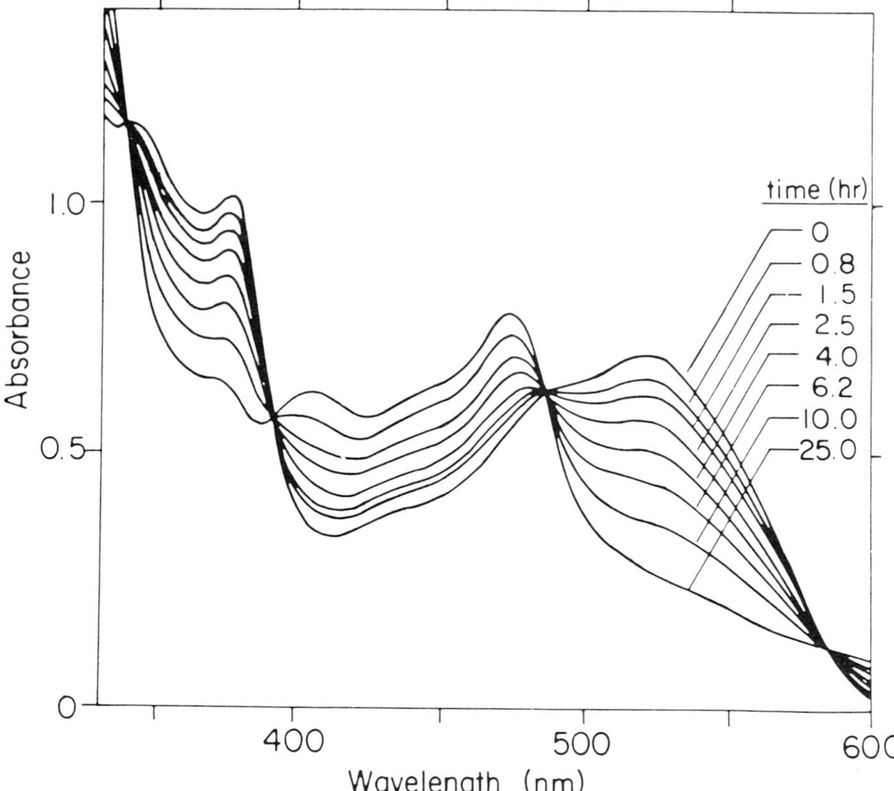

Figure 5: The visible spectrum for the anaerobic thermolysis of $AdoB_{12}$ in ethylene glycol showing 4 isosbestic points even without added TEMPO.

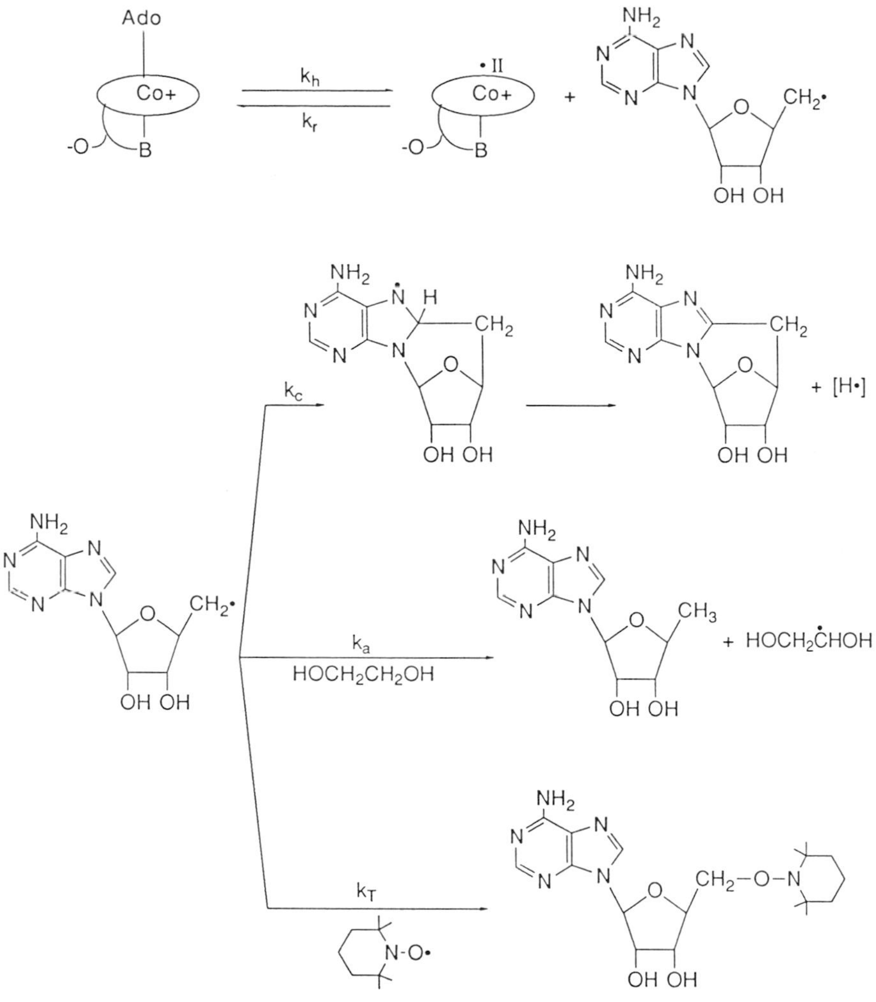

$$\frac{-d \ln AdoB_{12}}{dt} = k_h \left[\frac{k_c + k_a[HOCH_2CH_2OH] + k_T[TEMPO]}{k_c + k_a[HOCH_2CH_2OH] + k_T[TEMPO] + k_r[Co(II)B_{12}]} \right]$$

Figure 6: The kinetic scheme and derived rate law for $AdoB_{12}$ thermolysis in ethylene glycol.

two reactions which trap Ado·: the 'self–trapping' Ado· cyclization reaction to form 8,5'–anhydroadenosine after oxidative work–up (a H· is lost from the cyclized radical in this process; see Figure 6); and, to a lesser extent, H· abstraction by Ado· from the solvent, ethylene glycol. The use of the nitroxide trap is necessary, however, to obtain a mechanistically straightforward system where, for example, all the products can be characterized and quantified. Added nitroxide is also required to obtain a simple kinetic system exhibiting linear first-order kinetic plots.

The scheme by which the reaction kinetics were analysed is shown in Figure 6. Note that this work provides the first estimate for the rate constant for H· abstraction from ethylene glycol by a thermally generated Ado· [k_a(110°C) ≃ 7 x 10^3 $M^{-1}s^{-1}$].[12] Further studies of this reaction, its k_H/k_D values as a function of temperature [k_H/k_D(110°C) ≃ 6.7 ± 1.5],[12] and a comparison of the lower temperature k_H/k_D values to the unusually large enzymic isotope effects should prove most interesting and are in progress.

Because of the thermodynamic stability of the radical trap (TEMPO) towards Co(II) B_{12r} in ethylene glycol, it was possible to study the $AdoB_{12}$ homolysis kinetics as a function of added authentic Co(II) B_{12r}. The inverse linear dependence predicted for a Co–C homolysis is in fact observed, Figure 7, a result which provides the *first unequivocal evidence for the reversible thermal homolysis of the Co–C bond in AdoB$_{12}$ in the absence of any B$_{12}$–binding enzyme.*

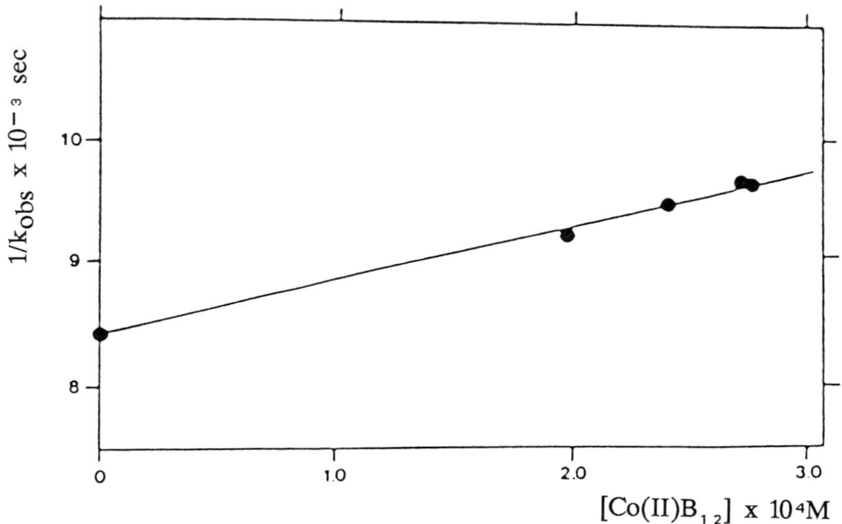

Figure 7: Plot of $1/k_{obsd}$ *vs* Co(II) B_{12r} providing the first direct demonstration of the $AdoB_{12}$ to Ado· + Co(II)B_{12r}· equilibrium in the absence of any B_{12}–binding protein.

Full details of these studies are available;[12] they include rate constants and activation parameters for the homolysis step of the axial benzimidazole base–on form of B_{12}, k_h, base–on. This work by Benjamin Hay includes a number of accomplishments noteworthy to anyone interested in the challenges and details of this area. These accomplishments include: a full error analysis optimizing the kinetic experiments (*eg* the wavelengths used) and the other conditions so that ± 3% precision data in over 50 kinctic runs was obtained (this level of precision was necessary in order to detect the small, *ca* 8% Co(II) B_{12r} dependence); *demonstration* for the first time that homolysis from the base–off form of $AdoB_{12}$ is negligibly slow (this finding is necessary to simplify the kinetics into a manageable form); the independent measurement of the temperature dependence of the base–on to base–off, axial–base equilibrium; and the use of this information to correct the k_{obsd} to yield the desired rate constants and activation parameters for the base–on form of $AdoB_{12}$ ($k_{h,on}$, $\Delta H^{\ddagger}_{h,on}$ and $\Delta S^{\ddagger}_{h,on}$, respectively).[12]

Table I. A Comparison of the Thermal Homolyses of $AdoB_{12}$ in Ethylene Glycol, $AdoB_{12}$ in pH = 7.0 H_2O, and $AdoCbi^+OH^-$ in pH =7.0 H_2O

Corrin Compound (solvent)	$k_{h,on}$ (sec^{-1})[a] (110°C)	$\Delta H^{\ddagger}_{h,on}$ (Kcal/mol)	$\Delta S^{\ddagger}_{h,on}$ (e.u.)	references
$AdoB_{12}$ (ethylene glycol)	2.7×10^{-4}	34.5 ± 0.8	14 ± 1	12
$AdoB_{12}$ (H_2O; pH=7.0)[b]	2.6×10^{-4}	33 ± 2	11 ± 3	13
$AdoCbi^+OH^-$ (H_2O; pH=7.0)	1.2×10^{-5}	37.5 ± 1.2	16 ± 3	18

(a) $k_{h,on} \equiv k_{homolysis,\ base\text{-}on}$

(b) Corrected for the observed small percentage of net heterolysis[13]

A second full paper on Dr Hay's analogous $AdoB_{12}$ thermal homolysis studies in* pH 7 H_2O has been published.[13] Significantly, homolysis also proceeds smoothly in *neutral* H_2O, providing a second clean system for study (4 isosbestic points in the visible spectrum; see reference 13 and Figure 4b therein). Complete product studies and excellent kinetic data and activation parameters were obtained (that are within experimental error of those obtained in ethylene glycol, see Table 1). Here, too, the temperature-dependent kinetic studies were corrected for the axial-base temperature dependence to yield the constants for base-on homolysis, $k_{h,on}$, $\Delta H^{\ddagger}_{h,on}$, and $\Delta S^{\ddagger}_{h,on}$. The results establish that there is no measurable solvent effect on $\Delta G^{\ddagger}_{h,on}$ *for the* $AdoB_{12}$ *homolysis reaction in pH 7* H_2O *vs ethylene glycol,* and only a 1.5 (~2.6) kcal/mol[12b,13] effect on $\Delta H^{\ddagger}$ despite the claim† that there should be a *ca* 6 kcal/mol difference.[14]

In a subsequent study by others, $AdoB_{12}$ has been thermolysed in *acidic*, pH 4.3 H_2O with added Co(II) cobaloxime ($Co[DMG]_2$), the latter purported to be an Ado· radical trap.[15] Hence, $AdoB_{12}$ has now been thermolysed under three sets of conditions, in ethylene glycol,[12] in pH 7 H_2O,[13] and in pH 4.3 H_2O.[15] We independently

---

* We note that the $AdoB_{12}$-binding portion of the protein is expected to have multiple H-bonding sites used to bind B_{12}, and that these are mimicked in part by protic solvents such as H_2O or ethylene glycol. However, one wonders if aromatic groups strategically placed at the active site could stabilize Ado·, substrate·, or product· radicals. For this reason, the thermolysis of $AdoB_{12}$ in solvents containing aromatic groups is of interest and under investigation (B_{12}'s solubility properties greatly limit the choices here however).

† This claim by others was made on the basis of studies of B_{12} *models* in H_2O *vs* ethylene glycol,[14] studies which fail to take into account solvent cage effects. The data was interpreted instead in terms of an inapplicable *gas phase* reaction coordinate diagram.[20] However, as discussed elsewhere,[20] the observed 6 kcal/mol difference most likely occurs largely in the enthalpy for diffusive *cage escape* ($\Delta H^{\ddagger}_d$)[20] and not in the 'solvent dependence of the activation enthalpy of the reverse recombination process'[14] (an example of the confusion that can result if one assumes and applies a gas phase reaction coordinate diagram). Moreover, cage effects will vary from system to system, especially for $AdoB_{12}$ *vs* '$AdoB_{12}$ models' that do not have even the same alkyl (*ie* with R ≠ Ado). These reasons are why the 6 kcal/mol effect found in the R-Co B_{12}-model studies would be expected to be different, *and in fact is different*,[12b,13] from the 1.5 kcal/mol effect *measured* for $AdoB_{12}$.

investigated several aspects of the reported pH 4.3 work and found the following results[13] which are at variance with the reported claims:[15]

(i) the added Co(II) cobaloxime ($Co[DMG]_2$) is not stable under the acidic reaction conditions but instead hydrolyses to $Co(H_2O)_6^{2+}$ (a result confirmed by the literature);[16]

(ii) adenine* and not Ado–$Co[DMG]_2$ is the final major product at pH 4.3 (77% at 85°C; 45% at 110°C);

(iii) some Co(III) B_{12a} is produced as an initial product rather than just Co(II) B_{12r} (in contrast to the clean ethylene glycol and pH 7 H_2O systems, no isosbestic points are seen in the pH 4.3 system; see Figure 4a of reference 13 and compare it to Figures 4b and 4c therein);

(iv) the observed reaction stoichiometry is not that reported of: $AdoB_{12}$ + $Co(II)[DMG]_2$ → Co(II) B_{12r} + $AdoCo[DMG]_2$ (see eqn 3 of reference 15); and

(v) the rate law for $AdoB_{12}$ Co–C cleavage in H^+ has a $[H^+]$ dependence (not unexpectedly, given the correct reaction stoichiometry).[13] This previously missed $[H^+]$ dependence may be important in explaining why the $\Delta S^\ddagger$ observed[15] at pH 4.3 differs (*ie*

---

* The adenine and Co(III) B_{12a} products observed at acidic pH values are those expected for $AdoB_{12}$ Co–C *heterolysis*. It was also shown that the adenine yield at pH = 4.0 was *unaffected* by 100 eq of TEMPO while the Ado• cyclization product was diverted to TEMPO trapped Ado-TEMPO in this same experiment.[13] Such results (plus control experiments demonstrating product stabilities)[13] rule out Ado• as a precursor to adenine. However, others have provided equally convincing evidence that $AdoB_{12}$ Co–C thermolysis can be suppressed by added [Co(II) B_{12r}] beyond a level expected for a parallel heterolysis pathway[17] (our analysis of the kinetics substantiates their claim that the direct heterolysis component can be no larger than the *demonstrated* suppression of the Co–C cleavage reaction). Although a mechanism has not yet been reported that will explain all of the available facts, we agree with the basic conclusion[17] that net (apparent) Co–C heterolysis can sometimes proceed, in selected systems under certain conditions, *via* a mechanism consisting of initial homolysis [followed, presumably, by electron transfer from available reductants like Co(II) B_{12r} or $Co(II)[DMG]_2$].

is not within experimental error)* from that observed in pH 7 H_2O (*vide infra*).

The problems with the pH 4.3 system serve to emphasize the difficulties in developing suitable radical traps (and thus the importance of the TEMPO nitroxide radical trapping method developed by Dr Smith[10]).

Overall, it is presently unclear as to exactly what reaction is being studied, and thus what the reported rate constants and activation parameters refer to, in the complicated pH 4.3/'Co[DMG]$_2$' system.[15] We look forward to the appearence of a full report on the pH 4.3 work[15,17] so that such uncertainties can be resolved.

Returning to the studies done at Oregon, a direct comparison of the rate of homolytic cleavage of the $AdoB_{12}$ Co–C bond *in vitro* to the rate of enzyme-assisted homolytic Co–C bond cleavage provided perhaps our most novel and most important result to date. The comparison shows, for example, that the enzymes diol dehydratase and ethanolamine ammonia lyase *accelerate homolytic cleavage of the $AdoB_{12}$ Co–C bond by a factor of $10^{12\pm1}$*. It is important that the B_{12}-dependent enzymes are able to do this, as otherwise the half-life for Co–C homolysis would be a biologically useless 22 years rather than its present few milliseconds! As discussed elsewhere,[12,18] this example of a 'unimolecular'[12,18] enzymic rate acceleration is as large as any on record. As such, it becomes an important research target in the active area of protein folding, especially with regard to the questions of:

(i) how protein folding is influenced by binding to such a large cofactor (one of Nature's more complex, non-polymeric natural products), and

---

* Given the complicated [H^+] dependence expected in the rate law, and the fact that $\Delta S^\ddagger$ at pH 4.3 is low and derived from a *negative* $\Delta S^\ddagger_{obsd}$, the near agreement between the $\Delta H^\ddagger$ values in pH 7.0 *vs* 4.3 solution (Δ4.4 kcal/mol, which is nearly within the combined experimental error of $\pm$3 kcal/mol) would seem to be accidental (*eg* may refer to an arbitrary choice of [H^+]). *A detailed study of the [H^+] dependence of the rate law is needed for the acidic pH system, including deconvolution of the data for the effects due to protonation of B_{12}'s axial base, but this has not been reported to date.*[15,17] Clearly, only the pH 7.0 results[13] (and those in ethylene glycol)[12] are reliable measures of $AdoB_{12}$'s Co–C homolysis activation parameters at this time.

(ii) how the protein–cofactor (plus protein–substrate) intrinsic binding energy is converted into catalysis by distortion of the Ado–B_{12} bond along the Co····C cleavage reaction coordinate.

A number of chemical models for the Co–C activation step have appeared, notably the corrin 'butterfly' conformational distortion theory. Lead references to the work of the groups of Glusker, Marzilli and Randaccio, Pratt, Schrauzer and Grate, Halpern, and others are available.[12,18] The interested reader is also referred to a discussion of the $10^{12\pm1}$ finding by Pratt.[1e] We have several projects nearing completion in this area that will be reported in due course, after which we are planning a comprehensive review of the explanations and chemical models to date for this intriguing $10^{12\pm1}$ enzymic rate acceleration.

Lastly, we have also reported[18] a study of Co–C thermolysis of $AdoCbi^{+}OH^{-}$,* work which was undertaken to test directly a sizeable chemical-model literature suggesting that the axial benzimidazole base in B_{12} is a key to accelerating Co–C cleavage (see reference 18 for lead references). The important result here is that the axial base appears *to play a more minor role than previously thought*, since its removal slows the Co–C homolysis process by only a factor of $\leqslant 10^{2}$. As discussed elsewhere, this result is fully consistent with the literature of the enzymes, which teaches that even axial base-free $AdoCbi^{+}$ is a reasonably effective cofactor.[18] This biochemical literature furthermore suggests that the axial base is more involved in *binding* of the B_{12} cofactor than in the Co–C cleavage step.

The highlights of the rate constants and activation parameters for the $AdoB_{12}$ and AdoCbi+ studies are provided in Table 1. These studies[12–14] provide the first and the only reliable chemical precedent for the long postulated, reversible thermal homolytic cleavage of the Co–C bond in Ado–B_{12}, including the finding of the enzymic $10^{12\pm1}$ rate enhancement.

---

* $AdoCbi^{+}OH^{-}$ is adenosylcobinamide, in which the appended axial 5,6-dimethylbenzimidazole base (see Figure 1) has been chemically removed (by cleavage at a P–O bond using $Ce(OH)_3$; see reference 18 for further details).

The AdoB$_{12}$ Co–C Bond Dissociation Energy

Since the enthalpy of activation for AdoB$_{12}$ Co–C bond homolysis, $\Delta H^{\ddagger}_{h,on}$, is now available, it is in principle possible to obtain a reasonable if not good estimate of AdoB$_{12}$'s Co–C bond dissociation energy (BDE) for the first time. Unfortunately there has been considerable confusion on exactly how to accomplish this, not just for AdoB$_{12}$ but also for all solution state metal–ligand bond thermolysis studies (recall that a BDE is defined in the gas phase where solvent effects of *all types* are absent). Much of the confusion stems from the failure of the oversimplified reaction coordinate model proposed in the literature.[19] Most significantly, the literature model fails to consider *solvent cage effects*[20] (the previously proposed model employs a *gas phase* reaction coordinate model[19] which, rigorously speaking, cannot apply in solution). In collaborative work with Professor Thomas Koenig at Oregon, an expert in radical cage chemistry, we have provided the proper equations and discussion necessary to obtain Co–C (and other) BDE's from solution thermolysis studies.[20] We are continuing our efforts at obtaining the proper solvent cage chemistry data and corrections necessary for a more rigorous determination of the BDE for AdoB$_{12}$. However, we have discussed elsewhere how the estimate of 30 ± 2 kcal/mol for AdoB$_{12}$'s Co–C BDE was obtained.[12,13]

3 THE SUBSTRATE REARRANGEMENT STEP: THE COBALT PARTICIPATION OR NON-PARTICIPATION QUESTION

The cobalt participation question, briefly defined earlier along with the presentation of Figure 3, has been the question most vigorously pursued by chemists interested in the B$_{12}$-dependent rearrangements.[21] A specific form of this question for the case of diol dehydratase is presented in Figure 8. Detailed therein are the Co non-participation (radical) pathway and the possible Co participation pathways, consisting of a Co(III) π-complex pathway and the two conceivable electron-transfer pathways. Two important background papers published in 1983 describe our synthetic[22a] and

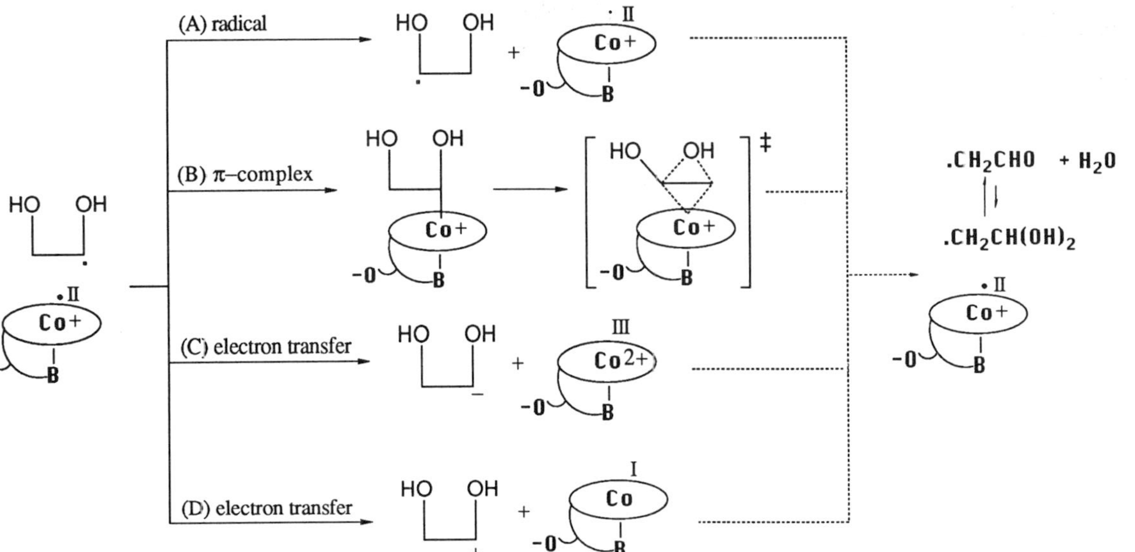

Figure 8: Cobalt non–participation, radical rearrangement mechanism (pathway A) and conceivable cobalt–participation rearrangement pathways (B–D). (The subsequent steps needed to get to the indicated products are not shown.)

a close B_{12} redox mimic,* our modified version of the so-called Costa B_{12} model.[23]

Presently, the most often cited mechanism for the diol dehydratase rearrangement step is the Co π-complex depicted in Figure 8, and presented in more detail in Figure 9. From reading the literature one gets the strong impression that this mechanism is 'accepted' by many workers[24a,b] despite the problems with this mechanism.[22] Hence, in practice this mechanism is the one that has been the focus of our more recent attention. (Such a limited focus is valid only because of our earlier work which provides evidence against the other mechanisms shown in Figure 8.[22])

Note that this scheme (Figure 9) hinges upon two key intermediates: the putative $HOCH_2CH(OH)$–Co B_{12} intermediate (that would result if the $HOCH_2CH(OH)\cdot$ and Co(II) B_{12} radicals combine in the enzyme); and the putative formylmethyl-B_{12} complex, Co–CH_2CHO, that is the ultimate product of the Co π-complex pathway (as well as all other Co-participation pathways we have been able to write; see the discussion of this point in references 22a and 22b). Hence important goals are the independent preparation of pure, unequivocally characterized samples of these two putative intermediates, followed by a study of their reactions and properties – are they consistent with the mechanisms written in Figure 9?

The needed studies were recently completed by Yun Wang for her Ph D thesis[25a] An initial communication on her findings has appeared[25b] and a full paper describing this work will also be

---

* An earlier paper crucial to our approach to the Co-participation question is a critical analysis of the literature of B_{12} models, followed by an electrochemical study of two of the best B_{12} models. The key finding is that the Costa B_{12} model is a preferred, $\pm$0.05 V mimic of B_{12}'s Co(III)/Co(II) and base-on Co(II)/Co(I) redox potentials.[23] This close correspondence with B_{12}'s Co(II)/Co(I) redox potential is the key reason why 'B_{12}-like' results were obtained[22,25] – results mirrored by our more recent study of the Co-participation question using the B_{12} cofactor itself.[25]

This work was also valuable in that it directed our efforts away from models and towards B_{12} itself – we learned that the differences greatly outnumber the similarities when one compares a wide range of B_{12} properties to even the best of the B_{12} models.[23] The reader who is especially interested in either B_{12} models or the Co-participation question is referred to this earlier work.

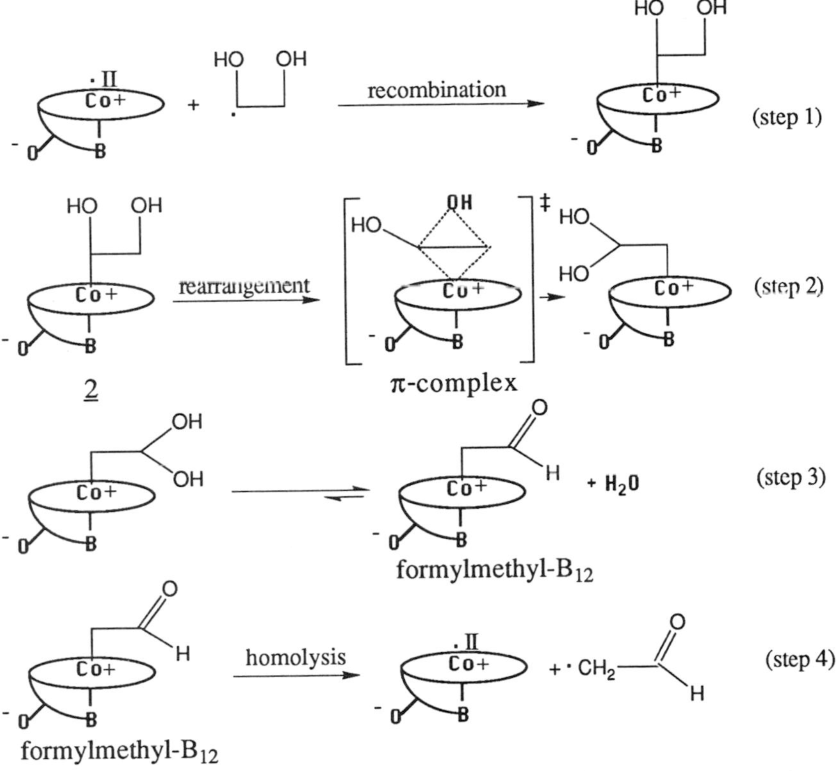

Figure 9: Hypothetical cobalt participation Co(II) π–complex pathway, illustrated for the specific case of diol dehydratase.

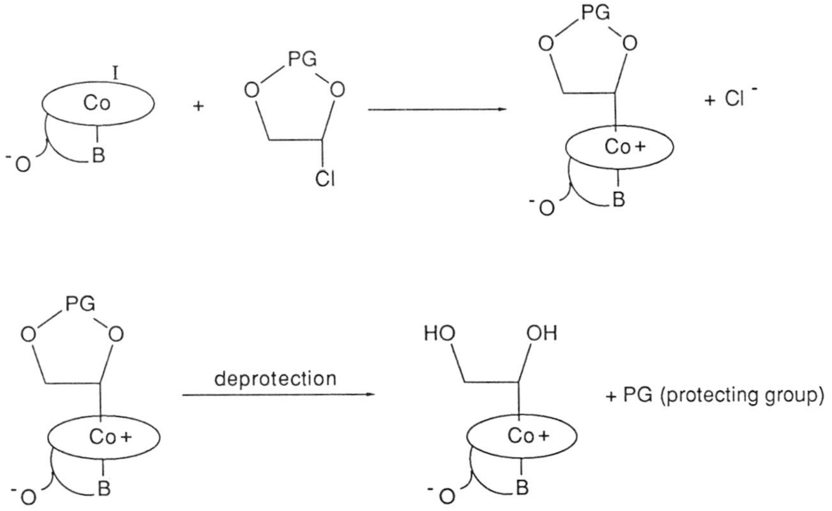

PG
Co + Cl → Co+ + Cl^-

deprotection
+ PG (protecting group)

Figure 10: The protected intermediate approach.

published.[25c] This work employed the B_{12} cofactor and our successful 'protected (diol) intermediate' approach* developed in the Costa B_{12}-model studies,[22] Figure 10.

The route used to synthesize a carbonate-protected form of the $HOCH_2CH(OH)$-Co B_{12} intermediate is shown in the top part of Figure 11, an approach that follows closely our previous studies with the Costa B_{12} model.[22] The resultant carbonate-protected intermediate is formed in modest but acceptable yield (33%), and was shown to be composed of two pure diastereomers by HPLC, mass spectroscopy, and NMR.[25] (Two diastereomers are formed since B_{12} is chiral and chloroethylene carbonate is racemic.) Interestingly, we have recently shown that a second product is formed in about the same yield as the desired β-isomer. This second isomer is a rare example of an α-isomer, the α prefix indicating that the alkyl group derived from chloroethylene carbonate is attached to *the α or 'bottom' side* of B_{12} (*ie* the side where the appended axial benzimidazole base normally binds).[26]

The products which result from the β-isomer, on removal of the carbonate *via* deprotection with CH_3O^-/CH_3OH, are most interesting (Figure 11, bottom half). *There are no detectable rearrangement products*, specifically no formylmethylcobalamin, no CH_3CHO, and no Co(II) B_{12r}. Instead, a non-enzymic redox side-reaction occurs to produce Co(I) B_{12s} and $HOCH_2CHO$. *This is exactly the reaction seen previously in our B_{12} model studies and thus predicted for these studies employing the B_{12} cofactor.*[22]

---

* We note that Professor Paul Dowd pioneered the approach of using B_{12}-bound substrates as probes of the rearrangement step.[21a] Unfortunately, however, only a few workers[21d] have done the necessary follow-up kinetic and mechanistic work necessary to take full advantage of this powerful approach, one that can simultaneously probe $[Co(III)]^+/R^-$, $[Co(II)]\cdot/R\cdot$, $[Co(I)]^-/R^+$, and Co-R participation pathways (R = substrate). Instead, one commonly finds statements in the literature that Co-participation from the Co-substrate precursor 'has been demonstrated' (as if it *must* have occurred for some unknown reason). This conclusion is common despite the general finding that the Co-substrate bond is *far too stable* to be a viable enzymic intermediate (in the absence of unprecedented[25b] enzyme effects), and despite the fact that photolysis is often required to *break* the Co-substrate bond before any rearrangement product is observed, implying the participation of R· intermediates. Moreover, often the *simplest or only precedent* for the observed rearrangement is *via* a free R· (or R^- or R^+) intermediate. Restated, these studies often interpret their results only in terms of their initial *hypothesis*.

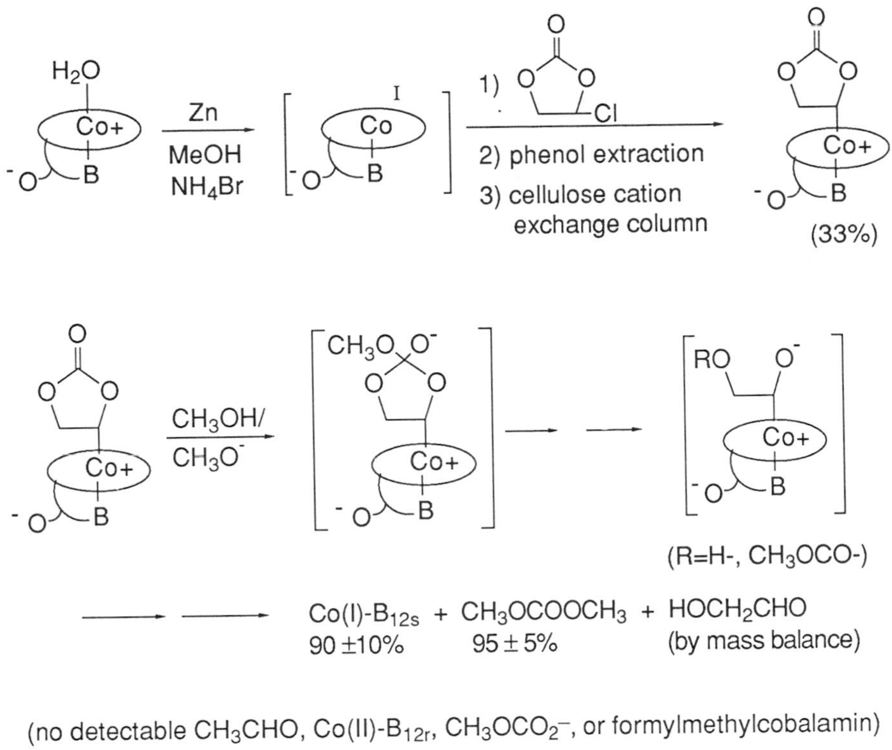

H_2O Co+ B $\xrightarrow{\text{Zn, MeOH, }NH_4Br}$ [Co(I) B] 1) (4-chloromethyl-1,3-dioxolan-2-one) 2) phenol extraction 3) cellulose cation exchange column → (carbonate)Co+ B (33%)

(carbonate)Co+ B $\xrightarrow{CH_3OH/CH_3O^-}$ [CH_3O, O^- adduct, Co+ B] → → [RO, O^-, Co+ B] (R=H-, CH_3OCO-)

→ → Co(I)-B_{12s} + $CH_3OCOOCH_3$ + $HOCH_2CHO$
90 ±10% 95 ± 5% (by mass balance)

(no detectable CH_3CHO, Co(II)-B_{12r}, $CH_3OCO_2^-$, or formylmethylcobalamin)

Figure 11: The synthesis of the carbonate-protected intermediate (top line) and the products that result from its CH_3OH/CH_3O^- deprotection. Note that *none* of the products expected for cobalt participation are observed.

Further, independent and corroborating evidence for our findings is provided by Meyerstein's pulse radiolysis work,[27] summarized in Figure 12. That work effectively extends our findings in basic solution to solutions as acidic as pH 3.1 – with identical findings and conclusions. *Cobalt–participation is not only unnecessary, it leads to a redox side reaction any time Co(II) B_{12r} and the strongly reducing ($E_{1/2}$ = ca −1.3 V(SCE))*[25b] *$HOCH_2CH(OH)\cdot$ radical come into contact!*

As part of our studies, the Co–CH_2CHO formylmethylcobalamin complex was synthesized, purified, and unequivocally characterized (work that resolved the remaining issues in an old literature

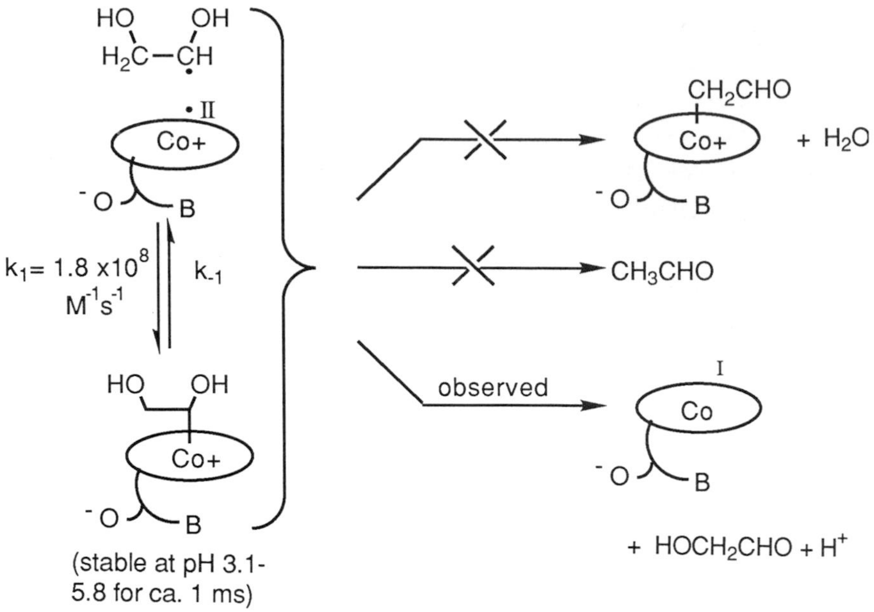

HO OH
$H_2C-\dot{C}H$
• II
Co+
^-O B
$k_1 = 1.8 \times 10^8$ $M^{-1}s^{-1}$
k_{-1}
HO OH
Co+
^-O B
(stable at pH 3.1-5.8 for ca. 1 ms)

CH_2CHO
Co+
^-O B
$+ H_2O$

CH_3CHO

observed
I
Co
^-O B
$+ HOCH_2CHO + H^+$

Figure 12: Meyerstein's pulse radiolysis results[27] in acidic, pH 3.1 water. These results are the same as those seen *via* the protected intermediate ,approach at basic pH.

controversy concerning its preparation and properties).[25] A control experiment showed that formylmethylcobalamin is *stable* to the CH_3O^-/CH_3OH carbonate deprotection conditions thereby *unequivocally ruling it out as an intermediate* in the methanolysis reaction in Figure 11. This, in turn, *rules out any and all Co–participation mechanisms that yield formylmethylcobalamin* – a significant finding since all cobalt-participation mechanisms that we or others have been able to write should yield at least some formylmethylcobalamin.[25] Moreover, when studied separately the formylmethyl complex was found to undergo net* thermolytic *heterolysis* in pH 7 H_2O to give Co(III) B_{12a} + CH_3CHO, consistent

---

* In light of recent reports,[17] this and all other apparent Co–C (and other M–L) heterolyses need to be examined for an inverse Co(II) B_{12r} (or M•) dependence as a test of a net (apparent) heterolysis reaction proceeding by a mechanism of initial homolysis followed by subsequent electron transfer.

with what others had reported previously, but for less well characterized samples of formylmethylcobalamin.[25a,c]

In other words, there is still no evidence that formylmethylcobalamin can undergo thermal homolysis as required by the Co-participation mechanism (Figure 9, step 4). Note also that the enzyme would fall into a *ca* 30 kcal/mol deep energy well (a reasonable guess at the minimum Co-CH_2CHO BDE) if it were to form this intermediate. This is exactly the type of step that enzymes evolve *to avoid!*. These and other arguments against the Co-participation mechanism in Figure 9 are discussed further elsewhere.[25b]

It is important to stress at this point that all of the available enzymic evidence argues that $HOCH_2CH(OH)$-Co B_{12} formation never occurs (step 1, Figure 9), and thus that cobalt does not participate (see the discussions provided elsewhere[22,25]). Firstly, all ESR-detected - *and kinetically competent* - Co(II) B_{12r} and substrate radicals are found to be at least 5 and probably 10Å apart based on extensive ESR studies and calculated fits to the observed spectra (these studies are from several research groups, and are all in general agreement). These ESR studies are summarized in Figure 13.[28] It

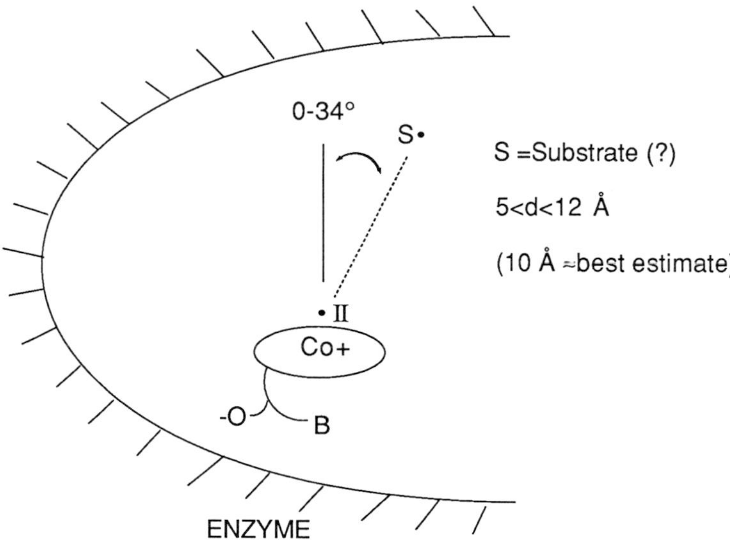

Figure 13: A schematic representation of the evidence against cobalt participation from the ESR studies of B_{12}-dependent enzymes.[28]

would seem that the enzyme is 'well aware' of the results pictured in Figure 13 – if the Co(II) and $HOCH_2CH(OH)\cdot$ radicals get too close,* a redox side reaction will ensue thereby ending the catalytic cycle.

Secondly, the extensive and elegant stereochemical studies of the reactions catalysed by diol dehydratase that were noted earlier provide compelling evidence for a pathway involving planar, sp^2 hybridized intermediates[7] – radical intermediates once the ESR and other spectroscopic evidence is also taken into account. These stereochemical studies *demand* some non-stereospecific pathway other than a (100% stereospecific) Co-participation one – a point discussed in greater detail elsewhere,[25b,1c] and a point which proponents of stereospecific Co-participation mechanisms[24] (*eg* Figure 9) must no longer ignore.[25b] Thirdly, the recent discovery of a non-B_{12} dependent diol dehydratase offers proof that a non-cobalt participation diol dehydratase reaction and mechanism† exists in Nature.[29]

To date, only the participation of protein-bound (*not free!*)[1c] radicals can explain *all* of the available chemical model, enzyme, stereochemical, and other evidence in the case of diol dehydratase.

Lastly, there is some well-known α-OH radical chemistry, Figure 14, that one might expect diol dehydratase to use to its advantage (for lead references, see references 1c, 22 and 30). The *ca* 10^5 increased acidity of an α-hydroxy radical (compared to an alcohol) and facile β-OH fragmentation are features that a protein equipped with a *basic* diol binding site could profitably employ (to bind a substrate-derived α-hydroxy radical and thereby to weaken (or dissociate completely) the β-OH group). The work of Golding and Radom (the lower half of Figure 14)[30] suggests that a second, adjacent, *acidic* diol

---

* Given the recent demonstrations of long-range electron transfers in proteins, the correct argument may be that the protein-assisted rearrangement of $HOCH_2CH(OH)\cdot$ is simply faster than electron transfer to Co(II) B_{12r} at whatever distance separates these two radicals.

† Since this protein has an Fe^{2+} dependence, one should qualify this statement by acknowledging the perhaps unlikely possibility of an Fe-dependent rearrangement mechanism.

OH-binding site in the protein would be ideal to assist further the *apparent* OH migration required by the enzyme stereochemical studies.[7] (This assistance could proceed *via* partial protonation,[1c,30] -O(H)...$^{+}$H-B^{-}, as Golding has previously suggested.[30]) In fact, since the enzymic stereochemical studies for propane diol, $CH_3CH(OH)CH_2OH$, require a 3-point attachment to the enzyme, *it is difficult to imagine how to bind propane diol except through the two OH groups* (plus a weaker interaction with the methyl group).

A

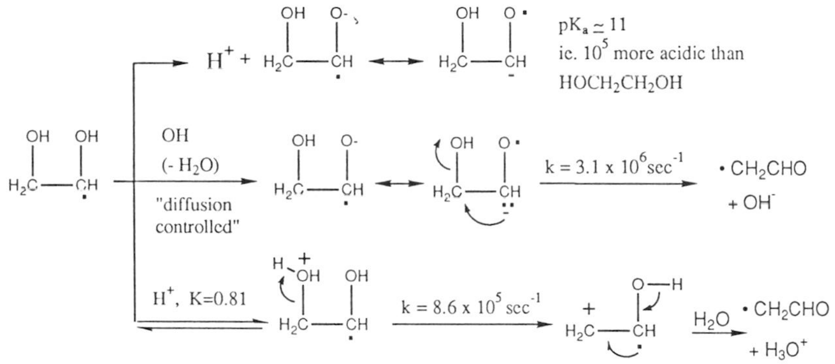

B

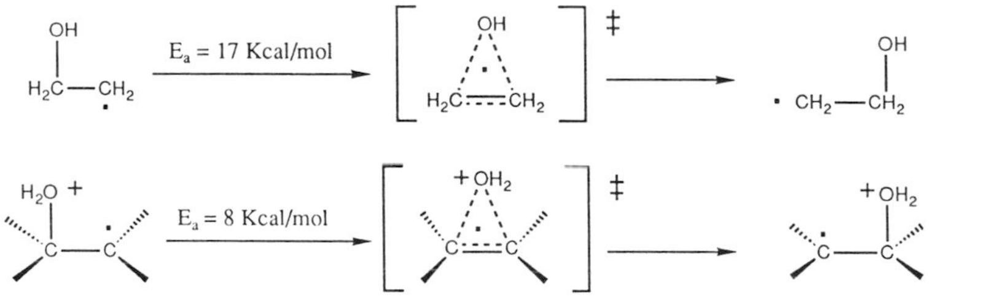

Figure 14: Chemical evidence suggesting that basic and acidic diol-binding sites may be present in diol dehydratase. (A) Well-known radical β-fragmentations that an enzyme could profitably employ by having a basic diol-binding site. (B) Computational studies of Golding and Radom demonstrating that protonation lowers the barrier to OH migration, work suggesting an acidic diol-binding site in diol dehydratase.

To summarize this section on the Co-participation or Co non-participation question:

(i) There is no chemical or biochemical evidence to date for Co-participation, and considerable evidence against it, in the case of diol dehydratase.*

(ii) The *protein*, not the B_{12} cofactor, must be controlling the substrate to product rearrangement step *via* tight binding of the radical intermediates.†

(iii) The finding of cobalt non-participation implies a resulting *simplification* of Coenzyme B_{12}-dependent mechanistic chemistry - Coenzyme B_{12} is nothing more (nor less!) than Nature's source of enzyme-bound Ado· radicals.

The mechanistic simplification and more limited role for the B_{12} cofactor is an extremely important finding in another respect. When viewed within the context of a larger body of information to be discussed in the next section, one is led to the speculative and unproven but exciting conclusion that the B_{12}-dependent enzymes will ultimately prove to be members of a widely distributed class of enzymes operating by a *radical chain mechanism.* In this mechanism, B_{12} and other cofactors operate as *radical chain initiators.* The interested reader is referred to Professor JoAnne Stubbe's chapter in this volume and also to her excellent discussions of this subject elsewhere.[33]

---

* If there is a case where Co-participation might yield some rate advantages, it is probably methylmalonyl-CoA mutase (*eg via* substrate carbanion formation). Whether or not this happens has been controversial.[21c,31] However, Halpern and Wollowitz have provided important studies showing that the substrate-derived radical can in fact rearrange[31a] (total yields of 1-9%), but only if the radical is given sufficient lifetime to rearrange. They conclude that providing relatively long radical lifetimes is probably one important role of the enzyme. Since they also find that the observed rate constants are at least 10^2 fold slower than the enzyme's k_{cat}, another probable role of the protein is to accelerate the radical rearrangement by $\geq 10^2$. (Others have failed to reproduce this same rearrangement[31b] or find it in only trace (*ca* 0.1%) yields.[31c] A detailed explanation for these discrepancies has not yet appeared, but one suspects radical lifetimes insufficient for rearrangement, due to competing reactions, in the cases where little to no rearrangement is seen.)

† Such tight binding comes as no surprise to an enzymologist - indeed, Professor Abeles, whose group discovered diol dehydratase,[32a] noted in 1972 that 'highly oriented' (not 'free') radicals are probably involved in diol dehydratase catalyzed reactions.[32b]

4 THE POSSIBILITY OF AN ADOB$_{12}$-INITIATED RADICAL CHAIN MECHANISM

What evidence exists for such a mechanism? What evidence contradicts such a mechanism? What does a working hypothesis of this mechanism look like? These obvious questions are just a few that come to mind; a brief discussion of them and a schematic representation of a possible radical chain mechanism follow.

The idea of a radical chain mechanism grew out of two observations: first, the observation of tritium wash-out (as ^{3}HOH) from C5' ^{3}H-labelled AdoB$_{12}$ [Co–C(^{3}H)$_{2}$–] into the solvent (water) in the presence of B$_{12}$-dependent ribonucleotide reductase (as described in 1973 by Tamao and Blakely,[34a] and then in 1975 by Sando, Blakely, Hogenkamp, and Hoffmann[34b]); and second, the observation of unusually high, *apparent* k_H/k_D and k_H/k_T isotope effects measured for B$_{12}$-dependent diol dehydratase and ethanolamine ammonia lyase (apparent k_H/k_T, respectively, of *ca* 125 and 160!).[35] In the 1975 paper,[34b] Sando *et al* postulate that an enzyme thiol site (Enz–SH) undergoes tritium exchange with C5' labelled Ado· by some unknown (and as written,[34b] chemically implausible) reaction. In 1979 Babior suggested that the most likely explanation for the large, apparent isotope effects '...is that there exists a hitherto unrecognized pool of enzyme-bound hydrogens which are in communication with the mobile hydrogens on the substrate and product.'[35] Next, Cleland, in a critical review of enzyme isotope effects published in 1982,[36] postulated the presence of a radical chain mechanism, one initiated by AdoB$_{12}$ serving '...as a source of radicals for the beginning of the reaction' and then operating *via* some 'completely unknown' group (which we now can postulate to be an enzyme '–XH' site). Cleland furthered this idea by showing quantitatively how a chain length of 9 would bring the apparent isotope effects seen with ethanolamine ammonia lyase down to the expected values.[36]

All these ideas were refined, clearly presented, and expanded upon in Stubbe's 1983 review[33a] on ribonucleotide reductases (her 1985[33b] and 1988[33c] reviews are also recommended). Therein AdoB$_{12}$ and the protein –X· site are seen as, and stated to be, the *chain initiator* and the *chain carrier*, respectively.

The 1983 Ph D thesis of McGee describes a reinvestigation of several types of isotope effects (*eg* substrate 'wash-in' and two types of 'wash-out' isotope effects)[37] while also using a more highly purified form of diol dehydratase.[5a] The results were again interpreted in terms of a Enz-XH H-transfer site. The data are puzzling, however, in that they apparently show a small, sub-stoichiometric use [*ca* 5% (±?)], rather than catalytic use, of this site for protio propane-1,2-diol substrate,* the remaining *ca* 95% (±?) of the H-transfers proposed to proceed *via* the traditional $AdoB_{12}$ C5' site - a finding that supports the classic statement that the C5' position is an 'obligatory' H-transfer site[6b] rather than supporting a radical chain mechanism. The reasons for this puzzling result† remain obscure (no publication of this work, other that McGee's Ph D thesis, has appeared).[37]

In 1985, Babior and co-workers provided additional evidence for the participation of a Enz-XH site in ethanolamine ammonia lyase,[38] and they suggested that -XH is a thiol. Furthermore, our own re-examination of the literature revealed two other observations that can be interpreted in terms of an -XH site and a radical chain mechanism. One is the report of 3-10% activity for methylmalonyl-CoA mutase apoprotein [that is even when 'no' (meaning 'no detectable') B_{12} cofactor is present].[39] Another

---

* For deuterated propane-1,2-diol, the Enz-XH site is calculated to participate in *ca* 12% of the turnovers, while for tritiated substrate the number rises to *ca* 33%.

† It is hard to believe that an Enz-XH site, apparently analogous to ones employed catalytically in other enzymes, is used *non-catalytically*, only 5% of the time in diol dehydratase (or, for that matter, even for only 9 turnovers in ethanolamine ammonia lyase). It will be important to reproduce some of these key isotope effects once cloned, over-produced, and even more highly purified forms of these enzymes become available. One specific reason for saying this is that different forms of diol dehydratase (apparently less pure, more proteolysed samples) gave *not less, but more* H abstraction from some other source.[37] This finding raises the serious concern that some of the H-transfers not involving the C5' position of $AdoB_{12}$ may be artifactual. Hence at present it is unclear which H-transfers involve (a) the putative Enz-XH site, or (b) are perhaps just indiscriminate H abstractions from random protein side-chains due to a 'loose,' partially proteolysed protein, one unable to control its normally tightly bound and highly oriented radical intermediates.

observation, one that is readily explainable by the Enz–XH site hypothesis, is the deuterium wash-out observed for non-B_{12} but SAM (S-adenosylmethionine) and pyridoxal phosphate-dependent lysine 2,3-amino mutase.[40] *The possibility that a radical chain, B_{12}/ribonucleotide reductase-like mechanism would extend all the way to this class of enzymes is an intriguing hypothesis, one worthy of careful experimental scrutiny – if only because of the unifying effect such a mechanistic finding would have.**

The best demonstration – a seemingly unequivocal demonstration – of the radical chain mechanism for a B_{12}-dependent enzyme is by Stubbe's group, using ribonucleotide reductase.[33c] Their work demonstrates that $AdoB_{12}$ can be the H-carrier at most 1 turnover out of 1000, that is a chain length of ≥1000 is indicated!

Clearly, there is enough evidence to warrant the construction of a schematic representation of the proposed radical chain mechanism as a *speculative, working hypothesis* (one intended to serve us until such time as an enzyme crystal structure becomes available). Such a schematic mechanism is presented in Figure 15.† Equally apparent, however, is the considerable effort needed in the area of the B_{12}-dependent rearrangements to put the radical chain mechanism#

---

* A noteworthy quote in this regard is Barker's insightful 1979 description of SAM as potentially 'a poor man's adenosylcobalamin.'[41]

† Note that this scheme is illustrated for diol dehydratase where it builds upon the evidence for $AdoB_{12}$ homolysis, against Co-participation in the rearrangement step, and where acidic and basic diol binding sites are proposed on the basis of the evidence cited earlier. This scheme is essentially that which we provided earlier,[1c] only upgraded by the addition of a protein –XH site. One can easily conceive of many changes that might occur once a crystal structure becomes available. To mention just one, chemical logic suggests that a protein operating by a radical chain mechanism might have *at least two substrate binding sites* so that a reactive –X• radical never has to sit and wait while product dissociates from, and fresh substrate is bound to, the enzyme.

\# It is worth noting that the problem discussed earlier with the last step of Figure 3 becomes a different problem, but also becomes somewhat less problematic, in a radical chain mechanism. Specifically, the difficult and poorly precedented last step, Ado–H + substrate• + $Co(II)B_{12r}$ → $AdoB_{12}$ + substrate–H (Figure 3), is replaced by a perhaps thermodynamically even less favourable last step involving a –X• radical. *However*, it is obviously better if the enzyme only has to accomplish this difficult step after a large number of productive catalytic cycles.

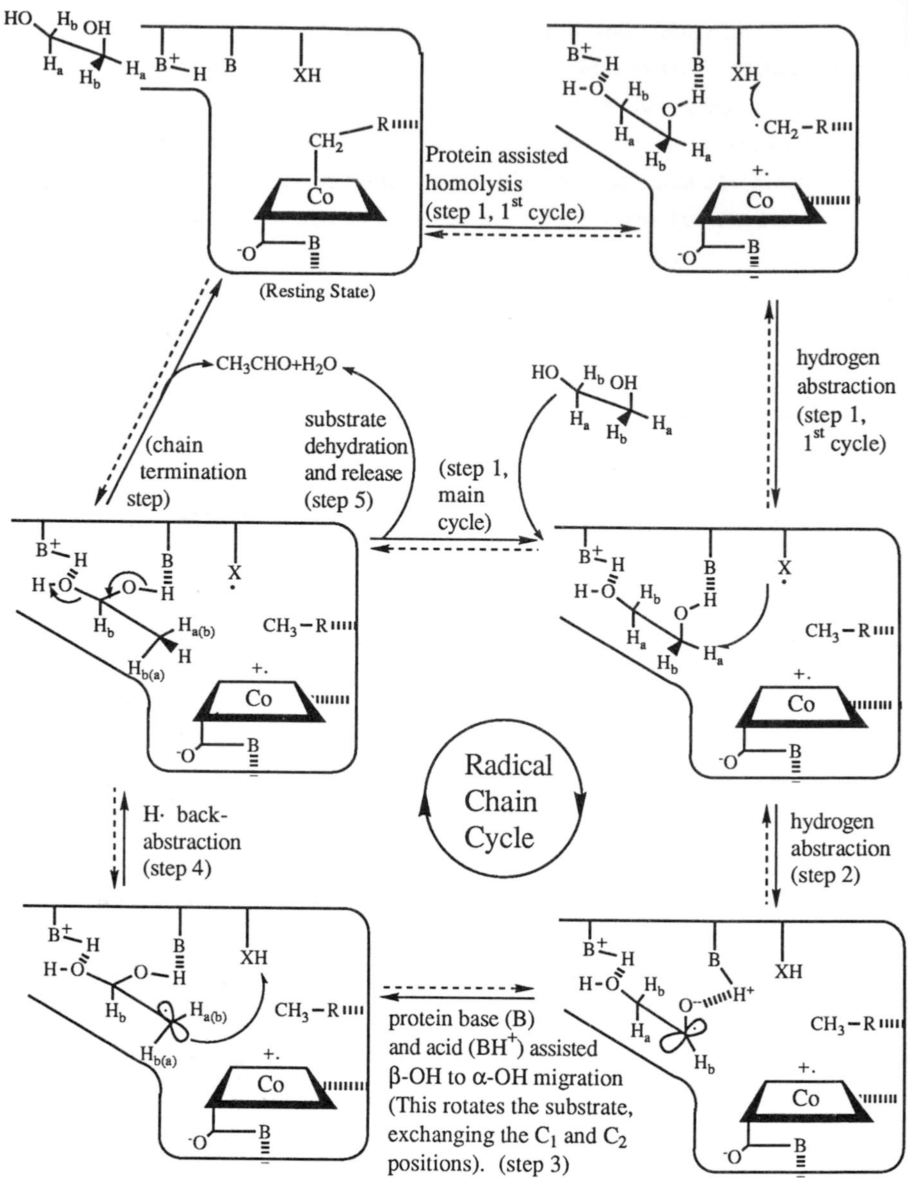

Figure 15: A schematic representation of the putative radical chain mechanism, intended as a working hypothesis.

on a firm footing or to provide experimental evidence against it and for better alternatives consistent with all the available evidence. Providing chemical evidence for the nature of the –XH site and for the proposed steps involving –X· (some of which include some apparent thermodynamic problems) are major goals of our own experimental efforts.

5 SUMMARY

To summarize, herein is described:

(1) Work which provided the first demonstration *in vitro* of the reversible, thermal homolytic cleavage of the Co–C bond for the cofactor, $AdoB_{12}$.
(2) A quantitativc comparison of the rate of Co–C bond cleavage in $AdoB_{12}$ to that in the holoenzyme, a comparison that reveals a previously unknown enzymic rate enhancement of $10^{12\pm1}$, as large a 'unimolecular' enzymic rate enhancement as is on record. This result also identifies the proteins that bind B_{12} as interesting targets for protein folding studies.
(3) Model studies of the Co-participation question for the diol dehydratase reaction, work that provides good evidence against cobalt-participation in the rearrangement step. This finding in turn implies a resulting simplification of mechanistic B_{12} chemistry, one that requires that the protein, and not the B_{12} cofactor, controls the rearrangement step and its stereochemistry. The finding that $AdoB_{12}$'s role is as a source of Ado· is consistent with evidence pointing towards a radical chain mechanism in which $AdoB_{12}$ functions as a radical chain initiator.
(4) A short summary of the evidence for and against a radical chain mechanism, followed by a schematic representation of this mechanism presented in the spirit of a working hypothesis.

It is clear that Nature has not yet revealed all of its secrets to the family of scientists interested in the B_{12}-dependent enzymes!

ACKNOWLEDGEMENTS

The author is indebted to Professor JoAnne Stubbe for insightful discussions over the past several years and to Ms Cheryl Garr for her help in constructing the Figures. This work was supported by NIH grant DK-26214.

REFERENCES

1 (a) 'B_{12}'; D Dolphin, ed, Wiley-Interscience: New York, 1982, Vols I and II; (b) 'Vitamin B_{12}': Proceedings of the Third European Symposium on Vitamin B_{12} and Intrinsic Factor, B Zagalak, and W Friedrich, eds; Walter de Gruyter, New York, 1979; (c) R G Finke, D A Schiraldi, and B J Mayer, *Coord Chem Rev*, 1984, 1; (d) 'Comprehensive B_{12}', Z Schneider and A Stroinski, eds; Walter de Gruyter: New York, 1987, Chapter 9; (e) J M Pratt, *Chem Soc Rev*, 1985, 161; (f) Many earlier reviews exist. References to reviews by Babior, Abeles and Dolphin, Golding and others are provided as the initial references in our publications, for example ref 1-17 of: R G Finke, W P McKenna, D A Schiraldi, B L Smith, and C Pierpont, *J Am Chem Soc*, 1983, **105**, 7592.

2 (a) D C Hodgkin, J Kamper, M MacKay, J Pickworth, K N Trueblood, and J G White, *Nature* (London), 1956, **178**, 64; (b) P G Lenhart and D C Hodgkin, *Nature* (London), 1961, **161**, 937.

3 R H Abeles, Proceedings Robert A Welch Foundation, *Conf Chem Res*, Vol XV, *Bio-Organic Chemistry and Mechanisms*, W O Milligan, ed, 1972, 95.

4 (a) E N Marsh, N McKie, N K Davis, and P F Leadlay, *Biochem J*, 1989, **260**, 345. Human liver L-methylmalonyl-CoA mutase has also been cloned, see; F D Ledley, M Lumetta, P N Nguyen, J F Kolhouse, and R H Allen, *Proc Natl Acad Sci USA*, 1988, **85**, 3518; (b) E N Marsh, S E Harding, and P F Leadlay, *Biochem J*, 1989, **260**, 353. See also: F Francalanci, N K Davis, J Q Fuller, D Murfitt, and P F Leadlay, *ibid*, 1986, **236**, 489;

(c) E N Marsh, and P F Leadlay, *ibid*, 1989, **260**, 339;
(d) E N Marsh, P F Leadlay, and P R Evans, *J Mol Biol*, 1988, **200**, 421.

5 (a) D E McGee, S S Carroll, M W Bond, and J H Richards, *Biochem Biophys Acta*, 1982, **108**, 547; D E McGee and J H Richards, *Biochemistry*, 1981, **20**, 4293; (b) K Tanizawa, N Nakajima, T Toraya, and H Tanaka, *Z Naturforsch C*, 1987, **42**, 353.

6 (a) See references 5 (a)–(d) in reference 1(c); (b) See references 6 (a)–(b) in reference 1(c).

7 (a) J Retey, A Umani-Ronchi, and D Arigoni, *Experientia*, 1966, **22**, 72; J Retey, A Umani-Ronchi, J Seibl, and D Arigoni, *ibid*, 1966, **22**, 502; (b) J Retey, 'Stereochemistry', Ch Tamm, ed; Elsevier Biomedical Press, Amsterdam, 1982, Chap 6; (c) R G Finke, D A Schiraldi, and B J Mayer, *Coord Chem Rev*, 1984, 1.

8 B M Babior, 'Vitamin B_{12}': Proceedings of the Third European Symposium on Vitamin B_{12} and Intrinsic Factor, B Zagalak and W Friedrich, eds; Walter de Gruyter, New York, 1979, p 461.

9 Lead references to Professor Halpern's extensive contributions in this area are: (a) J Halpern, *Polyhedron*, 1988, **7**, 1483 and references therein; (b) J Halpern, *Bull Chem Soc Jpn*, 1988, **61**, 13.

10 (a) B L Smith, Ph D Dissertation, University of Oregon, 1982; (b) R G Finke, B L Smith, B J Mayer, and A A Molinero, *Inorg Chem*, 1983, **22**, 3677.

11 (a) A Bakac and J H Espenson, *J Am Chem Soc*, 1984, **106**, 5197; R Blau and J Espenson, *ibid*, 1985, **107**, 3530; (b) J P Collman, L McElwee White, P J Brothers, and E Rose, *ibid*, 1986, **108**, 1332; (c) M K Geno and J Halpern, *ibid*, 1987, **109**, 1238; *idem*, *J Chem Soc Chem Commun*, 1987, 1052.

12 (a) R G Finke and B P Hay, *Inorg Chem*, 1984, **23**, 3041; (b) *idem*, *Polyhedron*, 1988, **7**, 1469.

13 B P Hay and R G Finke, *J Am Chem Soc* 1986, **108**, 4820.

14 (a) S C Stinson, *Chem Eng News*, 1985, **63(7)**, 28; (b) M K Geno and J Halpern, *J Chem Soc Chem Commun*, 1987, 1052.

15 J Halpern, S H Kim, and T W Leung, *J Am Chem Soc*, 1984, **106**, 8317.

16 See reference 13 for references 14 (a)–(e) and references 15 (a)–(j) therein.

17 J Halpern, *J Inorg Biochem*, 1989, **36**, 192 (An abstract corresponding to oral presentation E002).

18 B P Hay and R G Finke, *J Am Chem Soc* 1987, **109**, 8012.

19 (a) J Halpern, F T T Ng, and G L Rempel, *J Am Chem Soc*, 1979, **101**, 7124; (b) J Halpern, *Pure Appl Chem*, 1979, **51**, 2171. (c) See also references 24 (a)–(k) in reference 20.

20 (a) T W Koenig and R G Finke, *J Am Chem Soc*, 1988, **110**, 2657; (b) T W Koenig, B P Hay, and R G Finke, *Polyhedron*, 1988, **7**, 1499.

21 For lead references see: (a) P Dowd and R Hershline, *J Chem Soc Perkin Trans 2*, 1988, 61, and references therein; (b) A I Scott, J Kang, D Dalton, and S K Chung, *J Am Chem Soc*, 1978, **100**, 3603; (c) J H Grate, J W Grate, and G N Schrauzer, *ibid*, 1982, **104**, 1588; (d) R M Dixon, B T Golding, S Mwesigye-Kibende, and D N Ramakrishna Rao, *Phil Trans R Soc London B*, 1985, **311**, 531 and references therein; (e) P Moller and J Retey, *J Chem Soc Chem Commun*, 1983, 1343; M Fountoulakis and J Retey, *Chem Ber*, 1980, **113**, 650; (f) I I Merkelbach, H G M Becht, and H M Buck, *J Am Chem Soc* 1985, **107**, 4037; (g) Y Murakami and Y Hisaeda, *Pure Appl Chem*, 1988, **60**, 1363; Y Murakami, Y Hisaeda, J-I Kikuchi, T Ohno, M Suzuki, Y Matsuda, and T Matsuura, *J Chem Soc Perkin Trans 2*, 1988, 1237 and references therein; (h) See also reference 31 herein.

22 (a) R G Finke, W P McKenna, D A Schiraldi, B L Smith, and C Pierpont, *J Am Chem Soc*, 1983, **105**, 7592; (b) R G Finke, and D A Schiraldi, *ibid*, p7605.

23 C M Elliott, E Hershenhart, R G Finke, and B L Smith, *J Am Chem Soc*, 1981, **103**, 5558.

24 (a) See the discussion on pp 7592–7594, and references 12 and 14–16, contained in reference 22(a); (b) B B Wayland, S L Van Voorhess, and K L Del Rossi, *J Am Chem Soc*, 1987, **109**, 6513.

25 (a) Y Wang, Ph D Dissertation 'Synthesis, Properties, and Mechanistic Investigation of Two Putative Intermediates in the Coenzyme B_{12}-Dependent Diol Dehydratase Reaction,' University of Oregon, August, 1988; (b) Y Wang and R G Finke, *Inorg Chem*, 1989, **28**, 983; (c) Y Wang and R G Finke, submitted for publication.

26 Y Wang, P Fleming, and R G Finke, submitted for publication.

27 W A Mulac and D Meyerstein, *J Am Chem Soc*, 1982, **104**, 4124; *idem*, 1978, **100**, 5540.

28 See references 12(a)-(f) in reference 1(c).

29 (a) M G N Hartmanis and T C Stadtman, *Proc Natl Acad Sci USA*, 1987, **84**, 76; (b) Dr B Martin in the Finke research group is acknowledged for this point.

30 B T Golding and L Radom, *J Am Chem Soc*, 1976, **98**, 6331.

31 (a) S Wollowitz and J Halpern, *J Am Chem Soc*, 1984, **106**, 8319; *ibid*, 1988, **110**, 3112; (b) W M Best, A P F Cook, J J Russell, and D A Widdowson, *J Chem Soc Perkin Trans 1*, 1986, 1139; (c) U Aeberhard, R Keese, U-C Vogeli, W Lau, and J K Kochi, *Helv Chim Acta*, 1983, **66**, 2740.

32 (a) See p 118 of reference 3; (b) B Zagalak, P A Frey, G L Karabatsos, and R H Abeles, *J Biol Chem*, 1966, **241**, 3028.

33 (a) J Stubbe, *Mol Cell Biochem*, 1983, **50**, 25; (b) G W Ashley and J Stubbe, *J Pharmac Ther*, 1985, **30**, 301; (c) J Stubbe, *Biochemistry*, 1988, **27**, 3893.

34 (a) Y Tamao and R L Blakely, *Biochemistry*, 1973, **12**, 24; (b) G N Sando, R L Blakely, P C Hogenkamp, and P J Hoffmann, *J Biol Chem*, 1975, **250**, 8774.

35 B M Babior and J S Krouwer, *CRC Critical Rev Biochem*, 1979, **6**, 35 (see p 70).

36 W W Cleland, *CRC Critical Rev Biochem*, 1982, **13**, 385 (see p 416).

37 D E McGee, Ph D Dissertation, California Institute of Technology, 1983, Chapter 3.

38 R J O'Brien, J A Fox, M G Kopczynski, and B M Babior, *J Biol Chem*, 1985, **260**, 16131.

39 F Francalanci, N K Davis, J Q Fuller, D Murfitt, and P F Leadlay, *Biochem J*, 1986, **236**, 489.*

40 D J Aberhart, S J Gould, H-J Lin, T K Thiruvengadam, and B H Weiller, *J Am Chem Soc*, 1983, **105**, 5461.

41 J H Barker and T S Stadtman, 'B_{12}', D Dolphin, ed, Wiley-Interscience: New York, 1982, Vol II, pp 203-231 (see p 206 and reference 12).

42 (a) K Reynolds and J A Robinson, *J Chem Soc Chem Comm*, 1985, 1831; (b) G Brendelberger, J Retey, D M Ashworth, K Reynolds, F Willenbrock, and J A Robinson, *Angew Chem Int Ed Engl*, 1988, **27**, 1089.

* *Note added in proof*: Since authentic apoprotein is inactive (P Leadlay, private communication), this result implies the presence of only 3–10% of holoenzyme (*ie* enzyme containing $AdoB_{12}$).

A NEW PERSPECTIVE ON PYRIDOXAL- AND PYRUVATE-DEPENDENT ENZYMES FROM *IN VITRO* STUDIES

Ronald Grigg

School of Chemistry, University of Leeds, Leeds LS2 9JT, UK

1 INTRODUCTION

Over ten years ago[1] we proposed and demonstrated a new general type of prototropy (1)$\rightleftharpoons$(2), in X=Y–ZH systems in which the central Y atom possesses a lone pair of electrons:

$$X{=}\ddot{Y}{-}ZH \;(1) \quad \rightleftharpoons \quad X{=}\overset{+}{\underset{H}{Y}}{-}\overset{-}{Z} \;(2) \quad \text{1,2 - PROTOTROPY}$$

$$X{=}\ddot{Y}{-}ZH \;(1) \quad \rightleftharpoons \quad HX{=}Y{-}Z \;(3) \quad \text{1,3 - PROTOTROPY}$$

(atom numbering in (1): X = 3, Y = 2, Z = 1)

Thus, the long established 1,3-prototropy, (1)$\rightleftharpoons$(3), as illustrated for example by keto-enol, imine-enamine, nitroso-oxime, azo-hydrazo tautomerism, can in certain cases be accompanied by 1,2-prototropy.[2] This latter type of prototropy is found in imines,[3] hydrazones,[4] and oximes.[5,6] The importance of this type of prototropy is that it results in charge separation and leads to a novel type of 1,3-dipole (2). Such species were identified in 1963 by Huisgen as important precursors of 5-membered heterocycles *via* 1,3-dipolar cycloaddition reactions (Scheme 1).[7]

Now clearly the ease with which the 1,3-dipole (2) is formed will depend on the basicity of Y and the pK_a of the ZH group. These in turn will depend on the nature of the substituents on X and Z. These novel dipoles are usually present only in trace amounts and are normally not spectroscopically detectable. However, we can study

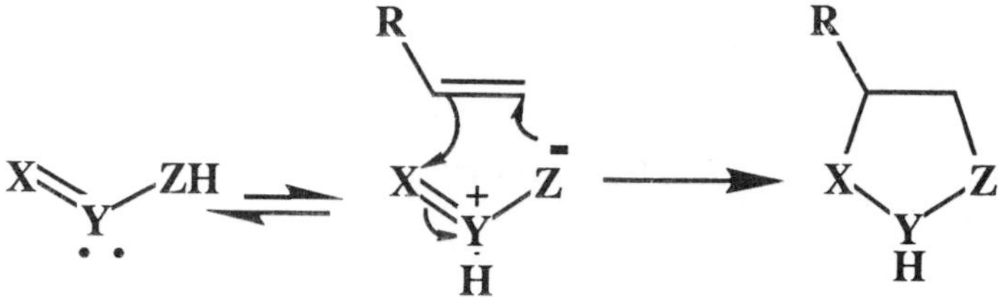

Scheme 1: Prototropy leading to a 1,3-dipole followed by 1,3-dipolar cycloaddition to an alkene

the effect of structure on the ease of dipole formation by measuring the rate of the cycloaddition reaction provided that the dipole-forming step is rate determining. This condition is met for imine cycloaddition reactions when the dipolarophile is a maleimide.[8] The trends observed are shown in Figure 1, and some rate data are collected in Figure 2.

As expected the trends in the rate of cycloaddition of imines to *N*-phenylmaleimide (NPM) show a decrease in rate as the basicity of the central nitrogen atom of the imine is decreased. A similar trend is observed as the pK_a of the ZH proton is increased. The rate data

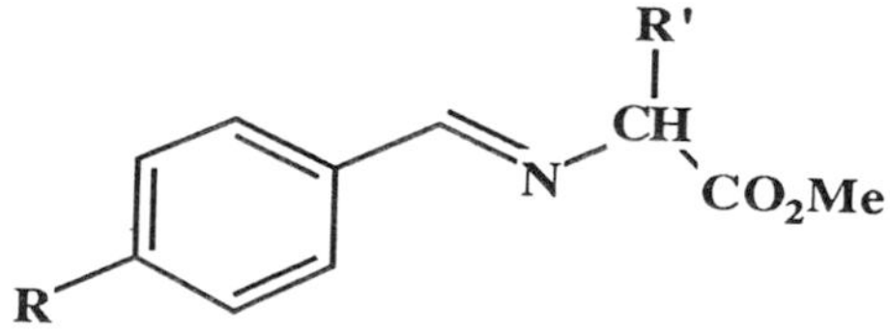

Rate decreases R = NMe_2 > OMe > Me > H > Cl > CN > NO_2

decreasing imine basicity →

Rate decreases R' = CO_2Me > Ph > Me

decreasing C–H acidity →

Figure 1: Trends in the rate of cycloaddition of imines to *N*-phenylmaleimide[8]

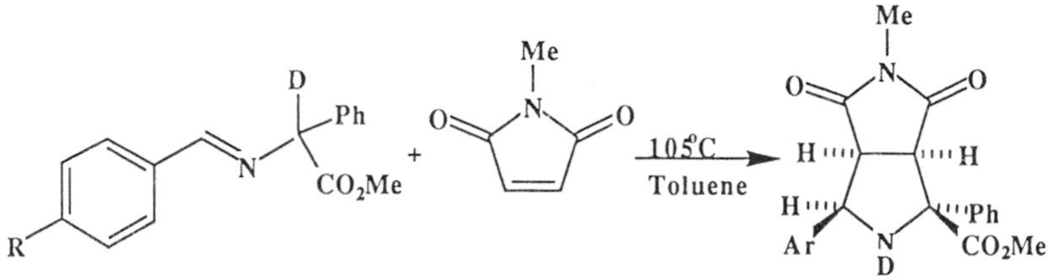

R	Rate Constant (s^{-1})	Isotope Effect
NMe_2	44.6×10^{-5}	1.21
OMe	7.8×10^{-5}	2.14
H	3.55×10^{-5}	2.70
CN	0.72×10^{-5}	2.75
NO_2	0.80×10^{-5}	2.17

Figure 2: Rate data for cycloaddition of imines to *N*-phenylmaleimide[8]

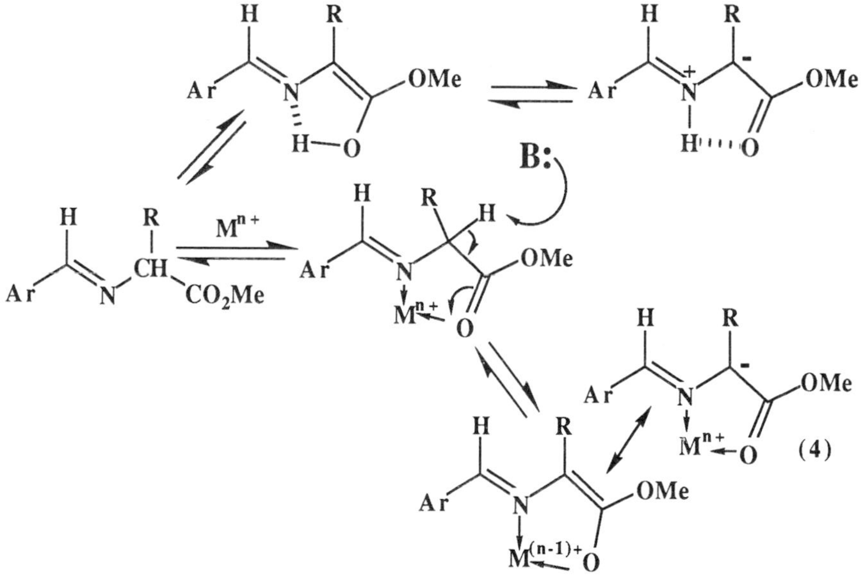

Scheme 2: Mechanistic factors influencing the stereochemistry of 1,3-dipoles

in Figure 2 illustrates that this trend, whilst not large, is nevertheless significant. A deuterium isotope effect is also observed, although the magnitude of this effect is noticeably less than that observed for the rate determining 1,3-prototropy in keto-enol tautomerism, which exhibits an isotope effect of *ca* 6. The cycloaddition produces a single cycloadduct stereoisomer and thus involves only one configuration of the dipole. Hence there appears to be some property inherent in the imine system which imparts a kinetic bias to one dipole. The simplest explanation of this observation is shown in Scheme 2.

The enolisation of the initial imine leads to an intermediate hydrogen-bonded enol which, upon proton transfer to nitrogen, yields the dipole with the configuration required by the observed stereochemistry of the NPM cycloadducts. The intermediate 5-membered hydrogen bonded enol in Scheme 2 may be rendered energetically more favourable by incorporation of a bridging water (5) or solvent molecule. 1,3-Dipole formation is catalysed by both weak Bronsted[9] and weak Lewis acids[3,9-12] and the use of such catalysis enables the cycloaddition reactions to be carried out at ambient temperature. In these catalysed cases the Lewis acid is considered to coordinate the imine nitrogen atom and/or the carbonyl oxygen atom of the ester substituent. Subsequent deprotonation affords a species which could exhibit the properties of a metallo-1,3-dipole **(4)** or a metallo-enolate **(5)**. Examples of both types of behaviour are found depending on the choice of Lewis acid. Lewis acids used successfully include salts of silver(I),[9,10] lithium,[9-11] zinc,[3,10] magnesium,[3,10] manganese(II),[3] cobalt(III),[3] and titanium (IV).[12] Dipoles generated from imines of optically active amino acids and their esters give rise to racemic cycloadducts. A wide range of electron withdrawing groups other than ester or carboxyl facilitate 1,2-prototropy in imines generating 1,3-dipoles.[13]

2 MODELS FOR TRANSAMINASES AND RACEMASES

The simple and facile method described above for generating 1,3-dipoles from imines of α-amino acids[14] and their esters and related compounds,[13] prompted consideration of possible biochemical processes in which such 1,3-dipoles might be implicated. Pyridoxal (vitamin B_6) phosphate-dependent enzymes occur widely and are responsible for the synthesis, racemisation, degradation, and interconversion of α-amino acids in living systems. These transformations are known to proceed *via* imine formation **(7)** and many of the mechanistic features of these processes are understood.[15] The initial reactive intermediates **(8)** or **(9)** are generated by cleavage of either bond (a) or (b) in **(7)** together with protonation of the pyridine nitrogen atom. Stereoelectronic control dictates that the breaking bond [bond (a) or (b) in **(7)**] is aligned with the pyridyl azomethine π-system.[16] Racemases and transaminases function by cleavage of bond (a) in **(7)** and lead us to consider, in the light of our experience with imines of α-amino acids and their esters, whether **(8)** might be more properly regarded as a 1,3-dipole with a proton residing on the imine nitrogen atom *ie* a novel type of azomethine ylide.[17] We therefore prepared a range of pyridoxal imines of

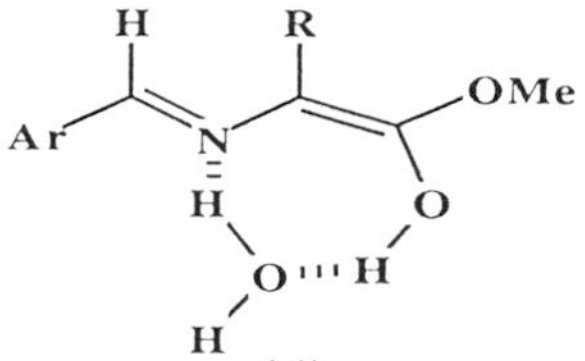

(6)

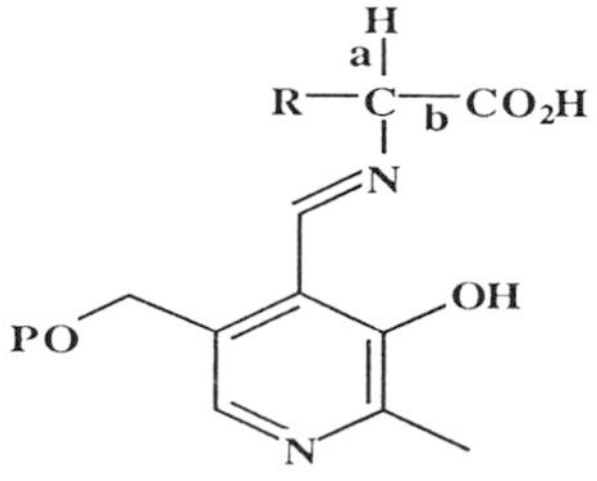

(7) $P = PO_3^{2-}$

(8) $P = PO_3^{2-}$

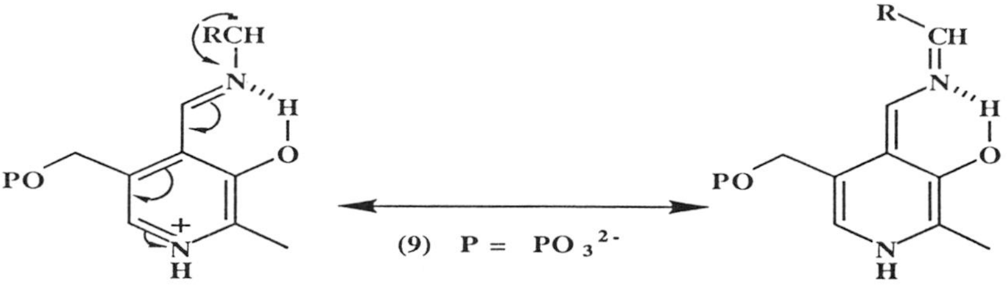

(9) $P = PO_3^{2-}$

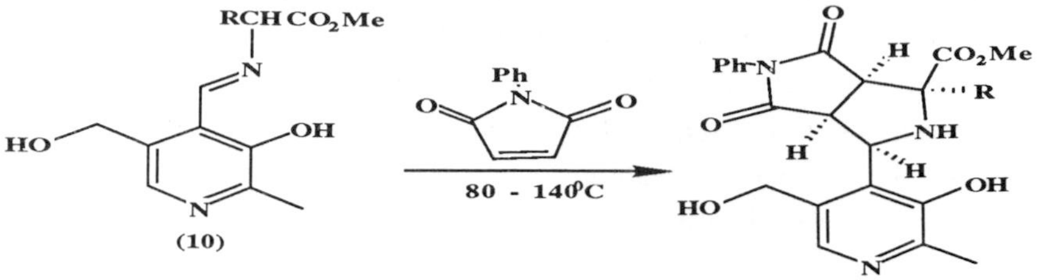

(10)

80 - 140°C

Solvents : MeCN, Toluene or Xylene

(11) R = H, Me, CH_2OH 3-Indolemethyl etc.

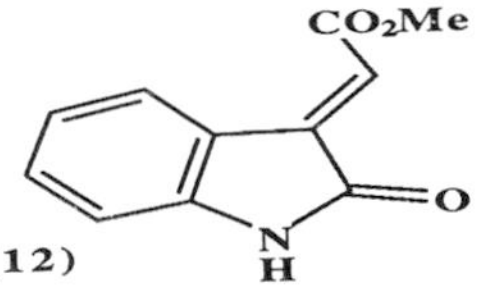

(12)

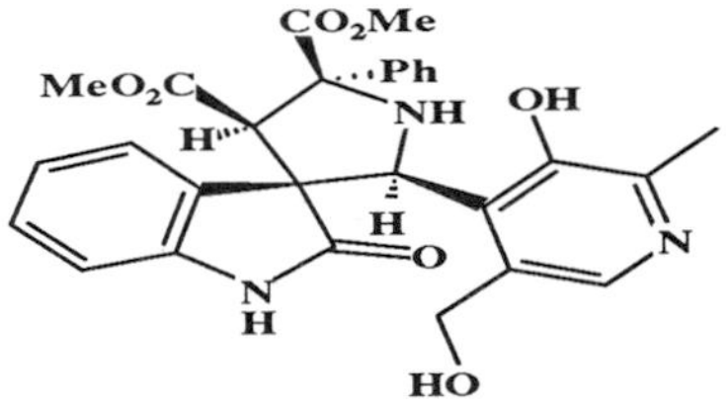

(13)

α–amino esters **(10)** and examined their potential as 1,3–dipole precursors. We were rewarded with a series of smooth, stereospecific cycloaddition reactions giving cycloadducts **(11)** in good to excellent yield.[18]

Other dipolarophiles can also be used as illustrated by the reaction (xylene, 130°C) of imine **(10**, R=Ph) with **(12)** to give **(13)** (59%).[18] Cycloadditions of this type suggest a possible new approach to the inhibition of pyridoxal enzymes. Thus, an α–amino acid incorporating a suitably positioned dipolarophile, such as **(14)**, might undergo an intramolecular cycloaddition as shown in Scheme 3.

The two imines in Scheme 3 do indeed undergo essentially quantitative cyclisation *via* the dipole on keeping at room temperature for 3 months.[19] Attempts to achieve the same process in a shorter time at elevated temperatures were successful with the *S*–allyl cysteinyl imine (MeCN solvent, 80°C, 4h).

In considering possible structures such as **(15)** for the 1,3–dipole derived from pyridoxal imines of α–amino acids, we became intrigued by the possibility that the bifurcated hydrogen–bonding might exert a special stabilising influence on the 1,3–dipole and hence promote its formation. We call this the 'proton sponge effect'. To test this idea we compared dipole generation from imines **(16)** and **(17)**. Imine **(16)** failed to form a dipole, but (17) in which bifurcated hydrogen bonding is possible did undergo dipole formation as evidenced by trapping experiments with maleimides.[20] However (17) gave rise to cycloadducts derived from both *syn* (18)- and *anti* (19)- dipoles showing that bifurcated hydrogen-bonding is not crucial to dipole formation.[20]

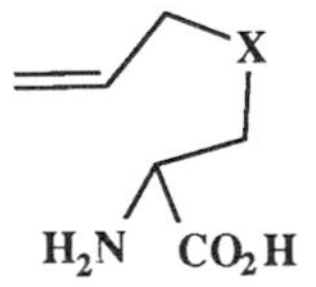

X
H_2N CO_2H

(14) X = CH_2 pentenylglycine

X = S S - allylcysteine

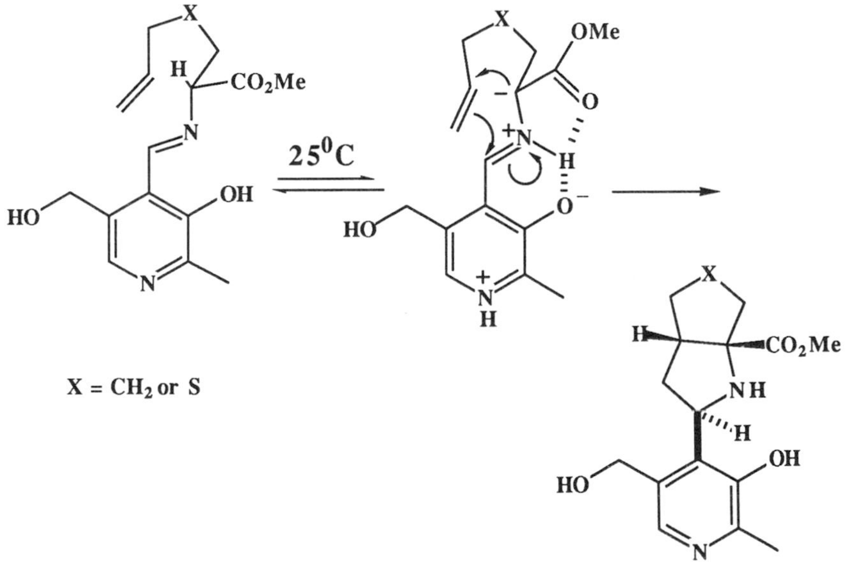

X
H
CO_2Me
N
OH
HO
N
$X = CH_2$ or S

25^0C

X
OMe
O
+N
H
O⁻
HO
+N
H

X
H
CO_2Me
NH
H
OH
HO
N

X = CH_2 90%

X = S 100%

Scheme 3: Intramolecular cycloaddition between a 1,3-dipole and pendant alkene

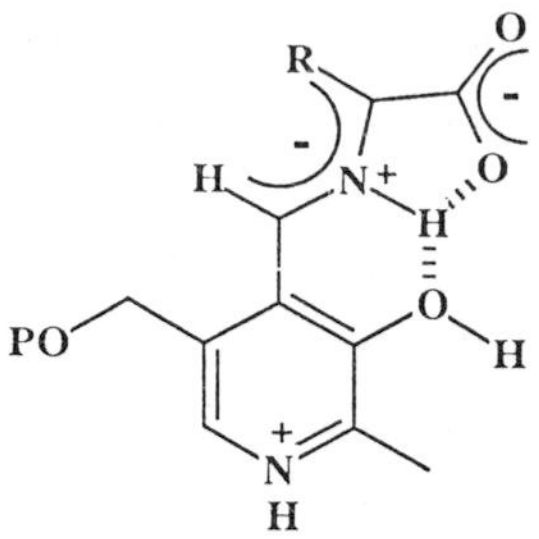

(15)

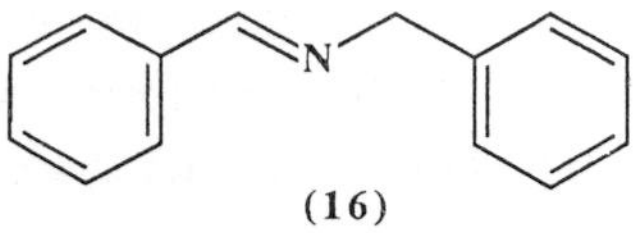

(16)

(17)

(18)

X = Y = OMe; X = OMe, Y = NMe_2 ; X = NMe_2 , Y = OMe

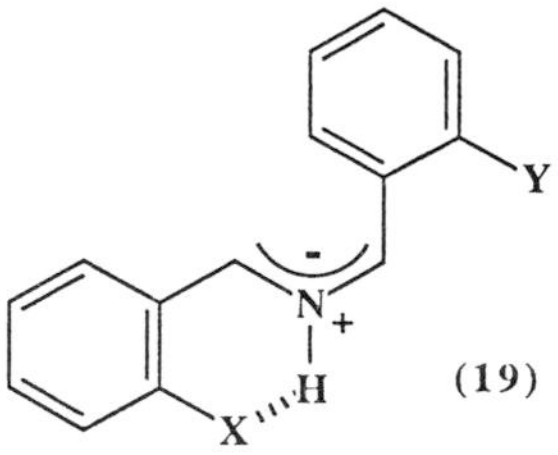

(19)

3 MODELS FOR PYRIDOXAL DEPENDENT DECARBOXYLASES

Enzymatic decarboxylations effected by pyridoxal enzymes (Scheme 4) are important biochemical processes which lead to the so-called biogenic amines:

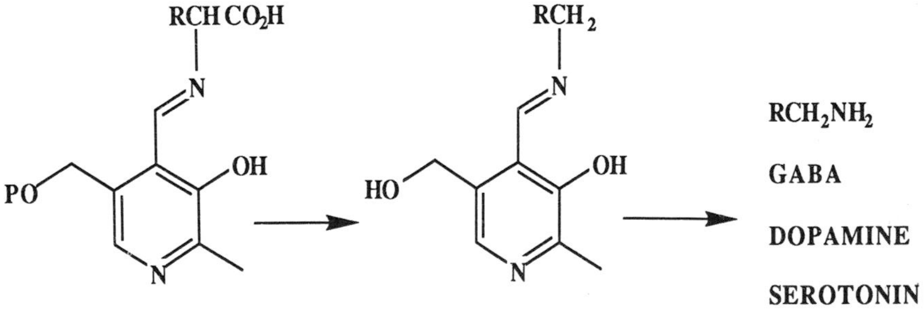

RCH_2NH_2

GABA

DOPAMINE

SEROTONIN

Strecker[21] was the first to observe the decarboxylation of α-amino acids (Strecker degradation) brought about by a carbonyl compound (alloxan). Recently we demonstrated that azomethine ylides are involved in the Strecker degradation. Extensive studies of the stereochemistry of the cycloadducts arising from trapping of these azomethine ylides with maleimides and other dipolarophiles, together with the influence of α-amino acid structure on the configuration of the azomethine ylide, lead us to propose that an intermediate oxazolidin-5-one is involved (Scheme 4).[22,23]

The stereochemistry of the azomethine ylide produced in the decarboxylative processes is controlled by the relative rates of the ring-chain equilibria, (20)$\rightleftharpoons$(21) and (22) (Scheme 4), and the stereospecific thermal cycloreversion steps generating the *syn*- and *anti*-dipoles.[23] Azomethine ylide formation is highly diastereoselective[23] and is exemplified by the reaction of pyridoxal,,

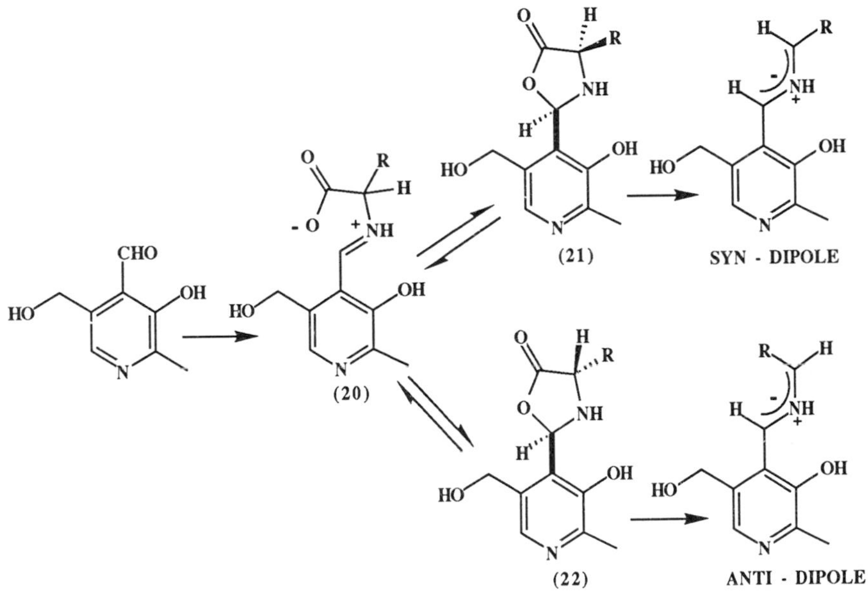

Scheme 4: Oxazolidin-5-one intermediates in the Strecker degradation leading to azomethine ylides from pyridoxal and an α-amino acid

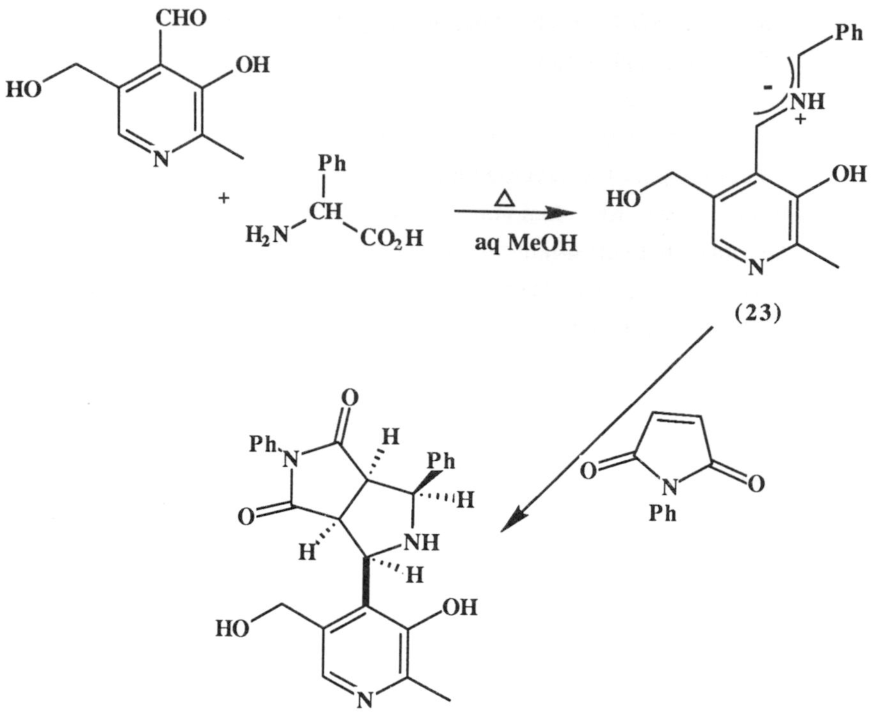

CHO
HO OH
N
Ph
+ CH
H_2N CO_2H
Δ
aq MeOH
Ph
NH
+
HO OH
N
(23)
O
Ph N H
Ph
H
O
H NH
H
HO OH
N
O N O
Ph

Scheme 5: Reaction of the azomethine ylide from pyridoxal and phenylglycine with *N*-phenylmaleimide

phenylglycine, and NPM in boiling aqueous methanol (Scheme 5). A single cycloadduct, derived from the *syn*-dipole **(23)**, is obtained in good yield. Other α-amino acids exhibit competition between generation of an azomethine ylide *via* 1,2-prototropy and decarboxylation. The former leads to cycloadducts which retain the carboxyl group of the α-amino acid.[22] These reactions are remarkably clean, despite the fact that pyridoxal, with its array of functionality, presents manifold opportunities for alternative reaction pathways.

These decarboxylative processes generating azomethine ylides can also be carried out with secondary α-amino acids and α,α-disubstituted amino acids. In addition, a wide range of aldehydes and ketones can replace pyridoxal.[22,23]

4 MODELS FOR PYRUVATE-DEPENDENT DECARBOXYLASES

The pyruvate-dependent decarboxylases,[24] although less ubiquitous than pyridoxal phosphate-dependent decarboxylases, are responsible for the decarboxylation of certain α-amino acids including histidine[25] and fulfil an important biochemical role. Snell[26] has suggested that they function in a manner similar to the pyridoxal-dependent decarboxylases (Scheme 6).

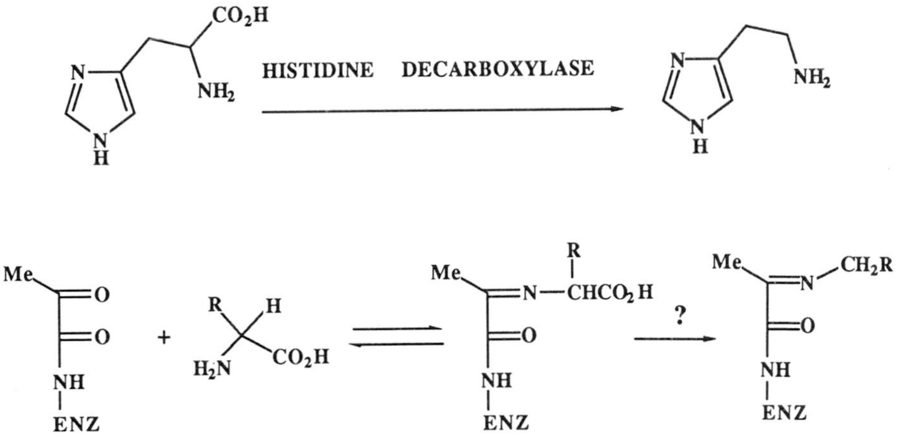

HISTIDINE DECARBOXYLASE

Scheme 6: Proposed mechanism of action of pyruvate-dependent decarboxylases

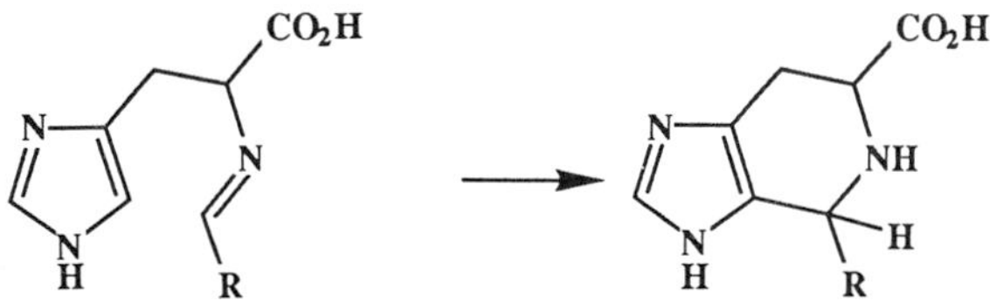

(25)

(26)

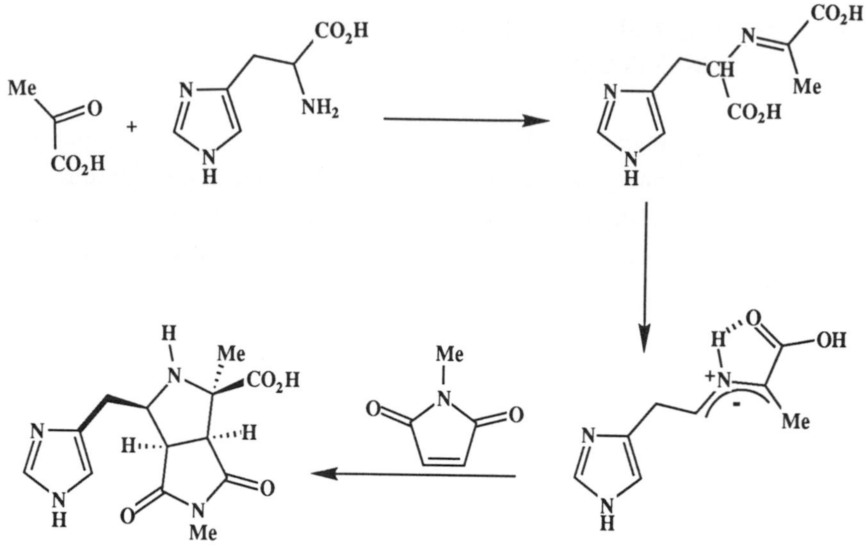

(24) 53%

(27)

Scheme 7: Reaction of histidine with pyruvic acid giving an azomethine ylide that reacts with *N*–phenylmaleimide to afford cycloadduct **(24)**

We were interested in probing the decarboxylation step, *in vitro*, for the presence of azomethine ylides. Heating a mixture of pyruvic acid, histidine and *N*–methylmaleimide (NMM) in dimethylformamide at 90°C for 2h afforded a single cycloadduct **(24)** (Scheme 7). This was despite the tendency of histidine imines **(25)** to undergo a facile cyclisation [**(25)** → **(26)**].

Cycloadduct **(24)** arises *via* regiospecific decarboxylation of the carboxyl group of the α–amino acid and the stereochemistry of **(24)** implicates dipole **(27)** as the intermediate. The reaction is general for pyruvic acid, and its esters and amides (Scheme 8). In all cases the reaction furnished a single cycloadduct **(30)** in good yield. The stereospecific formation of a single dipole suggests that an intermediate oxazolidin–5–one **(28)** (Scheme 8) is probably involved in these processes. It is considered that the product **(30)** arises from **(29)** *via* an *endo*–transition state. An alternative dipole **(31)**, adjudged less favourable energetically due to steric interactions of the RCO and R' groups, could give rise to **(30)** *via* an *exo*–transition state.

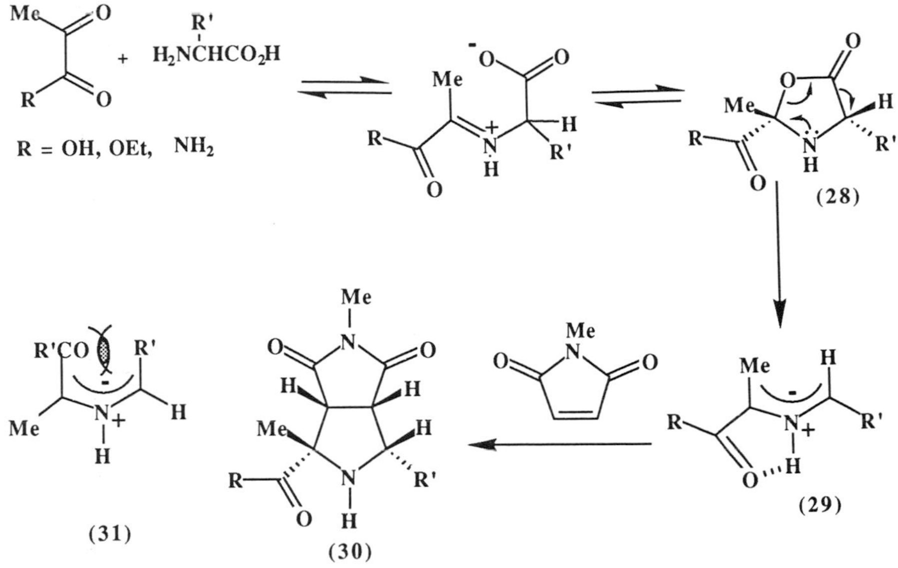

R = OH, OEt, NH_2

Scheme 8: Oxazolidin-5-one intermediate leading to an azomethine ylide from pyruvic acid and an amino acid

SUMMARY

Imines of α-amino acids and their esters react with pyridoxal, and with pyruvic acid derivatives, to give NH azomethine ylides either by 1,2-prototropy or by decarboxylation. The azomethine ylides are formed stereospecifically in most cases and can be trapped in cycloaddition reactions with dipolarophiles such as maleimides. It is suggested that pyridoxal phosphate- and pyruvate-dependent enzymes operate *via* azomethine ylides.

ACKNOWLEDGEMENTS

We thank ICI Colours and Fine Chemicals, SERC, Queen's University, Belfast and the University of Leeds for support.

REFERENCES

1 R Grigg, J Kemp, and N Thompson, *Tetrahedron Lett*, 1978, 2827.

2 R Grigg, *Chem Soc Rev*, 1987, **16**, 89.

3 K Amornraksa, D Barr, G Donegan, R Grigg, P Ratananukul, and V Sridharan, *Tetrahedron,* 1989, **45**, 4649 and earlier papers in this series.

4 R Grigg, M Dowling, M W Jordan, J Kemp, V Sridharan, and S Thianpatanagul, *Tetrahedron,* 1987, **43**, 5873.

5 W C Wildman and M R Slabough, *J Org Chem*, 1971, **36**, 3202; R Grigg and S Thianpatanagul, *J Chem Soc Perkin Trans 1*, 1984, 653; M H Norman and C H Heathcock, *J Org Chem.*, 1987, **52**, 226; A Hassner and R Maurya, *Tetrahedron Lett*, 1989, **20**, 2289; A Hassner, R Maurya, and E Mesko, *ibid,* 1988, **29**, 5313; A Padwa, U Chiacchio, D C Dean, A M Schosstall, A Hassner, and K S K Murthy, *ibid*, 1988, **29**, 4169.

6 See also: P Armstrong, R Grigg, and W J Warnock, *J Chem Soc Chem Commun,* 1987, 1325; P Armstrong, R Grigg, S Surendrakumar, and W J Warnock, *ibid,* 1987, 1327; R Grigg, M R J Dorrity, F Heaney, J F Malone, S Rajiviroongit, V Sridharan, and S Surendrakumar, *Tetrahedron Lett*, 1988, **29**, 4323; A Padwa and B H Norman, *ibid*, 1988, **29**, 2417.

7 R Huisgen, *Angew Chem Int Ed Engl,* 1963, **2**, 565 and 633.

8 R Grigg, H Q N Gunaratne, and J Kemp, *J Chem Soc Perkin Trans 1,* 1984, 41.

9 R Grigg and H Q N Gunaratne, *J Chem Soc Chem Commun,* 1984, 384; R Grigg, H Q N Gunaratne, and V Sridharan, *Tetrahedron,* 1987, **43**, 5887.

10 R Grigg, H Q N Gunaratne, J Kemp, P McMeekin, and V Sridharan, *ibid,* 1988, **44**, 557; D A Barr, G Donegan, and R Grigg, *J Chem Soc Perkin Trans 1*, 1989, 1550.

11 O Tsuge, S Kanewasa, and M Yoshioka, *J Org Chem*, 1988, **53**, 1384; *idem*, *Bull Chem Soc Jpn,* 1989, **62**, 869.

12 D A Barr, R Grigg, and V Sridharan, *Tetrahedron Lett*, 1989, **30**, 4727.

13 R Grigg, G Donegan, D Kennedy, H Q N Gunaratne, J F Malone, V Sridharan, and S Thiapatanagul, *Tetrahedron*, 1989, **45**, 1723.

14 K Amornraksa, R Grigg, H Q N Gunaratne, J Kemp, and V Sridharan, *J Chem Soc Perkin Trans 1*, 1987, 2285.

15 V C Emery and M Akhtar, in 'Enzyme Mechanisms', eds M I Page and A Williams, Royal Society Chemistry, 1987, Ch 18.

16 H C Dunathan, *Proc Natl Acad Sci USA,* 1966, **55**, 712; J R Fisher and E H Abbot, *J Am Chem Soc*, 1979, **101**, 2781.

17 R Grigg and J Kemp, *Tetrahedron Lett*, 1978, 2823; P Armstrong, D T Elmore, R Grigg, and C H Williams, *Biochem Soc Trans,* 1986, 404.

18 R Grigg, S Thianpatanagul, and J Kemp, *Tetrahedron,* 1988, **44**, 7823.

19 W P Armstrong and R Grigg, *Tetrahedron*, 1989, **45**, 7581.

20 R Grigg, P McMeekin, and V Sridharan, unpublished observations.

21 A Strecker, *Liebigs Ann Chem,* 1862, **123**, 363.

22 M F Aly, R Grigg, S Thianpatanagul, and V Sridharan, *J Chem Soc Perkin Trans 1*, 1988, 949.

23 R Grigg, S Surendrakumar, S Thianpatanagul, and D Vipond, *ibid*, 1988, 2693; R Grigg, J Idle, P McMeekin, S Surendrakumar, and D Vipond, *ibid*, 1988, 2703; H Ardill, R Grigg, V Sridharan, and S Surendrakumar, *Tetrahedron*, 1988, **44**, 4953; H Ardill, R Grigg, V Sridharan, and J F Malone, *J Chem Soc Chem Commun,* 1987, 1926.

24 R B Wickner, C W Tabor, and H Tabor, *J Biol Chem,* 1970, **245**, 2132; C W Tabor and H Tabor, *Adv Enzymol Relat Areas Mol Biol*, 1984, **56**, 251; J M Williamson and G M Brown, *J Biol Chem*, 1979, **254**, 8074; R Scandurra, V Consalvi, L Politi, and C Gallina, *ibid*, 1988, **263**, 192; T Warner and E Dennis, *ibid,* 1975, **250**, 8004; E Hawnot and E Kennedy, *Proc Natl Acad Sci USA*, 1975, **72**, 112.

25 W D Riley and E E Snell, *Biochemistry*, 1968, **7**, 3520; *idem, ibid,* 1970, **9**, 1485; E Snell, *J Biol Chem*, 1981, **256**, 687.

26 P A Recsei and E E Snell, *Ann Rev Biochem,* 1984, **53**, 357.

27 R Grigg, D Henderson, and A J Hudson, *Tetrahedron Lett*, 1989, **30**, 2841.

CHEMICAL SYNTHESIS AND ENZYMIC RECYCLING OF THE INOSITOL PHOSPHATES

David C Billington

Merck Sharp and Dohme, Neuroscience Research Centre, Terlings Park, Eastwick Road, Harlow, Essex CM20 2QR, UK

1 INTRODUCTION

Communication between cells is essential for the maintenance of life in complex organisms. A large range of chemical messengers is used for intracellular communication in mammalian systems, and these signals may be classified as hormones, neurotransmitters *etc*, depending on their function. These chemical messengers are received by specific receptors on the surface of the target cell, and the signal is then converted into a response within the cell by a transduction system.[1]

One class of receptors contains or is closely linked to an ion channel, which spans the plasma membrane. Stimulation of the receptor leads to a direct effect on the ion flux across the membrane. A second class of receptors are enzymes embedded in the membrane. These enzymes, tyrosine kinases, are activated on binding the transmitter molecule, and phosphorylate intracellular target proteins, leading to the overall cellular response. A third class of receptors have no intrinsic activity as ion channel control systems or enzymes. These receptors rely on the production of intracellular second messengers to exert their effects. The receptors are coupled to the enzyme or ion channel which they control *via* a class of proteins known as G-proteins.[1] Activation of the receptor causes its associated G-protein to release guanosine diphosphate (GDP) and bind guanosine triphosphate. In this form the G-protein regulates the activity of the enzyme or ion channel which is the next link in the chain. The classical example of this type of receptor system produces the second messenger cyclic adenosine monophosphate (cAMP) from adenosine triphospate *via* the G-protein regulated enzyme adenylate cyclase. In recent years overwhelming evidence has been obtained for a new and widely distributed second messenger system, involving the hydrolysis of inositol phospholipids (see Figure 1).

A detailed discussion of the current state of the inositol phosphate pathway is outside the scope of this chapter. The biology and chemistry of this system have been the subject of recent reviews.[2,3] Briefly, the receptor controlled hydrolysis of the phospholipid phosphatidylinositol 4,5-bisphosphate (PIP2) leads to the release of the two second messengers inositol 1,4,5-triphosphate (1,4,5-IP3) and diacylglycerol (DAG) (Figure 1). 1,4,5-IP3 binds to specific receptors on the endoplasmic reticulum, and causes the release of calcium from intracellular storage sites. A complex metabolic cycle then converts 1,4,5-IP3 into free inositol, which is used for the resynthesis of PIP2. 1,4,5-IP3 is either sequentially dephosphorylated *via* inositol 1,4-bisphosphate (1,4-IP2) and inositol 1- and 4-phosphates (I-1-P and I-4-P) to inositol, or phosphorylated to inositol 1,3,4,5-tetrakis-phosphate (1,3,4,5-IP4). The latter compound is dephosphorylated *via* a similar sequence of events to give free inositol. 1,3,4,5-IP4 may have a role as a secondary messenger in its own right by regulating the influx of calcium into the stimulated cell.

The need for authentic samples of the intermediates in the inositol cycle, both for identification purposes, and to allow isolation of the enzymes involved in their metabolism, led us to develop efficient syntheses of a number of these complex metabolites.

2 SYNTHESES OF INOSITOL PHOSPHATES

The enantiomers of inositol 1-phosphate were prepared *via* resolution of the known alcohol (1), Scheme 1.[4] Conversion of (1) into the diastereomeric camphanate esters (2) and (3) with (*R*)-(−)-camphanic acid chloride, followed by separation on silica gel and subsequent ester hydrolysis gave (+)- and (−)-(1). Phosphorylation with diphenyl chlorophosphate followed by transesterification using the anion of benzyl alcohol gave the dibenzyl phosphates. Hydrogenolysis over palladium cleaved the benzyl esters and ethers, to give the enantiomers of inositol 1-phosphate (4) and (5), isolated as the crystalline cyclohexylammonium salts. This use of camphanate esters

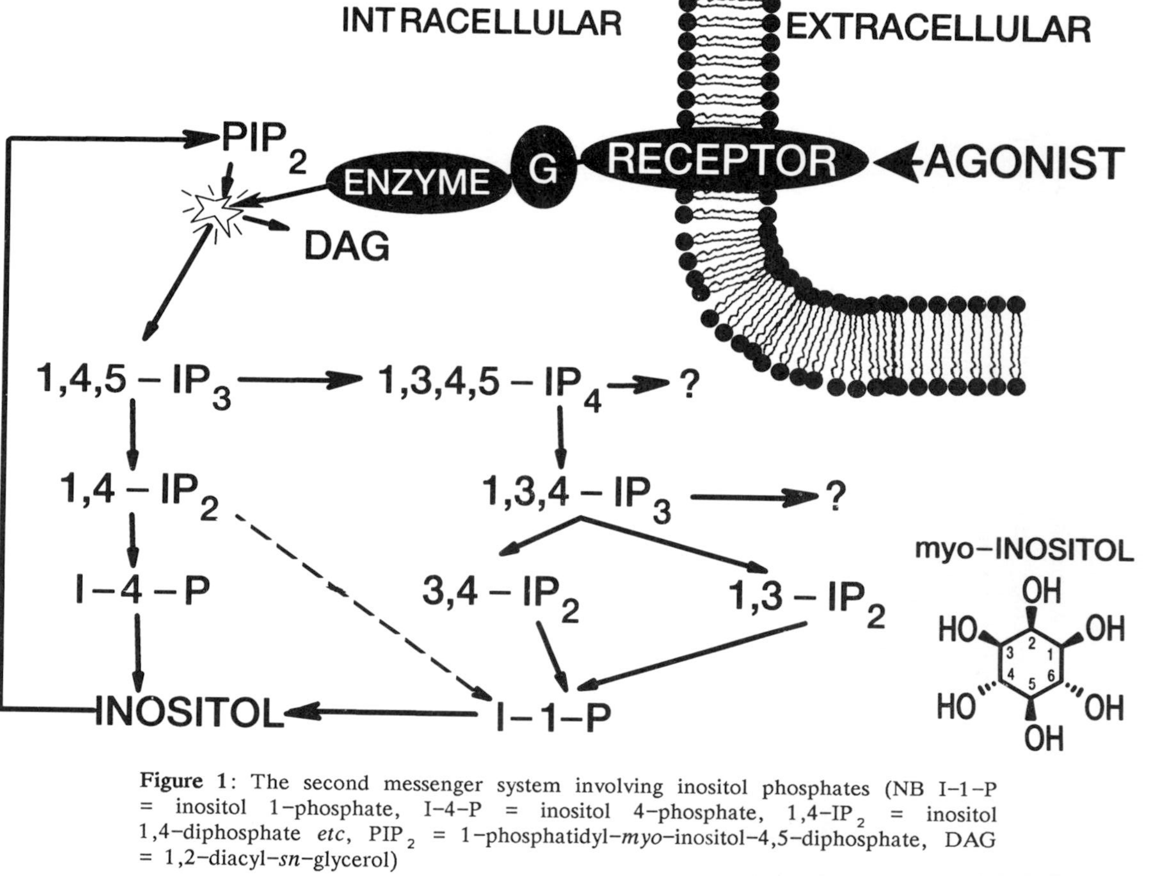

Figure 1: The second messenger system involving inositol phosphates (NB I–1–P = inositol 1–phosphate, I–4–P = inositol 4–phosphate, 1,4–IP_2 = inositol 1,4–diphosphate *etc*, PIP_2 = 1–phosphatidyl–*myo*–inositol–4,5–diphosphate, DAG = 1,2–diacyl–*sn*–glycerol)

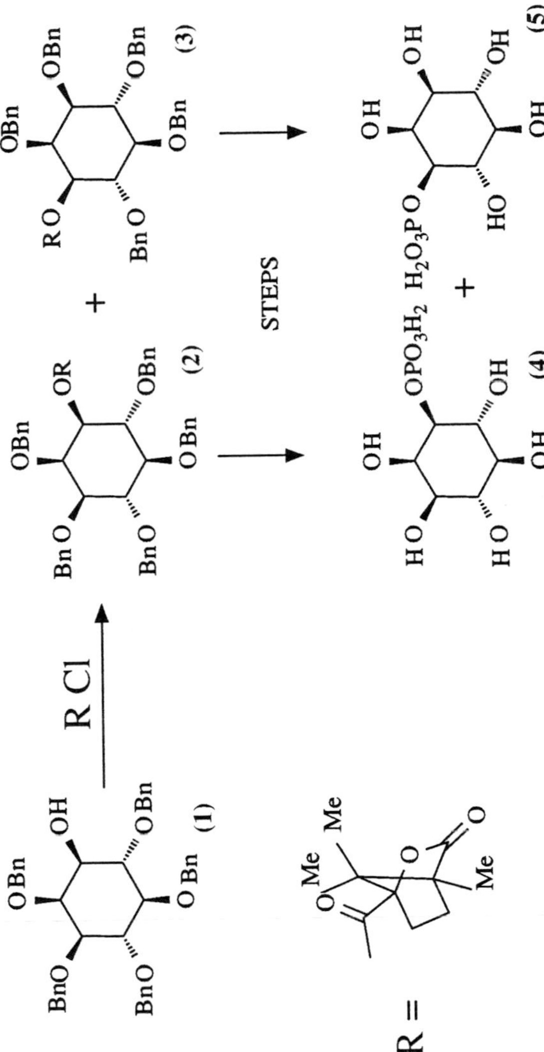

Scheme 1: Synthesis of the enantiomers of inositol 1–phosphate

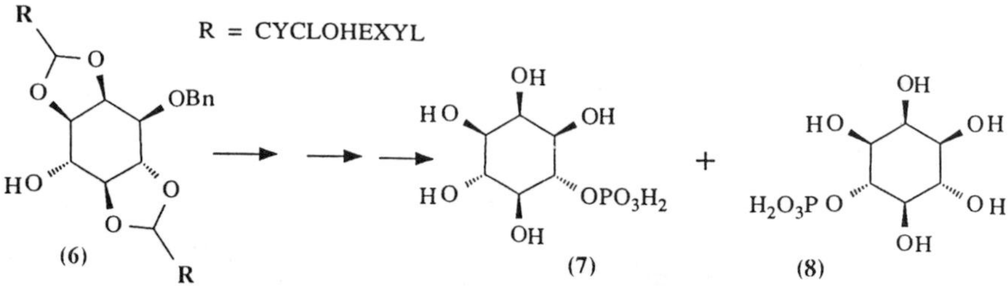

R = CYCLOHEXYL

(6) (7) (8)

Scheme 2: Synthesis of enantiomers of the inositol 4-phosphate

for the resolution of protected inositol derivatives has proven to be a generally applicable method. The enantiomers of the protected derivative (**6**) (Scheme 2) may be readily prepared by this technique, and then converted into the enantiomers of inositol 4-phosphate (**7**) and (**8**).[5]

Conventional syntheses of the inositol phosphates often use readily prepared cyclic acetals, such as (**6**) as intermediates.[6] These intermediates are however *not* suitable for the preparation of inositol phosphates having 1,3-substitution patterns, inositol 1,3,4,5-tetrakis-phosphate for example. We explored the possibility of selective protection of the orthoformate (**9**) (Scheme 3), which is readily

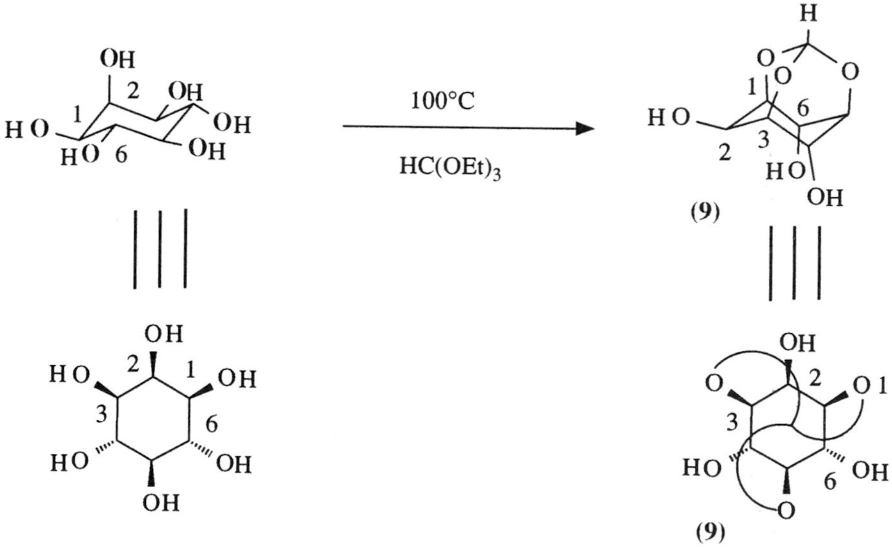

100°C

$HC(OEt)_3$

(9)

Scheme 3: Protection of *myo*-inositol at the 1,3,5-positions

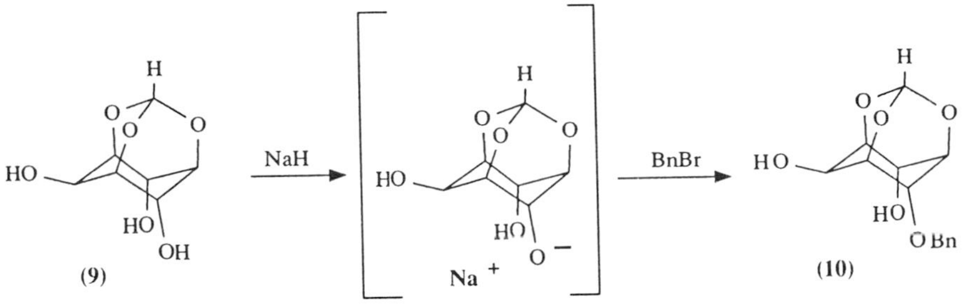

Scheme 4: Selective benzylation of orthoformate (9)

available from *myo*-inositol. This provides an intermediate protected at the 1,3,5-positions, and has reversed the normal axial/equatorial relationships of the remaining free hydroxyl groups at the 2,4,6-positions.

Reaction of orthoformate (9) with t-butyldimethylsilyl chloride and base gave a gross mixture of products. In contrast, reaction with sodium hydride in dimethylformamide (DMF), followed by alkylation with benzyl bromide gave a single monoalkylated product (10) (Scheme 4) in 80% yield.[7] The NMR spectrum of this compound clearly indicated that it was a 4-substituted derivative (dissymmetric spectrum showing 6 signals for the ring protons; 2-substitution gives symmetric NMR spectra, with C-1 = C-3 and C-4 = C-6, showing 4 signals for the ring protons). This selective reaction is presumably

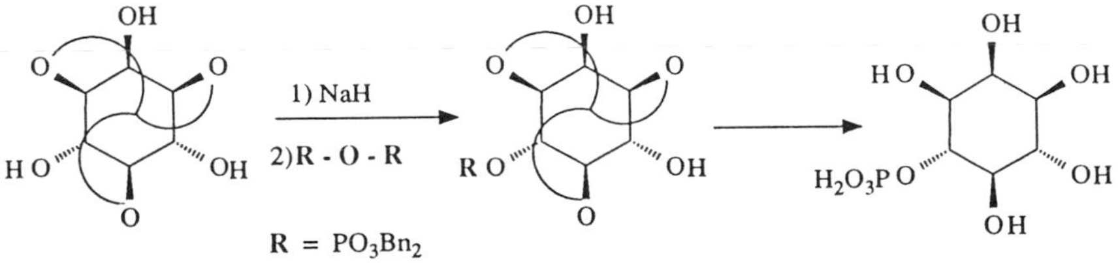

Scheme 5: Conversion of orthoformate (9) into *rac*-inositol 4-phosphate

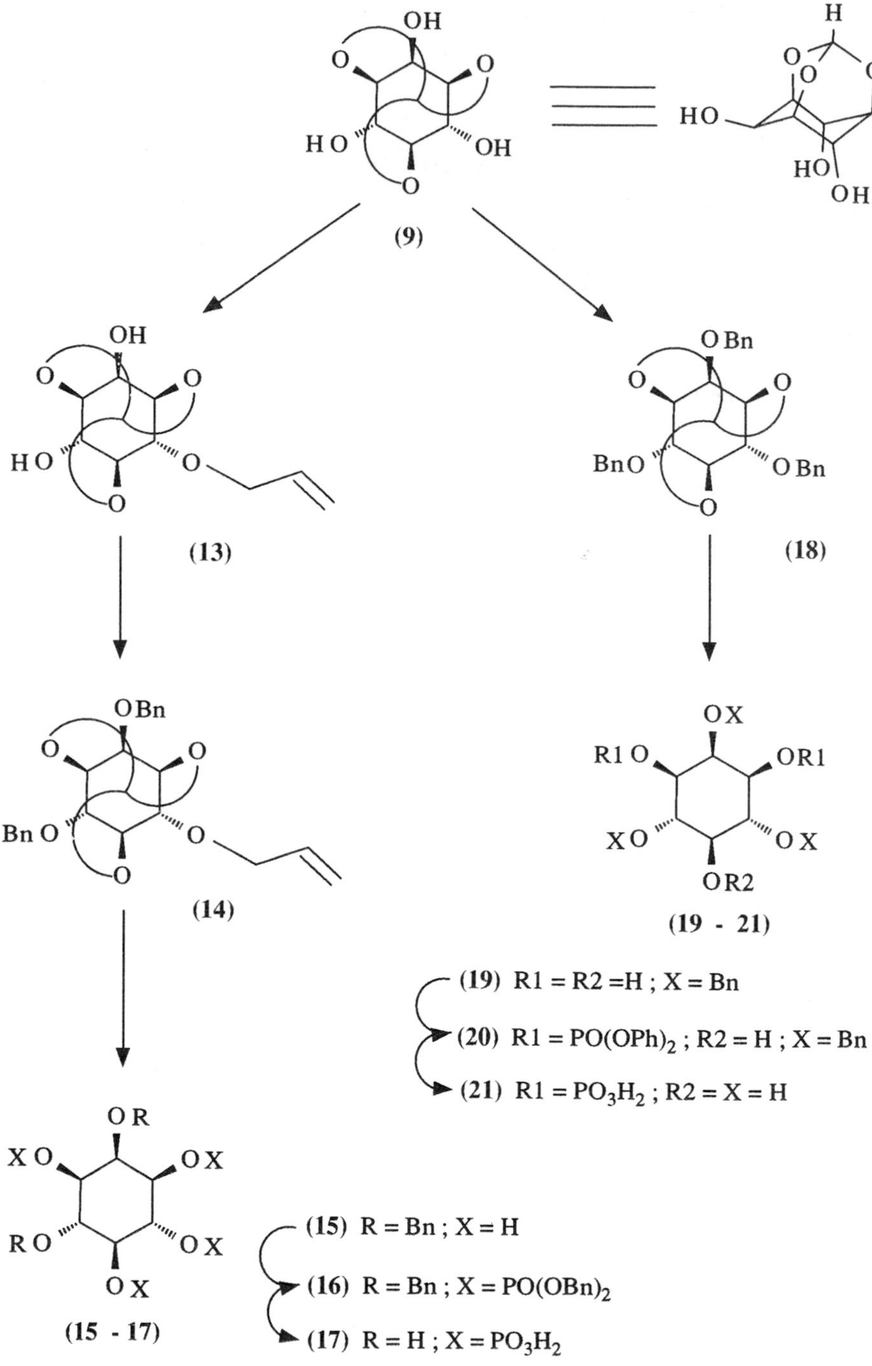

(9)

(13)

(18)

(14)

(19 - 21)

(19) R1 = R2 = H ; X = Bn

(20) R1 = $PO(OPh)_2$; R2 = H ; X = Bn

(21) R1 = PO_3H_2 ; R2 = X = H

(15 - 17)

(15) R = Bn ; X = H

(16) R = Bn ; X = $PO(OBn)_2$

(17) R = H ; X = PO_3H_2

Scheme 6: Synthesis of inositol 1,3,4,5-tetrakis-phosphate

due to the fact that chelation controlled alkylation occurs under these conditions. A range of alkylating agents gave similar results. We have exploited this selectivity in the synthesis of a number of inositol phosphates.[7,8]

Reaction of the anion of orthoformate (9) with tetrabenzyl pyrophosphate in DMF (Scheme 5) gave a 72% yield of alcohol (11). Hydrogenolysis of (11) followed by acidic hydrolysis gave racemic inositol 4-phosphate (12) in high yield. This represents an extremely short and efficient synthesis of *rac*-(12).

Allylation of the anion of (9) (Scheme 6) gave (13) in good yield. Benzylation of the remaining free hydroxyl groups gave (14). Isomerisation of the allyl group to the enol ether, followed by acidic hydrolysis of the enol ether and orthoformate gave the tetraol (15). Phosphorylation of the anion of (15) with tetrabenzyl pyrophosphate in tetrahydrofuran proceeded readily in the presence of a catalytic quantity of imidazole or 18-crown-6, giving the fully benzylated tetrakis-phosphate (16). Hydrogenolysis over palladium cleaved all of the protecting groups and gave racemic inositol 1,3,4,5-tetrakis-phosphate (17), isolated as its cyclohexylammonium salt.

Benzylation of orthoformate (9) gave the tris-benzyl derivative (18) (Scheme 6). Hydrolysis of the orthoformate then gave triol (19). Phosphorylation with diphenyl chlorophosphate gave a mixture of 1,3- and 1,5-bis-phosphorylated materials. The desired 1,3-isomer (20) was obtained pure after one recrystallisation from diethyl ether-light petroleum. Attempts to transesterify (20) led to decomposition. The alcohol (20) was directly deprotected using lithium/liquid ammonia, to give inositol 1,3-bisphosphate (21), isolated as its cyclohexylammonium salt.

3 ENZYMIC REACTIONS OF INOSITOL PHOSPHATES

Using the above substrates, the enzymes of the inositol recycling pathway have been isolated and characterised. It has been found that a single enzyme is responsible for removal of the 5-phosphate group of both 1,4,5-IP3 and 1,3,4,5-IP4 (giving 1,4-IP2 and 1,3,4-IP3, respectively). Metabolism of 1,4-IP2 proceeds almost entirely *via* I-4-P in the brain and in other tissues examined, in contrast to previous literature reports. A single enzyme removes the 4-phosphate from both 3,4-IP2 and 1,3,4-IP3. The enzyme responsible for the hydrolysis of I-4-P accepts both enantiomers of the substrate almost equally well. This enzyme also cleaves both enantiomers of I-1-P at essentially the same rate. Nature therefore exercises considerable economy in the number of enzymes used in this recycling pathway.

ACKNOWLEDGEMENTS

The author wishes to acknowlege the valuable contributions of Drs R Baker, J R Huff, J J Kulagowski, S J deSolms, and J P Vacca, and of Mr I M Mawer.

REFERENCES

1 L R Rawls, *Chem Eng News,* 1987, 26.
2 R Michell, *Nature (London)*, 1986, **319**, 176; J Altman, *Nature (London),* 1988, **331**, 119; C W Taylor, *Trends Pharmacol Sci,* 1987, **8**, 79.
3 D C Billington, *Chem Soc Revs,* 1989, **18**, 83.
4 D C Billington, R Baker, J J Kulagowski, and I M Mawer, *J Chem Soc Chem Commun,* 1987, 314.
5 D C Billington, R Baker, I Mawer, J J Kulagowski, J P Vacca, S J deSolms, and J R Huff, *Tetrahedron*, 1989, **45**, 5679.
6 S Ozaki, Y Kondo, H Nakihara, S Yamaoka, and Y Watanabe, *Tetrahedron Lett*, 1987, **28**, 4691; Y Watanabe, H Nakihara, M Bunya, and S Ozaki, *ibid*, p 4179; J P Vacca, S J deSolms, and J R Huff, *J Am Chem Soc*, 1987, **109**, 3478.
7 D C Billington and R Baker, *J Chem Soc Chem Commun,* 1987, 1011.
8 D C Billington, R Baker, I M Mawer, J J Kulagowski, J P Vacca, S J deSolms, and J R Huff, *J Chem Soc Perkin Trans 1*, 1989, 1423.

MECHANISM-BASED INHIBITORS AND SITE-DIRECTED MUTANTS AS PROBES OF THE MECHANISM OF RIBONUCLEOTIDE REDUCTASE

Shi-Shan Mao, Madeleine I Johnson, J Martin Bollinger, C Hunter Baker, and JoAnne Stubbe

Department of Chemistry, Massachusetts Institute of Technology, Cambridge, MA 02139, USA

1 INTRODUCTION

Ribonucleotide reductases (RDPRs) play a central role in DNA biosynthesis by catalysing the conversion of nucleoside diphosphates (NDPs) into deoxynucleoside diphosphates (dNDPs) concomitant with dithiol oxidation (equation 1):

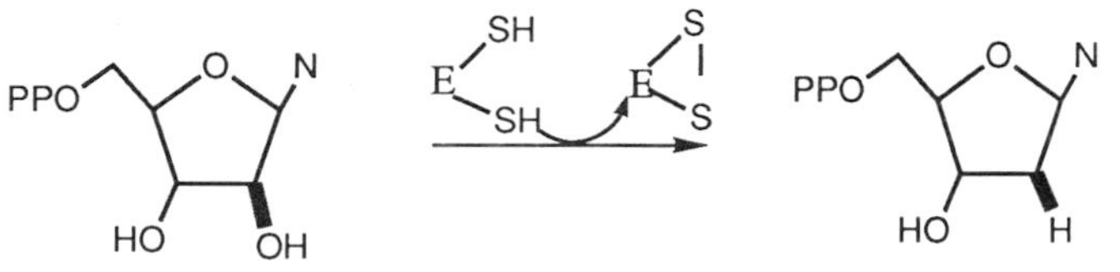

eqn 1

The enzyme from *Escherichia coli* is composed of two subunits each of which is made up of two equivalent protomers. B_1 ($\alpha\alpha'$) binds the NDP substrates, contains two sets of redox active thiols per protomer and contains the binding sites for the allosteric regulators. Recent evidence from Que's laboratory suggests that B_2 ($\beta\beta$) is composed of one binuclear iron centre and tyrosyl radical per protomer.[1] The active site is at the interface between the subunits.[2,3] Whether both sites are active remains to be established. It is intriguing that in the past few years reductases have been characterized from a variety of sources each with a novel cofactor. The protein from *Lactobacillus leichmannii* utilizes adenosylcobalamin (AdoCbl) and evidence implicates protein radical involvement in catalysis.[2] The enzyme from

Bacillus ammoniagenes contains a unique manganese centre whose spectroscopic properties in comparison with model inorganic compounds suggest a binuclear manganese centre–protein radical motif.[4] Very recently, Reichard's laboratory has uncovered yet an additional reductase from anaerobic *E coli* which promises to reveal yet another novel cofactor.[5] What all of these reductases possess in common is a way to generate an organic radical to initiate their chemistry.

Our efforts have focused on unraveling the mechanism of the tyrosyl radical- and AdoCbl-dependent enzymes. Consideration of non-enzymic reactions described by Walling and Johnson,[6] Gilbert and Norman,[7] and von Sonntag and collaborators[8] suggests that the ribonucleotide reductase reaction could be accomplished by the series of transformations indicated in Figure 1. The major thesis is that a protein radical X· mediates a hydrogen atom abstraction from the 3'-carbon of ribose yielding a 3'-radical adjacent to the bond being cleaved. Loss of the 2'-hydroxyl, following its protonation by a redox active thiol, would result in a cation radical. The resulting 'formyl methyl radical equivalent' could then be reduced by multiple electron transfer reactions. The penultimate step, reduction of a 3'-ketodeoxynucleotide *via* a disulfide radical anion has no chemical precedent to our knowledge. The final step is re-oxidation of the protein residue (XH) to regenerate protein radical X· concomitant with deoxynucleotide production.

Studies using [3'-^{2}H] and [3'-^{3}H]NDPs support the thesis that the 3'-carbon-hydrogen bond of the substrate is cleaved and that the hydrogen abstracted from the 3'-position in starting material is returned to the same position in the product.[2] However, all efforts using stopped flow kinetics methods to detect reduction of the tyrosyl radical, which should be accompanied by loss in absorbance at 410 nm and increase in absorbance at 270 nm, have thus far failed. While the proposed mechanism (Figure 1) is consistent with all available data, no evidence in support of radical intermediates or a correlation between 3'-carbon-hydrogen bond cleavage and tyrosyl radical reduction are presently available with the normal substrates.

Our laboratory has therefore taken two approaches to obtain evidence for substrate derived radical intermediates and the direct involvement of the tyrosyl radical in the chemistry. The first has

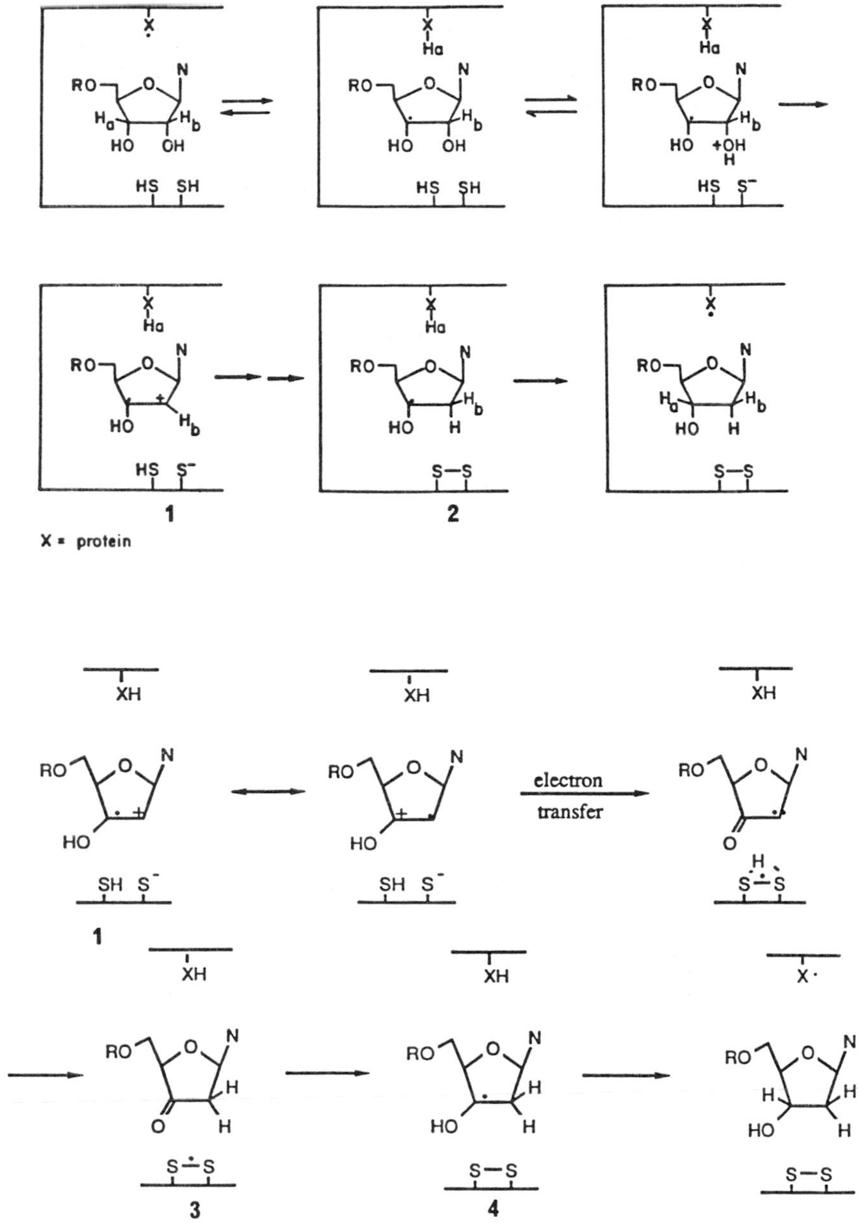

X = protein

Figure 1 : Pathway proposed for the conversion of a nucleoside diphosphate into a deoxynucleoside diphosphate by ribonucleotide reductase (X· is a protein-bound radical, RO= diphosphate, N = nucleotide base).

utilized mechanism-based inhibitors, that is modified substrates, and the second has utilized site-directed mutagenesis, that is, modified proteins. Both have proven informative and as discussed subsequently have provided evidence for radical intermediates. In addition, both approaches after further chemical studies, should help elucidate the function of the tyrosyl radical.

While the emphasis in the present paper will be on use of modified proteins to elucidate mechanism, a brief summary of results from studies with the mechanism based inhibitor 2'-azido-2'-deoxyuridine 5'-diphosphate (N_3UDP) is required to understand the results with the mutant proteins.

2 INTERACTION OF RIBONUCLEOTIDE REDUCTASE WITH 2'-AZIDO-2'-DEOXYURIDINE 5'-DIPHOSPHATE

In the past few years interaction of isotopically labeled N_3UDPs [β-^{32}P, 5'-^{3}H, 3'-^{3}H, 1'-^{2}H, 2'-^{2}H, 3'-^{2}H, 4'-^{2}H] with reductase have been investigated in detail (Figure 2). The enzyme is inactivated in a single turnover which is accompanied by production of one equivalent of N_2, uracil and inorganic pyrophosphate (PPi). The B_1 subunit of the enzyme is stoichiometrically labelled with the sugar moiety remaining and the entire process is initiated by 3'-carbon-hydrogen bond cleavage. In addition, the tyrosyl radical on the B_2 subunit is destroyed and a new nitrogen-centred radical is produced. These results, while not understood in detail from a mechanistic viewpoint, provide the first evidence for radical intermediates in a RDPR-catalysed reaction. Kinetic studies were then undertaken to address the major thesis that 3'-carbon-hydrogen bond cleavage is accompanied by tyrosyl radical reduction. To test this hypothesis, [3'-^{2}H]N_3UDP was prepared and the rate loss of the tyrosyl radical (monitored by a change in absorbance at 410 nm) and the rate of inactivation of the enzyme were monitored. Comparison of studies with [3'-^{1}H] and [3'-^{2}H]N_3UDP indicated that both measurements are accompanied by an isotope effect as would be predicted by the mechanism in Figure 2. However, two puzzling features remain to be explained. First, the kinetics of tyrosyl radical loss are biphasic with only an isotope effect being observed on the

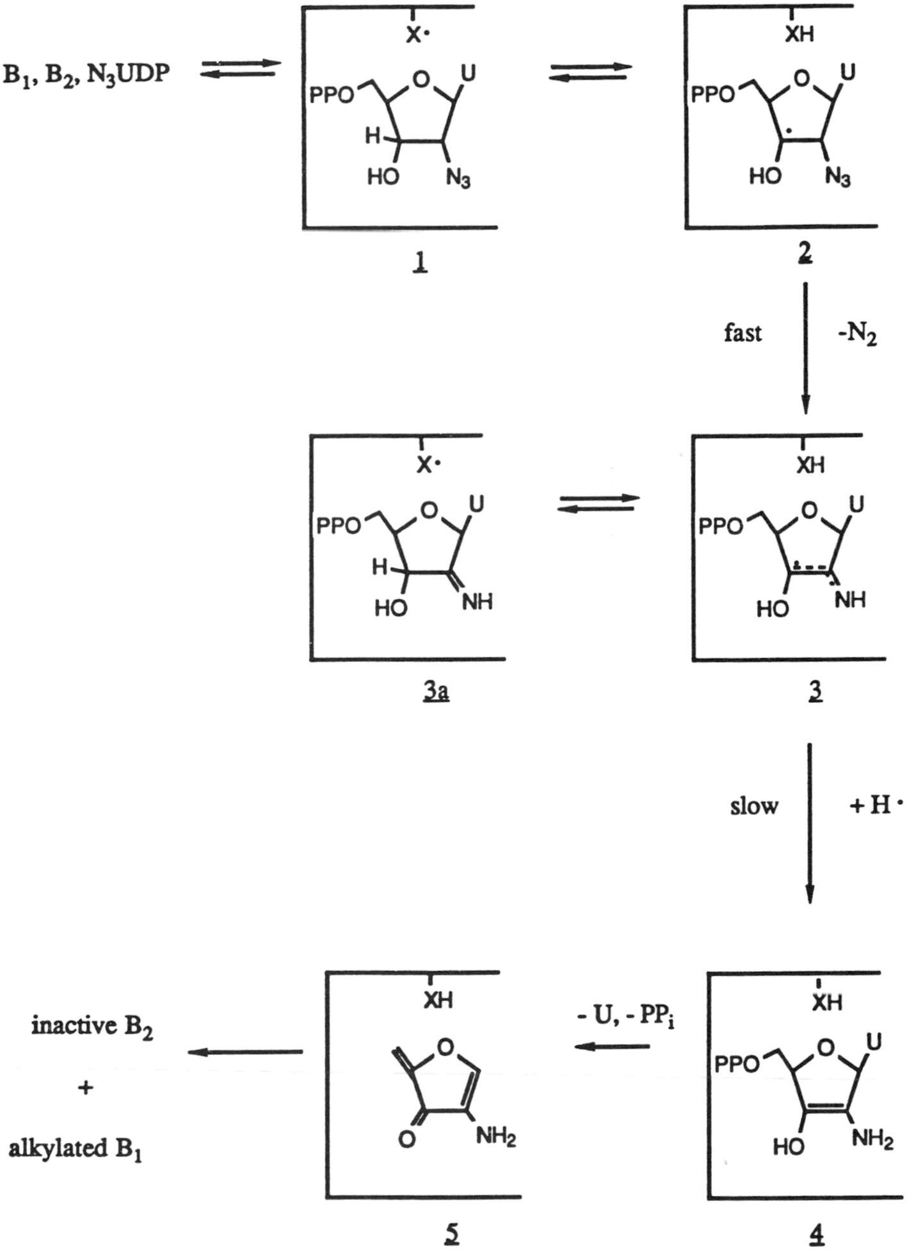

B_1, B_2, N_3UDP ⇌ [1: X·, PPO, O, U, H, HO, N_3] ⇌ [2: XH, PPO, O, U, HO, N_3]

2 → (fast, $-N_2$) → 3

[3a: X·, PPO, O, U, H, HO, NH] ⇌ [3: XH, PPO, O, U, HO, NH]

3 → (slow, + H·) → 4

inactive B_2 + alkylated B_1 ← [5: XH, O, O, NH_2] ← ($-$ U, $-$ PP_i) ← [4: XH, PPO, O, U, HO, NH_2]

Figure 2a: Pathway proposed for the inactivation of ribonucleotide reductase by 2'–azido–2'–deoxyuridine 5'–diphosphate.

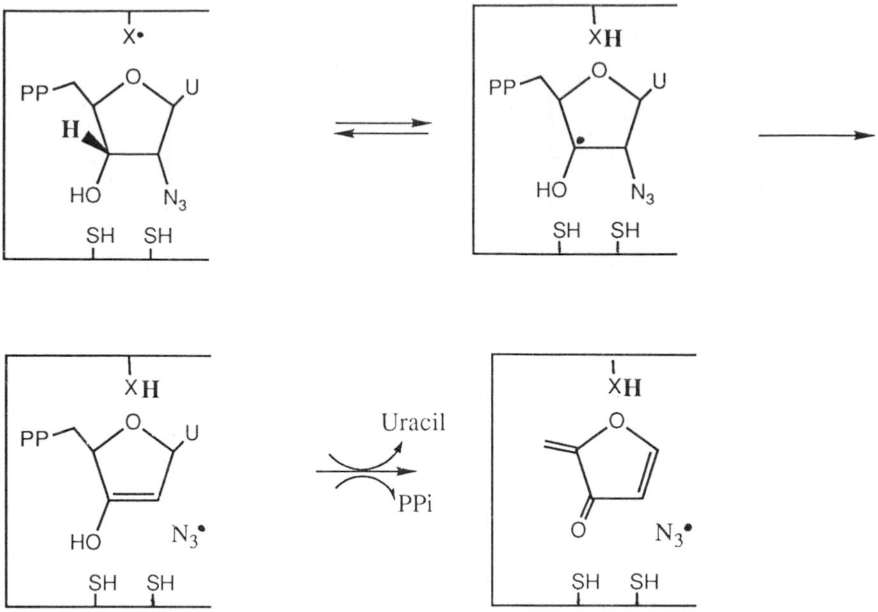

Figure 2b: Alternative mechanism of inactivation of ribonucleotide reductase by N_3UDP based on recent evidence.

rapid phase. The recent proposal of Que and coworkers[1] that suggests fully active enzyme should contain 2 tyrosyl radicals per B_2 and the fact that the maximum number of tyrosyl radicals observed to date is 1.5/B_2[1,9] might account for the biphasic nature of radical loss. A second puzzling feature of the inactivation of RDPR by N_3UDP is that the rate loss of the tyrosyl radical is slower than the rate of inactivation.[10] To accommodate this unexpected observation, one must postulate that the tyrosyl radical is regenerated (Figure 2a, **3** ⟶ **3a**) subsequent to its initial reduction. The tyrosyl radical in **(3a)** (Figure 2a) would be spectroscopically indistinguishable from that observed in the resting enzyme and hence the rate of its initial disappearance would be slower than it actually is. A number of other puzzling features of this scheme remain to be elucidated:

1 The EPR signal of the nitrogen centred radical has not yet been assigned a satisfactory structure;
2 There is no good chemical precedent for why a radical β to an alkyl azide should result in rapid loss of nitrogen gas;
3 The structure of the sugar moiety which covalently modifies B_1 remains to be established.

Recently [2'-^{13}C, 2'-$^{15}N_3$]UDP has been prepared and investigated by EPR spectroscopy. Very preliminary results suggest no difference between this spectrum and that produced by [2'-^{12}C, 2'-$^{15}N_3$]UDP in contrast to what would be predicted in Figure 2a. An alternative to this mechanism is shown in Figure 2b.

While this system needs to be examined in more detail, the studies of the interaction of N_3UDP with RDPR have provided the first clear evidence that RDPR can generate radical intermediates during turnover.

3 SITE-DIRECTED MUTANTS OF THE B_1 SUBUNIT OF RIBONUCLEOTIDE REDUCTASE

The second approach to unraveling mechanism has involved the preparation of site-directed mutants (cysteine → serine) of the B_1 subunit of RDPR. While this approach is questionable in the absence of structural data, our previous protein modification studies made predictions which we felt were experimentally testable using this methodology. We reasoned that if we could remove one or both of the thiols involved in reduction of the substrate, then the rate constants for various steps along the reaction pathway might be sufficiently altered to permit observation of tyrosyl radical reduction *via* stopped flow kinetics. The first problem encountered was identifying the target cysteines. There are 22 cysteines per B_1 subunit (11 per protomer). Efforts to use a number of mechanism-based inhibitors have thus far failed to identify active site residues due to the lability of the inhibitor-enzyme complex to peptide mapping procedures. The following method was therefore developed to locate the redox active thiols. RDPR was pre-reduced with dithiothreitol (DTT), the reductant removed and the enzyme incubated with cytidine diphosphate (CDP) in the absence of any external reductant. The active site thiols should be oxidized and hence protected from alkylating agents. Denaturation of the enzyme in the presence of iodoacetamide, reduction of active site thiols with DTT and alkylation with [^{14}C]-iodacetamide should alkylate all remaining thiols. Removal of the iodoacetamide should modify only the active site thiols. Tryptic digest of the RDPR has allowed isolation of the modified peptides and elucidation of the thiols to be targeted by mutagenesis.[11] Technically, these are difficult experiments. The results revealed, under single turnover conditions, that 3 dCDPs are produced accompanied by formation of approximately six carboxamidomethylated cysteines. Peptide mapping proved difficult but revealed two major radiolabeled peptides and a third minor peptide. Peptide sequencing established that *ca* 65% of the radioactivity resided within the C-terminal peptide of B_1: C_{754}-E-G-G-A-C_{759}-P-I-K. Peptide(s) containing the remaining 35% of modified cysteines were less tractable to analysis. 60% of this 35% was found within the peptide Q-F-S-S-C_{225}-V-L-I-E-C_{230}-G-P, while the remaining peptide found contained residues 443-453 and presumably cysteine 462.

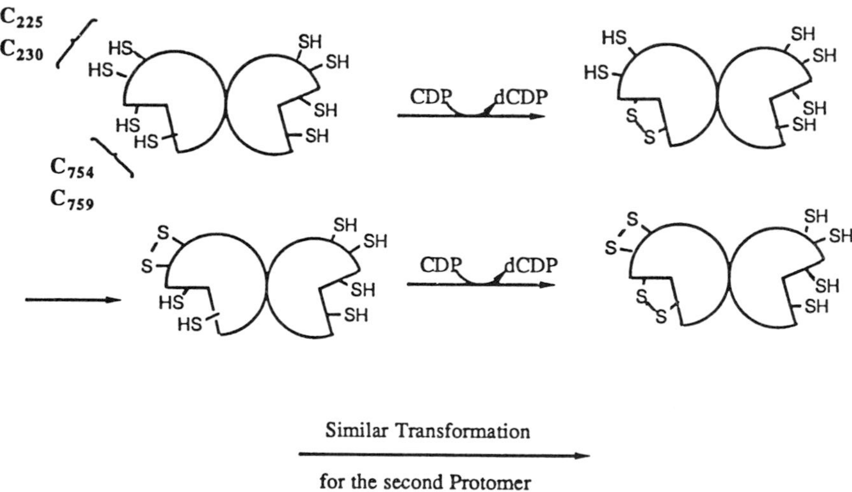

Figure 3: Postulated role of cysteines in ribonucleotide reductase.

The following model was proposed to accommodate the observation of three dCDPs and multiple labeled peptides (Figure 3). We postulated that cysteines 225 and 230 served as a conduit of electrons between the active site cysteines 754 and 759 of B_1 and thioredoxin. The thesis that 754 and 759 were at the active site was based on a 'homologous' peptide isolated from a similar experiment conducted with the adenosylcobalamin-dependent reductase: C-E-S-G-A-C-K-I.[11] In addition, of the eight sequences of B_1 reported, all possess two C-terminal cysteines although the sequence and motif, $C-(X)_2-C$ *versus* $C-(X)_4-C$ are different. This hypothesis predicts that mutants C754S or C759S should be unable to reduce CDP to dCDP and that mutants C225S and C230S should be unable to transfer electrons between thioredoxin and B_1. Both of these predictions can be examined biochemically.

We have recently prepared a number of mutants of RDPR using methods developed by Tabor and Richardson[12] for the cloning of toxic genes and methods of Taylor *et al*[13] for mutagenesis. The mutants can be purified to homogeneity using a dATP affinity column, suggesting that the allosteric binding sites are intact. A problem with this method is that there are no *E coli* host cells which lack the *nrd*A gene required for B_1 production. Therefore, all of the mutants discussed below contain contaminating wild type B_1 (*ca* 2 to 3%).

The properties of the C754S, C759S and C754S and C759S double mutant are summarized in Table 1. All three mutants have very low activity, 3 to 7% of the wild type enzyme prepared by identical procedures, using thioredoxin (TR), thioredoxin reductase (TRR) as reductant. We estimate that most of the observed activity can be accounted for by contaminating wild type enzyme. Efforts are underway to alleviate this problem using molecular biology methods. These results were in accord with our expectation that C754 and C759 are within the active site. When these mutants, however, were examined with DTT as a reductant in place of TR and TRR, dramatically different results were observed. The single mutants C754S and C759S had identical activity to that of the wild type in the presence of 10 mM DTT, while the double mutant was eight times more active and possessed 70% of wild type activity using TR/TRR! These results strongly implicate the C-terminal cysteines as the

Table 1: Activity of mutant B_1 subunits

MUTATION	Rate of dCDP production ($nmol\ min^{-1}\ mg^{-1}$)	
	TR/TRR	DTT
Wild-type B1	470	42
C754S	18 (4% wt)	40
C759S	35 (7% wt)	46
double mutant C754S & C759S	14 (3% wt)	330
C225S	13 (3% wt)	1.2
C230S	554	
double mutant C225S & C230S	10.9(2.3% wt)	3.4

mediators of electrons between thioredoxin and the active site of B_1, in contrast to our original postulate. In retrospect, the available sequence information is also consistent with this role. *L leichmannii* ribonucleotide reductase is reduced by *E coli* thioredoxin, which provides an explanation for the observed sequence homology of cysteine containing peptides. Furthermore, the T4 phage reductase, which is not reduced by *E coli* thioredoxin, lacks sequence homology in the C–terminal region, but the homology is highly conserved in the region of both C225 and C462. Thus C754 and C759 probably function in the role we originally attributed to C225 and C230.

Similar studies have been conducted on the C225S and C230S mutants (Table 1). The C225S RDPR has 3% the activity of the wild–type enzyme in both the TR/TRR and DTT assays. The observed dCDP is again consistent with the expected levels of contaminating wild type enzyme. In contrast, the C230S mutant has 120% the activity of the wild type enzyme in the TR/TRR assay! One can conclude therefore that C230 is *not* one of the thiols involved in direct substrate reduction. In addition, comparisons of the eight available sequences of B_1 indicate *no* conservation of this

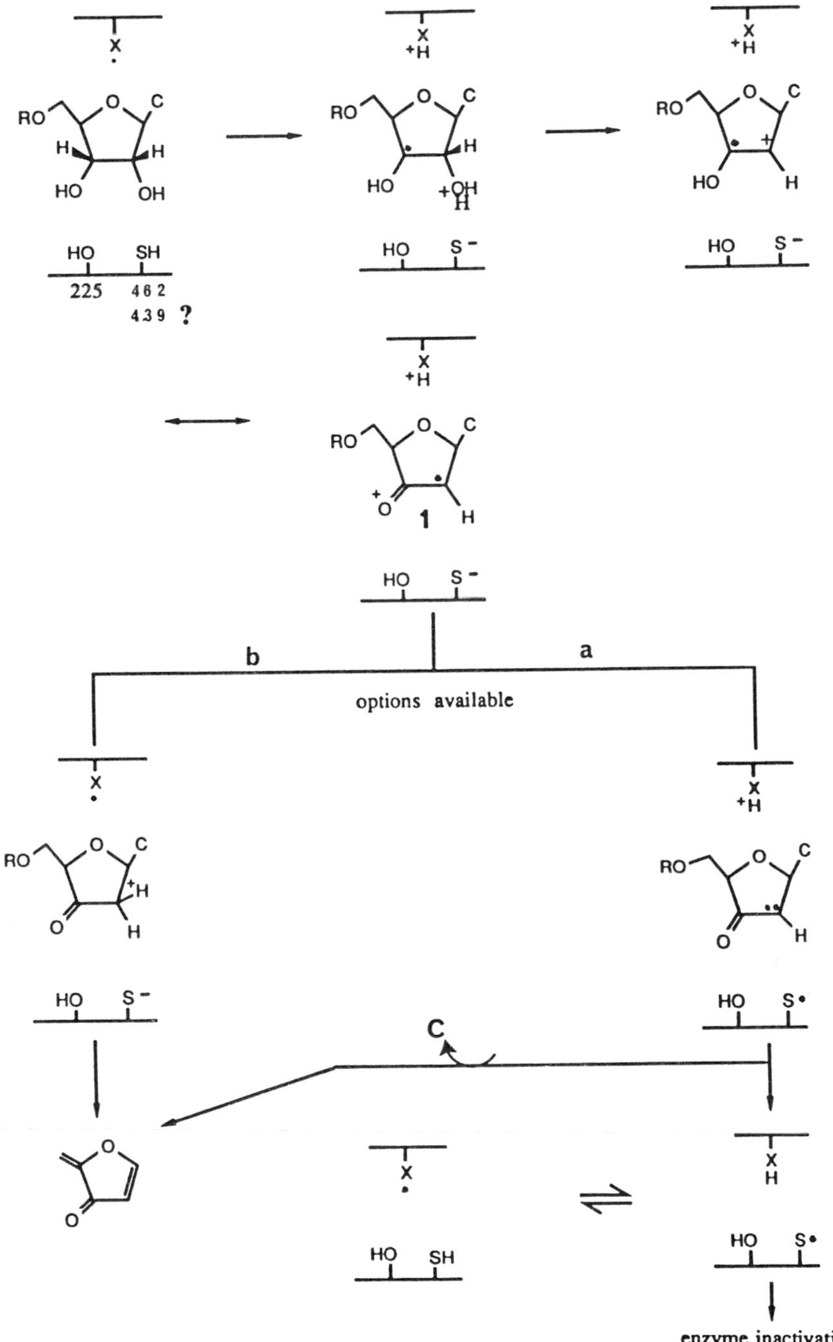

Figure 4: Pathway proposed for the degradation of a nucleoside diphosphate effected by ribonucleotide reductase in which cysteine 225 has been mutated to a serine.

residue. The B_1 subunit of the T4 phage reductase possesses an alanine in this position. In retrospect, Nature has in fact done a mutagenesis experiment for us. In contrast, C225 is conserved in all known sequences, as are C439 and C462.

Further analysis of the C225S mutant has revealed some remarkable properties. Analysis of the products and kinetics of their production indicated that in addition to dCDP, cytosine is produced. The kinetics of dCDP production are linear for 30 min and the rates of production with DTT are 10% that observed with TR/TRR. These observations are consistent with dCDP production resulting from contaminating wild type activity. The K_m value for CDP is identical to that for the wild type RDPR consistent with this hypothesis. In contrast, the kinetics of cytosine production are biphasic indicating that the enzyme is inactivating itself.[14] The observation of cytosine release concomitant with enzyme inactivation is remarkably similar to previous studies with mechanism-based inhibitors ClUDP and N_3UDP.[2] Incubation of [^{14}C]CDP and C225S-B_2 indicates that inactivation is accompanied by covalent modification of the protein. Furthermore, monitoring of the tyrosyl radical absorbance at 410 nm reveals time-dependent disappearance. The kinetics of tyrosyl radical loss (50% loss in 5 min) are at least biphasic, reminiscent of its loss when the wild type B_1 is incubated with N_3UDP. Recent studies utilizing [3'-^{3}H]UDP reveal that an analogous process to that observed with CDP occurs and that ^3H_2O is produced. 3'-Carbon-hydrogen bond cleavage therefore appears to be a prerequisite for cytosine release and enzyme inactivation.

Even more amazing is the unexpected observation that during this process the B_1 subunit (C225S) is cleaved into two pieces (26 and 61.5 kd) in a time-dependent, substrate-dependent, B_2-dependent process. Sequence analysis reveals that the small piece, 26 kd, is the *N*-terminal end of B_1. The cleavage is thus occurring in the vicinity of the mutation.

A model to account for these observations and data available from studies with mechanism-based inhibitors is presented in Figure 4. Studies are underway to test the validity of this proposal. Amazingly, a single substitution of a cysteine with a serine converts the normal substrate (NDP) into a potent mechanism-based inactivator in which both the B_1 and B_2 subunits are extensively modified.

Table 2: Results from one turnover experiments

MUTATION	NUMBER OF DCDP/B1
Wild-type	2.6 ±0.3
C754S	1.2 ±0.1
C759S	0.9 ±0.1
double mutant C754S & C759S	1.1 ±0.1

We therefore postulate that C225 is one of the key residues involved in substrate reduction. Very recently several additional mutants have been prepared (C439S and C462S) and assays have been completed. C462S is rapidly inactivated in the presence of TR/TRR and inactivation is accompanied by cytosine release and no protein cleavage. We currently favour the thesis based on our original peptide mapping experiments[11] and these data that C462 is the other active site cysteine required for substrate reduction.

Finally, to avoid the problem of contaminating wild type enzyme's interference with assay of the mutant proteins, one turnover experiments have been performed. The amount of dCDP produced from pre-reduced enzyme after a 20 min incubation in the absence of external reductant is indicated in Table 2. Results indicate that with the wild type enzyme *ca* 3 dCDPs are produced, whereas with the C754S and C759S mutants only a single dCDP is observed. The kinetics of dCDP production are presently being measured using rapid quench methods. The C225S mutant also produces 0.7 equivalents of dCDP after 30 min incubation. The rates of dCDP production must be established.

The model we presently favour to account for these results is shown in Figure 3. The basic tenets of the original model have not changed significantly. However, the residues assigned to carry out the various functions, that is, active site reduction and shuttling of reducing equivalents have changed substantially. We now favour the interpretation that C754 and C759 interact with the TR and are most easily reduced. The active site residues we believe to be involved in substrate reduction are C225 and C462. The mutagenesis studies have clearly eliminated C230 from consideration for this role. Extensive chemical studies with both the single and double active site mutants should lead to a better understanding of the mechanism by which the 3'-carbon-hydrogen bond of substrate NDP is cleaved, the involvement or non-involvement of the tyrosyl radical on B_2, and the role of the redox active thiols.

REFERENCES

1 J B Lynch, C J Garcia, E Munck, and L Que, *J Biol Chem*, 1989, **264**, 8091.
2 G W Ashley and J Stubbe, *Pharmacol Ther*, 1985, **30**, 301.
3 P Reichard and A Ehrenberg, *Science*, 1983, **221**, 514.
4 A Willing, H Follmann, and G Auling, *Eur J Biochem*, 1988, **170**, 603.
5 M Fontecave, R Eliasson, and P Reichard, *Proc Natl Acad Sci USA*, 1989, **86**, 2147.
6 C Walling and R A Johnson, *J Am Chem Soc*, 1975, **97**, 2405.
7 B C Gilbert, J P Larkin, and R O C Norman, *J Chem Soc Perkin Trans 2*, 1972, 794.
8 M S Aklaq, H P Schumann, and C von Sonntag, *Carbohydr Res*, 1987, **164**, 71.
9 M J Bollinger, unpublished results.
10 J Stubbe, *Adv Enzymol Relat Areas Mol Biol*, 1989, **63**, 349.
11 A I Lin, G W Ashley, and J Stubbe, *Biochemistry*, 1987, **26**, 6905.
12 S Tabor and C Richardson, *Proc Natl Acad Sci USA*, 1985, **82**, 1074.
13 J W Taylor, J Ott, and F Eckstein, *Nucleic Acids Res*, 1985, **13**, 8765.
14 S-S Mao, M I Johnston, J M Bollinger, and J Stubbe, *Proc Natl Acad Sci USA*, 1989, **86**, 1485.

MOLECULAR RECOGNITION IN β-LACTAMASES

Michael I Page and Andrew P Laws

Department of Chemical and Physical Sciences, Huddersfield Polytechnic, Huddersfield HD1 3DH, UK

1 INTRODUCTION

There are two families of enzymes which specifically recognise the β-lactam antibiotics, such as the penicillins **(1)** and cephalosporins **(2)**. The β-lactamases (penicillin amido β-lactam hydrolase EC 3.5.2.6 and EC 3.5.2.8) catalyse the hydrolysis of the β-lactam to yield biologically inactive products, equations 1 and 2:

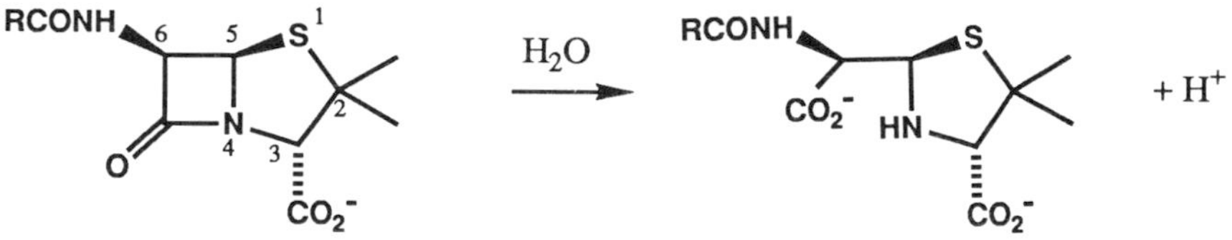

$\xrightarrow{H_2O}$ $+ H^+$

1

eqn 1

$\xrightarrow{H_2O}$ $+ H^+$

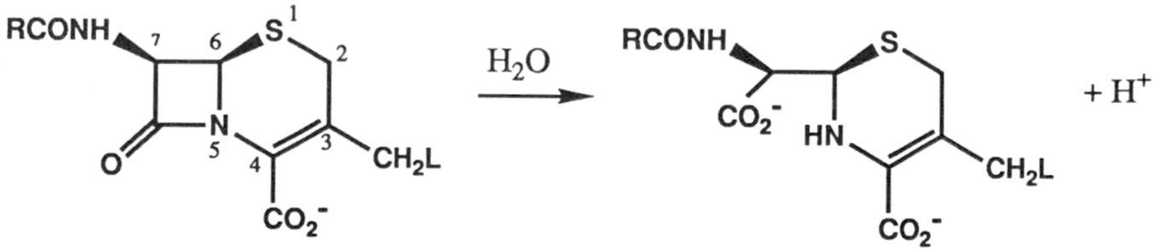

2

eqn 2

Plasmid–mediated formation of β–lactamases by bacteria is largely responsible for the resistance of many bacteria to the lethal action of β–lactam antibiotics.[1] The inhibitory target for β–lactam antibiotics that kill bacteria are the transpeptidase enzymes, and these represent the second class of enzymes that interact with β–lactams. The intermediates formed between transpeptidases and the β–lactam antibiotics are relatively stable and are the inhibited form of the enzyme preventing cell wall synthesis.[2]

2 β–LACTAMASES

Gram–positive bacteria excrete a β–lactamase in the culture medium during growth, but they also have a different β–lactamase bound to the membrane.[3] In Gram–negative bacteria, β–lactamases are often located in the periplasmic space and appear to be indistinguishable from the membrane bound enzyme.[4] There has been much discussion on the evolutionary relationship between β–lactamases and the transpeptidases. The biosynthesis of penicillin involves the intermediate formation of a linear tripeptide, and although the β–lactamases may be regarded as a type of peptidase, no acyclic peptide has been shown to be a substrate for these enzymes. However, monocyclic, as well as bicyclic β–lactams can be substrates for β–lactamases.[5] Similar to peptidases, there are different types of enzymes which catalyse the hydrolysis of β–lactams. Mechanistically these are serine and zinc enzymes, similar to, say, α–chymotrypsin and carboxypeptidase, respectively. Based on their primary structures there have been three classes recognised, A, B and C.[6] Class B is

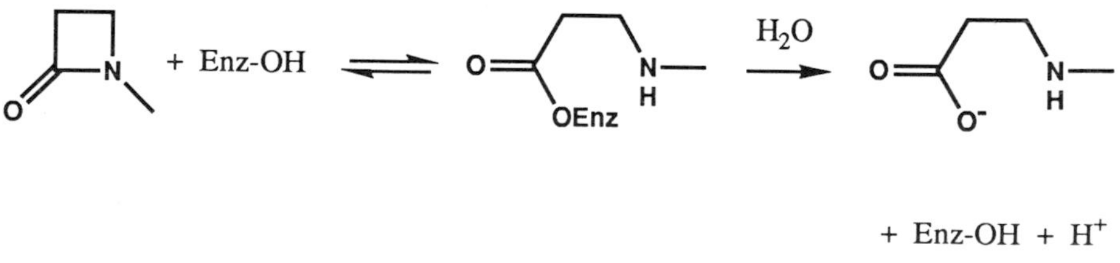

+ Enz-OH ⇌ O=C(OEnz)–CH₂–CH₂–NH– —H_2O→ O=C(O⁻)–CH₂–CH₂–NH–

+ Enz-OH + H^+

Scheme 1: The acylation–deacylation mechanism for serine β–lactamases distinguished by the requirement of zinc ions for activity, whereas classes A and C, although structurally distinguishable, act by using an active site serine.[7] The hydrolysis of penicillins and cephalosporins catalysed by serine β-lactamases is thought to occur by a two step acylation-deacylation mechanism (Scheme 1). In the serine proteases the active site histidine is apparent from the pH-rate profile and shows a pK_a of 7. The pH-rate profile for the serine β-lactamases shows a very broad bell-shaped pH dependence which is apparently independent of substrate. The classic kinetic model for describing this is given in equation 3:

$$k_{cat}/K_m = k_{max}/(1 + H^+/K_1 + K_2/H^+) \qquad \text{(eqn 3)}$$

where the enzyme has two ionising groups with ionisation constants K_1 and K_2 for the dissociation of EH_2 and EH_1, respectively. In the active form of the enzyme one group is in its unprotonated form and the other is in its protonated form, *ie* EH. The pH-independent second order rate constant for the reaction of EH and the substrate is given by k_{max}. The apparent pK_a's for the *Bacillus cereus* β-lactamase A are 5.6 and 8.6. It has not been possible to deduce whether these acid/base groups on the enzyme are mechanistically important.

3 REACTIVITY OF β-LACTAMS TO NUCLEOPHILES

Until recently, a basic tenet of the chemistry of β-lactam antibiotics had been that their presumed enhanced reactivity was due either to strain in the four membered ring or to reduced amide resonance. It is now generally accepted that the chemical reactivity of β-lactam antibiotics is not unusual and that there is little evidence to support the importance of strain or reduced amide resonance as major factors in determining either their biological or chemical reactivity.[8] The four membered ring does not open readily and the strain energy of the ring does not significantly facilitate C-N bond cleavage.[9]

The attack of nucleophiles on the β-lactam carbonyl carbon to form the tetrahedral intermediate **(3)** is reversible and expulsion of the attacking group Nu often occurs faster than cleavage of the

β-lactam C–N bond.[8,10] If the attacking atom of the nucleophile is bonded to an ionisable proton then general base catalysis is required to facilitate the reaction by removing this proton. Cleavage of the β-lactam C–N bond is also facilitated by general acid catalysis as in structure **(4)**.[8,10] Despite the assumed high reactivity of β-lactams and the release of strain that occurs upon ring opening, cleavage of the β-lactams involves similar mechanistic features to the hydrolysis of peptides.

The alcoholysis of penicillin occurs through the alkoxide of the alcohol and there is no evidence for the reaction proceeding by general base catalysed addition of the neutral alcohol. The rate limiting step for the alcoholysis is breakdown of the tetrahedral intermediate **(4)**.[11] The tetrahedral intermediate **(5)**, formed by attack of the β-lactamase serine on penicillin, is therefore likely to require a general base to remove the proton from oxygen and probably a general acid to protonate the β-lactam nitrogen before the acyl enzyme **(6)** can be formed. In chymotrypsin these two functions are carried out by the same histidine residue. It is possible that with β-lactamase these two proton transfers are carried out by two separate residues **(7)**.

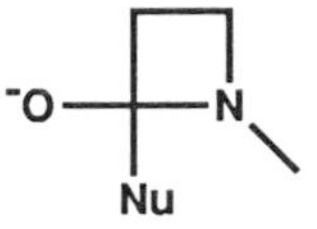

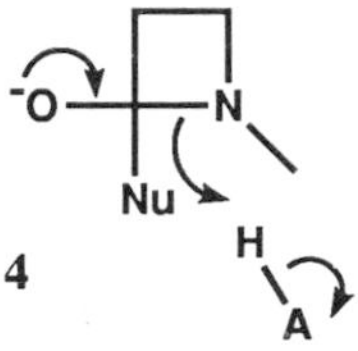

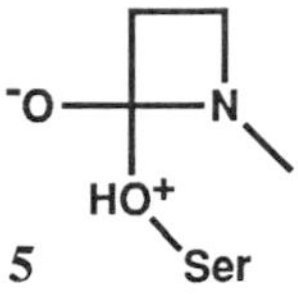

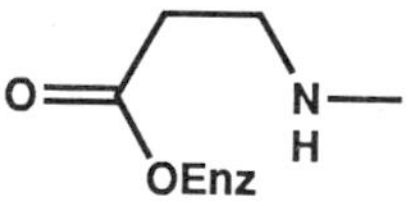

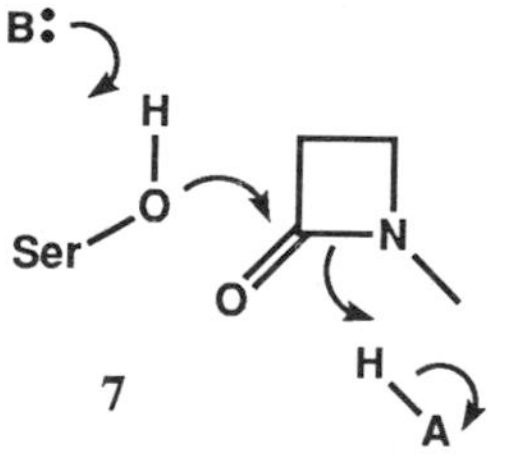

4 MECHANISMS OF REACTIONS OF β–LACTAM ANTIBIOTICS WITH β–LACTAMASES

The solvent isotope effect (SIE) on k_{cat}/K_m for the acylation of the classic serine protease α–chymotrypsin, is near unity, whereas it is between 2 and 3 for the deacylation of acylchymotrypsins.[12] The solvent isotope effect on k_{cat}/K_m for the β–lactamase–catalysed hydrolysis of β–lactam antibiotics is about unity for a variety of substrates, with individual effects on k_{cat} and K_m varying between 1 and 2. Although the SIE on k_{cat} is indicative that proton transfer occurs in the rate limiting step, it is not of the magnitude to allow this conclusion to be drawn firmly. Chemical modification studies have so far been unsuccessful in showing unequivocally the presence of an essential group, other than serine, in any β–lactamase. It is still not clear therefore, whether the ionisable groups indicated by the pH rate profiles are simply required for binding or have a mechanistic role.

Recently it has been shown that β–lactamase I from *B cereus* is inactivated by a water–soluble carbodi–imide and it is considered that this is mainly due to the conversion of the carboxy group of glutamic acid–168 into an amide.[13] It is not unambiguous that this carboxy group has a catalytic function because glutamic acid–168 is not conserved in all β–lactamases, and its replacement by aspartic acid by mutagenesis has little effect on the enzyme's activity.[14] On the other hand, replacement of the conserved glutamic acid–166 by glutamine gives enzyme with little or no activity.[14] The pH dependence of k_{cat}/K_m for the *B cereus* β–lactamase I–catalysed hydrolysis of cephaloridine is bell–shaped and indicates ionisable groups on the enzyme of pK_a 5.60 and 8.64. The lower pK_a could correspond to a carboxy group in a slightly non–polar environment.

Crystallographic studies of the class A β–lactamases have been in progress for many years with little result until very recently.[15] It has been suggested, on the basis of the crystal structure of the β–lactamase from *Staphylococcus aureus*, that a lysine residue acts as a general base catalyst.[16] This seems extremely unlikely because the lysine amino group will almost certainly be protonated at neutral pH.

A commonly accepted assumption about the molecular recognition by enzymes of β–lactam antibiotics is the necessity of an anionic group, such as carboxylate, at C3 in penicillins **(1)** and at C4 in cephalosporins **(2)**. It is also usually assumed that good substrates or inhibitors of β-lactamases also require the presence of a carboxylate, or other anionic residue. This prime recognition site is presumed to have a complementary positively charged residue on the β-lactamase enzyme.

Small ring lactones are constrained to the *E* configuration for C-O-CO and it is possible that lactones may be good mimics of carboxylate residues. This appears to be true for the cephalosporin lactone **(8)**, which is a good substrate for β-lactamase I.[17] Lactone **(8)** is a 50-fold better substrate than an analogous cephalosporin **(2)** with a free carboxylate group at C4. This could be due to an enhanced reactivity of the lactone or to improved binding energy between the substrate and β-lactamase.

The effect of changes in substrate structure on enzyme catalytic activity are often used to identify specific binding sites between parts of the enzyme and substrate. Changes in substrate structure can induce different intrinsic 'chemical' effects such as inductive, resonance and steric ones as well as different extrinsic intermolecular interactions between the substrate and enzyme. An important step towards understanding enzyme catalysis is to separate these effects.[18] One method of determining the 'chemical' effect on structural changes in β-lactam antibiotics on intrinsic reactivity is to determine the susceptibility of the β-lactam towards nucleophilic attack. The magnitude of these effects will depend upon the nature of the nucleophile and comparisons of non-enzyme and enzyme catalysed processes may be complicated by there being different rate limiting steps or even mechanisms for the two processes. Despite these reservations, we choose to use the hydroxide ion catalysed hydrolysis of β-lactams as a measure of the intrinsic effects of substituents.

For the lactone **(8)**, the ratio of k_{cat}/K_m for β-lactamase I to the second order rate constant for hydroxide ion catalysed hydrolysis is 3×10^4. *The rate enhancement for hydrolysis brought about by the enzyme is as great for the lactone (8) as it is for cephalosporins with a carboxylate group at C4.* This indicates that either the C4-carboxylate in cephalosporins is not a primary and necessary recognition site for substrates to show high reactivity with β-lactamases or that the lactone residue contributes a similar binding energy to that of the carboxylate.[18] The former explanation could be the reason for the lower reactivity of cephalosporins compared with penicillins in their β-lactamase catalysed hydrolysis.

The carbonyl group of esters, lactones and carboxylic acids is highly polarised, but it is difficult to quantify this in terms of 'atomic charge'. For example, despite the traditional view of resonance in carboxylic acids and carboxylate anions, it has recently been calculated[19] that the change in relative negative charge on the oxygens upon ionisation of carboxylic acids is only 0.1 to 0.2.

Small ring lactones must adopt the *E* conformation, which for esters is estimated to be between 16 and 35 kJ mol^{-1} less stable than the *Z* isomer.[20] The dipole moment of *Z*-methyl formate is only 1.98D and considerably smaller than the 4.60D for the *E* rotamer.[21] The high value for the dipole moment is similar to the 3.90D and 4.2D reported for γ-butyrolactone.[22] It therefore appears that the two oxygen atoms of the lactone in the cephalosporin **(8)** carry considerable negative charge and could interact with a suitably placed positive charge or a dipole of an α-helix in the enzyme **(9)**. This cannot occur with the *Z* conformation of esters **(10)** and any favourable binding energy between the enzyme and the *E*-conformation of the ester is insufficient to compensate for the unfavourable conformational change (**10** → **11**) as the methyl esters of penicillins and cephalosporins are poor substrates.[17]

In principle, further mapping of the molecular recognition between β-lactams and their substrates is possible with reversible, non-covalent inhibitors. However, until recently no effective inhibitors were known, which could exploit the recently reported *X*-ray structure of a β-lactamase.[16] The methyl ester of (3*S*,5*R*,6*R*)-benzylpenicilloate **(12)**, but not other epimers, is a competitive reversible inhibitor of β-lactamase I from *Bacillus cereus*.[23] It is not an inhibitor of the zinc enzyme, β-lactamase II.[23] Lactamase I does not catalyse either hydrolysis or epimerisation of the ester **(12)**. The inhibition constant, K_i, is 6.70 x 10^{-4} and 6.06 x 10^{-4} M using cephaloridine and benzylpenicillin as the substrates, respectively. The pK_a of the thiazolidine nitrogen conjugate acid, IH, of the methyl ester **(12)** is 3.80. The inhibition constant, K_i, is apparently pH independent between pH 5 and 8. Over this pH range the inhibitor exists predominantly in its free base form, although the concentration of its conjugate acid changes by a factor of 10^3. The simplest interpretation of the pH independence of K_i is that the inhibitor binds in its free base form, **(12)**, to the active form of the enzyme, EH. Other binding modes would be possible if the pK_as of the inhibitor were severely modified upon binding to the enzyme compared with the pK_as in aqueous solution.

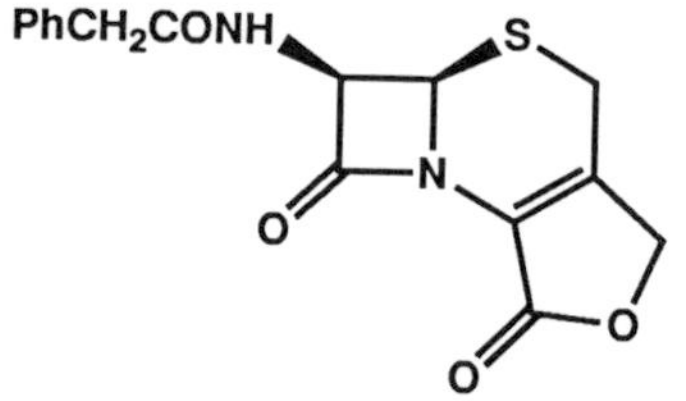

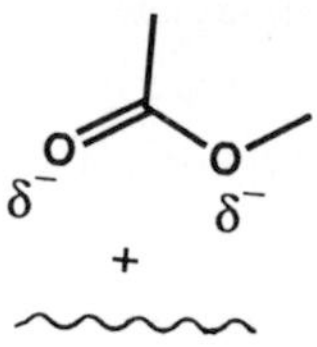

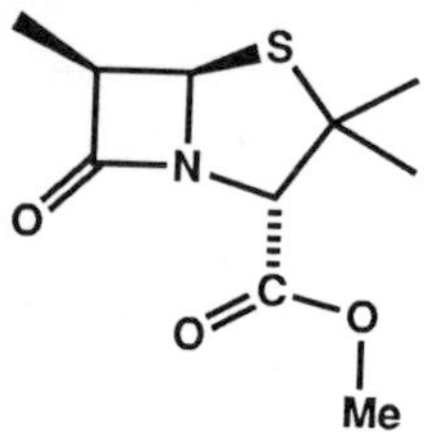

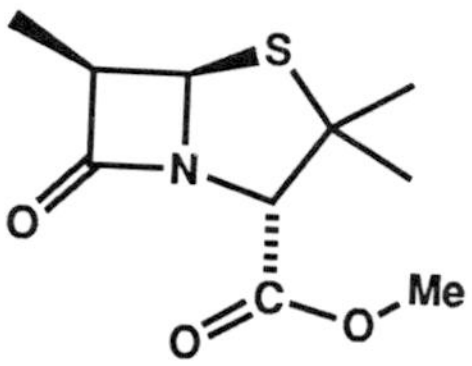

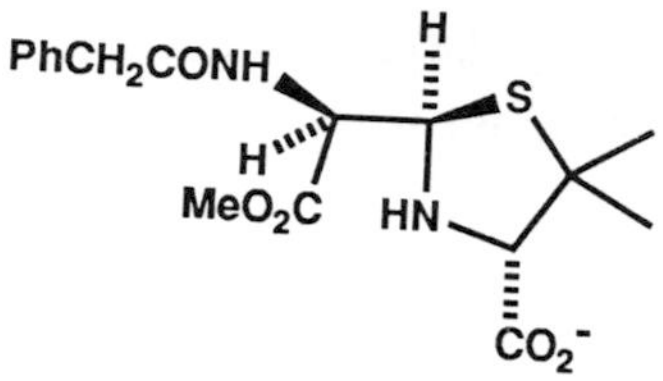

The hydrolysis product of benzylpenicillin, benzylpenicilloate, is a very weak inhibitor of β-lactamase and it is difficult to measure K_i accurately. Using cephaloridine as a substrate the inhibition constant was found to be $>5 \times 10^{-2}$ M. The increased binding of the ester **(12)**, relative to benzylpenicilloate, suggests that the C6 carboxylate residue of the latter may be near an anionic site on the enzyme. The recent *X*-ray crystallographic study[16] of the β-lactamase from *S aureus* suggests that the active site contains, in addition to the catalytically important serine 70 residue, lysine and glutamate residues (Scheme 2). Because the free base form rather than the protonated form of the inhibitor binds to the enzyme this may be indicative that the thiazolidine nitrogen acts as a proton acceptor rather than donor with respect to the protein. This suggests hydrogen bonding to the 73-Lys-NH_3^+ (Scheme 2), which would also be compatible with this group acting as a general acid catalyst in the protonation of the β-lactam nitrogen, thus facilitating C-N bond cleavage.

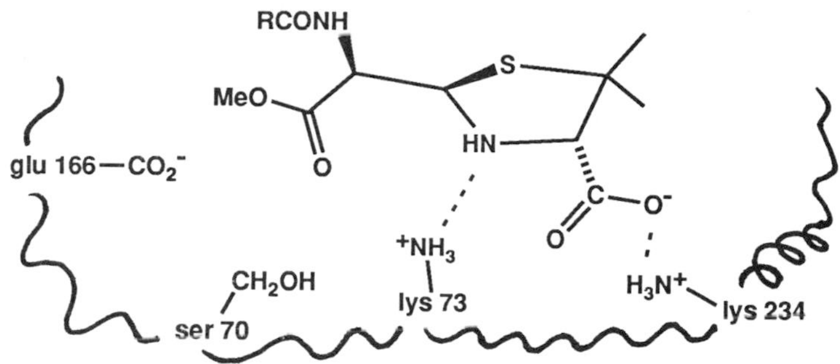

Scheme 2: A possible mode of binding at the active site of serine β-lactamase

Preliminary NOE NMR experiments of the inhibitor **(12)** bound to β-lactamase indicate that the benzene of the C6 benzylamido side chain is close to the C2 methyl groups of the thiazolidine residue. This rather surprising but interesting observation indicates that the inhibitor is bound in a U-shape conformation.

Other recognition sites in β-lactamase which have been investigated are the C6 and C7 side chains in penicillins, **(1)**, and cephalosporins, **(2)**, respectively, and the C3 residues in cephalosporins, **(2)**.[24,25,26]

B cereus β-lactamase I catalyses the hydrolysis of 6-alkylpenicillins with values of k_{cat}/K_m which are at least 50-fold greater than that shown by 6-aminopenicillanic acid and which increase with increasing chain length, reaching a maximum with hexylpenicillin.[24] However, the binding energy of the alkyl group is weak, only 1.46 kJ mol^{-1} per methylene residue, so although there appears to be a recognition site for the amido group there is no specific pocket in β-lactamase I for the recognition of hydrophobic residues in the 6-acylamido side chain of penicillins.

Kinetic parameters have also been determined for the *B cereus* β-lactamase I and β-lactamase II catalysed hydrolysis of a series of cephalosporins substituted in the 7-position.[25] As expected, there is no significant dependence of the rate of the base catalysed hydrolysis upon the nature of the side chain substituent. However, for β-lactamase I k_{cat}/K_m varies over 10^5, whereas for β-lactamase II the variation with substituents is only 10-fold. For alkyl substituents, k_{cat}/K_m increases with chain length and passes through a maximum, which, for β-lactamase I occurs with the undecyl derivative. There is again no evidence for a significant cavity in either enzyme to host aromatic residues. An ionised carboxylate residue on the side chain significantly reduces reactivity with β-lactamase I, which may indicate that the side chain is located near a negatively charged residue in the enzyme.[25]

Finally, the second order rate constants for the alkaline hydrolysis of 3-thiol substituted cephalosporins are independent of the pK_a of the thiol over a pK_a range of 9, which shows that if there is a leaving group at C3' it is expelled after the β-lactam ring is opened and the expulsion of the leaving group does not enhance the rate of β-lactam C-N bond fission.[26] The zinc enzyme β-lactamase II is about a 100-fold better catalyst than the serine enzyme β-lactamase I for the hydrolysis of the same cephalosporin. The second order rate constant k_{cat}/K_m for both β-lactamase enzymes shows no dependence on the nature of the substituent at C3' which is not explicable by the different chemical reactivity of the cephalosporins. There is thus no evidence for a significant recognition site in either enzyme for the C3' substituent and no evidence of the enzyme facilitating departure of the C3 leaving group.[23]

REFERENCES

1 J M T Hamilton-Miller and J T Smith, eds, 'β-Lactamases', Academic Press, New York, 1979.

2 J M Frère, *CRC Crit Rev Microbiol,* 1985, **11**, 299.

3 K Johnson, C Duez, J M Frère, and J M Ghuysen, *Methods Enzymol,* 1975, **43**, 687; J B Nielsen and J O Lampen, *Biochemistry,* 1983, **22**, 4652.

4 P Mantsala and I Suominen, *Acta Chem Scand Ser B,* 1981, **35**, 567.

5 R F Pratt, E G Anderson, and I Odeh, *Biochem Biophys Res Comm,* 1980, **93**, 1266.

6 R P Ambler, *Philos Trans Roy Soc London Ser B,* 1980, **289**, 321; B Jaurin and T Grundstrom, *Proc Natl Acad Sci USA,* 1981, **78**, 4897.

7 V Knott-Hunziker, S Petursson, S G Waley, B Jaurin, and T Grundstrom, *Biochem J,* 1982, **207**, 315; J Fisher, R L Charnas, S M Bradley, and J R Knowles, *Biochemistry,* 1981, **20**, 2726; S A Cohen and R F Pratt, *Biochemistry,* 1980, **19**, 3996.

8 M I Page, *Acc Chem Res,* 1984, **17**, 144; M I Page, *Adv Phys Org Chem,* 1987, 165.

9 M I Page, P Webster, L Ghosez, and S Bogdan, *Bull Soc Chim Fr,* 1988, 272; M I Page, P Webster, S Bogdan, B Tremerie, and L Ghosez, *J Chem Soc Chem Commun,* 1986, 1039.

10 M I Page and P Proctor, *J Am Chem Soc,* 1984, **106**, 3820; N P Gensmantel and M I Page, *J Chem Soc Perkin Trans 2,* 1979, 137.

11 P Proctor, A Davies, and M I Page, unpublished results.

12 M L Bender, C E Clement, F J Kézdy, and H D Heck, *J Am Chem Soc,* 1964, **86**, 3680; M L Bender and F J Kézdy in 'Proton Transfer Reactions', eds E Caldin and V Gold, Chapman and Hall, London, 1975, p 385.

13 C Little, E L Emanuel, J. Gagnon, and S G Waley, *Biochem J,* 1986, **240**, 215; S G Waley, *ibid,* 1975, **149**, 547.

14 P J Madgwick and S G Waley, *Biochem J,* 1987, **248**, 657.

15 B Samraoui, B J Sutton, R J Todd, P J Artymiuk, S G Waley, and D C Phillips, *Nature (London),* 1986, **320**, 378; J A Kelly, O Dideberg, P Carlier, J P Wery, M Libert, P Moews, J R Knox, C Duez, C Frarpont, B Joris, J Dusart, J M Frère, and J M Ghuysen, *Science,* 1986, **231**, 1429.

16 O Herzberg and J Moult, *Science,* 1987, **236**, 694.

17 M I Page and A Laws, *J Chem Soc Perkin Trans 2,* 1989, 1577.

18 M I Page in 'The Chemistry of Enzyme Action', ed M I Page, Elsevier, Amsterdam, 1984, p 1.

19 M R Siggel, A Streitweiser, Jr, and T D Thomas, *J Am Chem Soc,* 1988, **110**, 8022.

20 C E Blom and H Gunthard, *Chem Phys Lett,* 1981, **84**, 367; S Rushkin and S H Bauer, *J Phys Chem,* 1980, **84**, 306; R Huisgen and H Ott, *Tetrahedron,* 1959, **6**, 253; W C Closson, P J Orenski, and B Goldschmidt, *J Org Chem,* 1967, **32**, 3160.

21 K Wiberg and K E Laidig, *J Am Chem Soc,* 1987, **109**, 5935.

22 L H L Chia, H H Huang, and Y F Wong, *J Chem Soc B,* 1970, 1138; I Wallmark, M H Krackov, S H Chu, and H G Mautner, *J Am Chem Soc,* 1970, **92**, 4447.

23 M Jones, S C Buckwell, M I Page, and R Wrigglesworth, *J Chem Soc Chem Commun,* 1989, 70.

24 S C Buckwell, M I Page, and J L Longridge, *J Chem Soc Perkin Trans 2,* 1988, 1809.

25 S C Buckwell, M I Page, S G Waley, and J L Longridge, *J Chem Soc Perkin Trans 2,* 1988, 1815.

26 S C Buckwell, M I Page, S G Waley, and J L Longridge, *J Chem Soc Perkin Trans 2,* 1988, 1823.

MOLECULAR STUDIES ON THE BIOSYNTHESIS OF ENTEROBACTIN

Christopher Walsh, Jun Liu, and Frank Rusnak

Department of Biological Chemistry and Molecular Pharmacology, Harvard Medical School, Boston, MA 02115, USA

1 INTRODUCTION

Organisms require iron to live and when enterobacteria such as *Escherichia coli* are in iron-deficient environments, they turn on a metabolic pathway (Figure 1a) to biosynthesize an iron-chelating, catechol-containing molecule called enterobactin (also known as enterochelin). This is exported from the cell to scavenge for Fe^{3+}, which it does very successfully, with an estimated K_F of 10^{52} M^{-1}. The ferric enterobactin complex is then taken back into the bacterial cell and the iron is released from the siderophore either by Fe^{3+} to Fe^{2+} reduction or by enzymic partial disassembly of the macrocyclic trilactone ring of enterobactin. The *ent* genes, in enterobactin biosynthesis (Figure 1a), the *fep* genes (in iron permeation) and the *fes* gene ('ferric enterobactin' esterase) are clustered together in several regulons on the *E coli* chromosome (Figure 1b). They are transcriptionally regulated by intracellular Fe^{2+} levels by the metalloregulatory Fur (Fe uptake regulation) protein that is a DNA-binding protein that recognizes specific 'iron boxes' in the 5' upstream regions of these genes.[1,2]

The seven step pathway of enterobactin biosynthesis involves enzymes encoded by the *entA-G* genes (Figure 1a) and the first step involves utilization of the metabolite chorismate, a key common precursor to the essential aromatic amino acids phenylalanine, tyrosine and tryptophan, and the folate coenzymes. Diversion of chorismate flux to isochorismate is achieved by isochorismate synthase (*entC*). Isochorismate next undergoes carboxyvinyl ether hydrolysis (*entB*) to the *trans*-dihydrodiol 2,3-dihydro-2,3-dihydroxybenzoate, in turn a

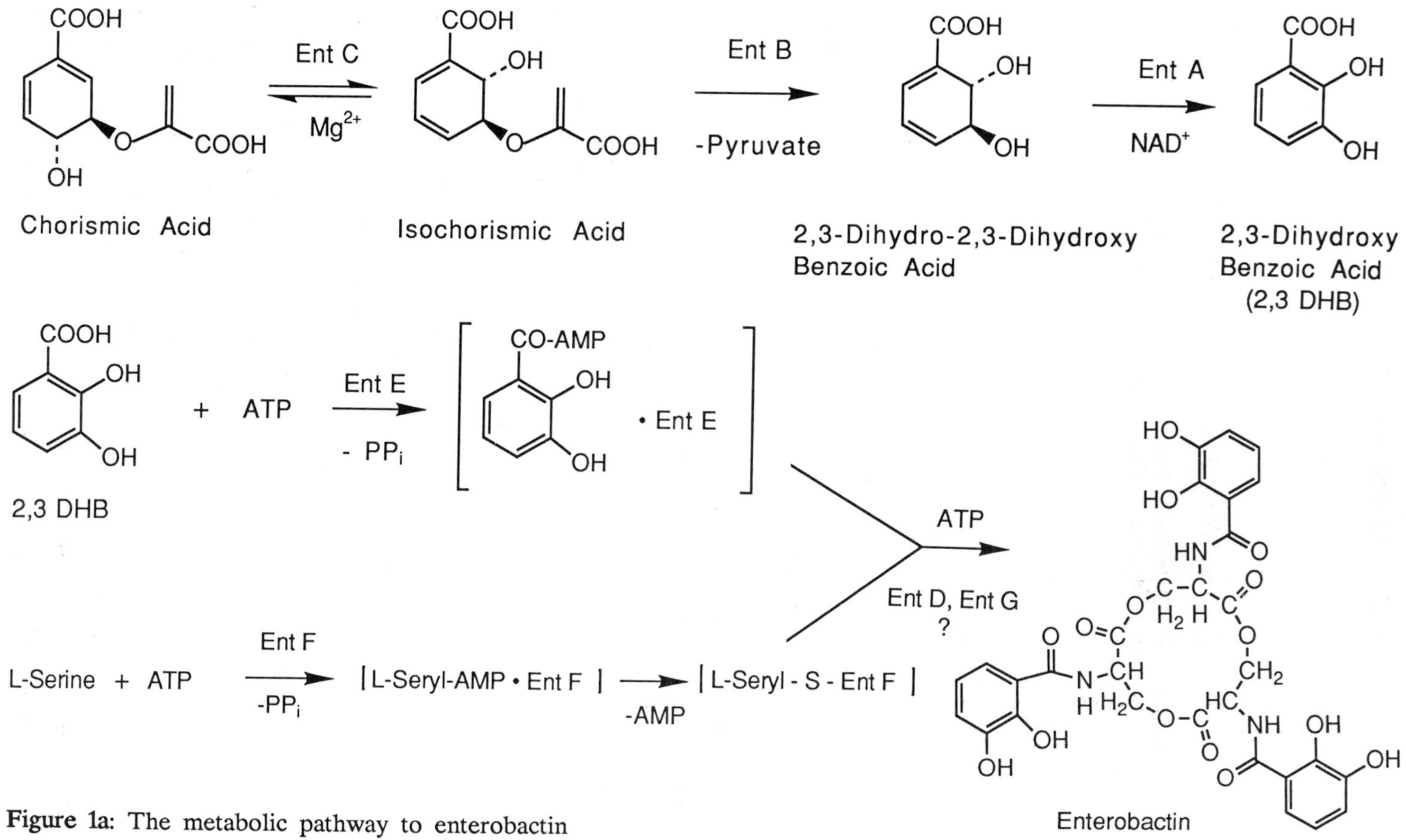

Chorismic Acid

Ent C

Mg^{2+}

Isochorismic Acid

Ent B

-Pyruvate

2,3-Dihydro-2,3-Dihydroxy Benzoic Acid

Ent A

NAD^+

2,3-Dihydroxy Benzoic Acid (2,3 DHB)

2,3 DHB + ATP

Ent E

$- PP_i$

[CO-AMP • Ent E]

L-Serine + ATP

Ent F

$-PP_i$

| L-Seryl-AMP • Ent F |

-AMP

| L-Seryl - S - Ent F |

ATP

Ent D, Ent G ?

Enterobactin

Figure 1a: The metabolic pathway to enterobactin

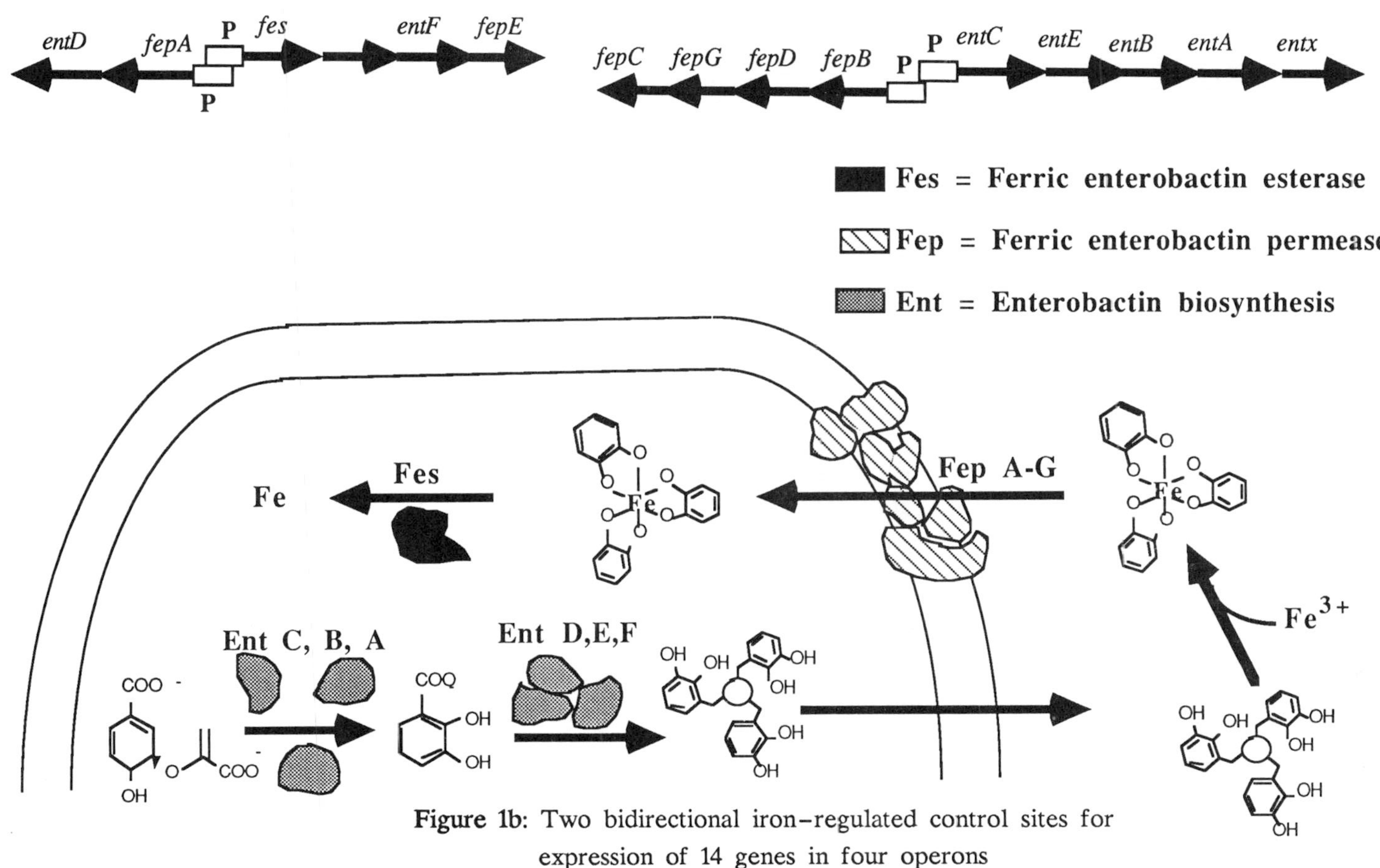

Figure 1b: Two bidirectional iron-regulated control sites for expression of 14 genes in four operons

substrate for an aromatizing dihydrodiol dehydrogenase (*entA*), which constructs the catecholic moiety in 2,3-dihydroxybenzoate, the iron-chelating moiety. The remaining steps are concerned with activation of the carboxyl groups of 2,3-dihydroxybenzoate and L-serine in preparation for amide and ester bond formation, respectively. Three dihydroxybenzoyl-serine amide linkages and three intramolecular serine ester linkages lead to the cyclic trilactone product in which the three catecholic moieties can engage in octahedral coordination of Fe^{3+} (Figure 2). In the ferric enterobactin complex the chirality of the L-serine units in the trilactone scaffolding imposes a Λ helicity at the coordinated Fe^{3+} locus.[3]

2 THE EntC, TrpE, AND PabB ENZYME FAMILY

The chorismate to isochorimate conversion is analogous to the transformation of chorismate to anthranilate (*trpE*, in tryptophan biogenesis) and to *para*-aminobenzoate (*pabB*, in folate biosynthesis), both in mechanistic type (Figure 3) and in structural homology between *entC, trpE,* and *pabB* genes and corresponding enzymes. While EntC releases the dihydroisochorismate product, TrpE holds onto the corresponding amino analogue of isochorismate and carries out aromatization to yield anthranilate by expulsion of pyruvate. The PabB enzyme complex utilizes the corresponding 4-aminodihydro species and presumably generates it during catalysis from an as yet uncharacterized earlier intermediate. The folate and tryptophan pathways are targets for antibacterial and/or herbicidal agents and an understanding of the structure/function issues in the family of enzymes in Figure 3 may lead to improved inhibitor design.

3 CLONING, EXPRESSION, AND PURIFICATION OF EntC, E, B, AND A

The study of the enterobactin pathway needs the enzymes, none of which had been significantly purified, and access to such molecules as chorismate, isochorismate, and 2,3-dihydro-2,3-dihydroxybenzoate. These molecules have been synthesized by Berchtold and colleagues at

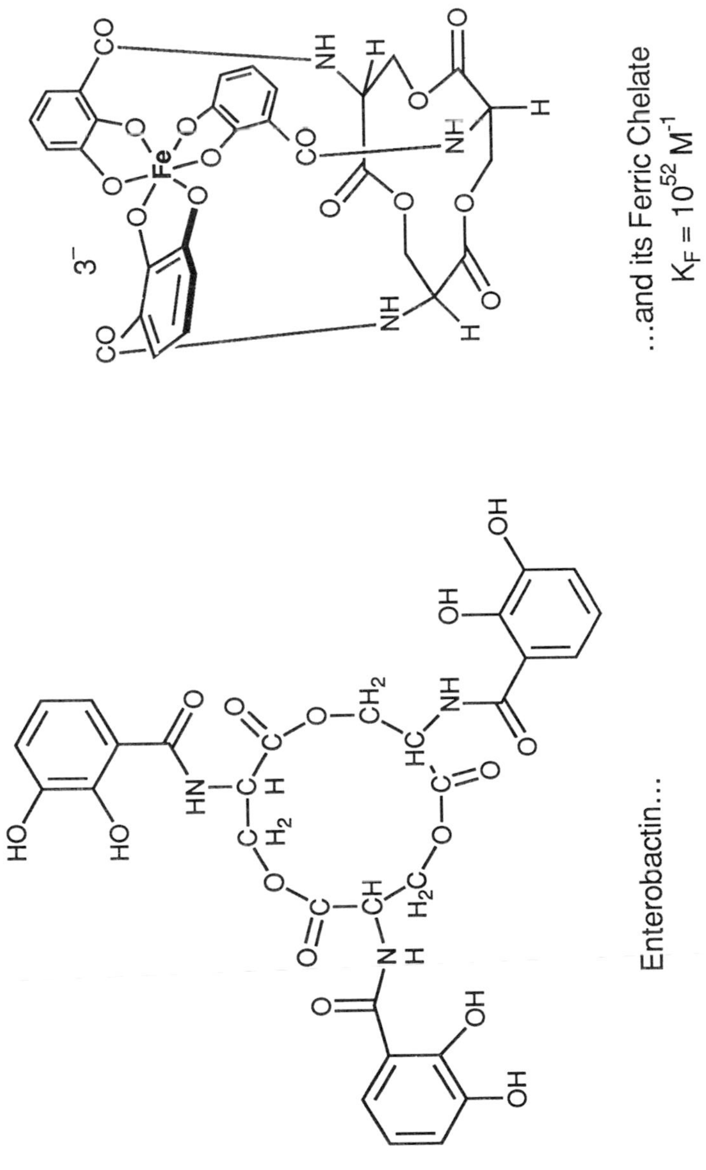

Enterobactin...

...and its Ferric Chelate

$K_F = 10^{52}\ M^{-1}$

Figure 2: The structure of enterobactin and its ferric chelate

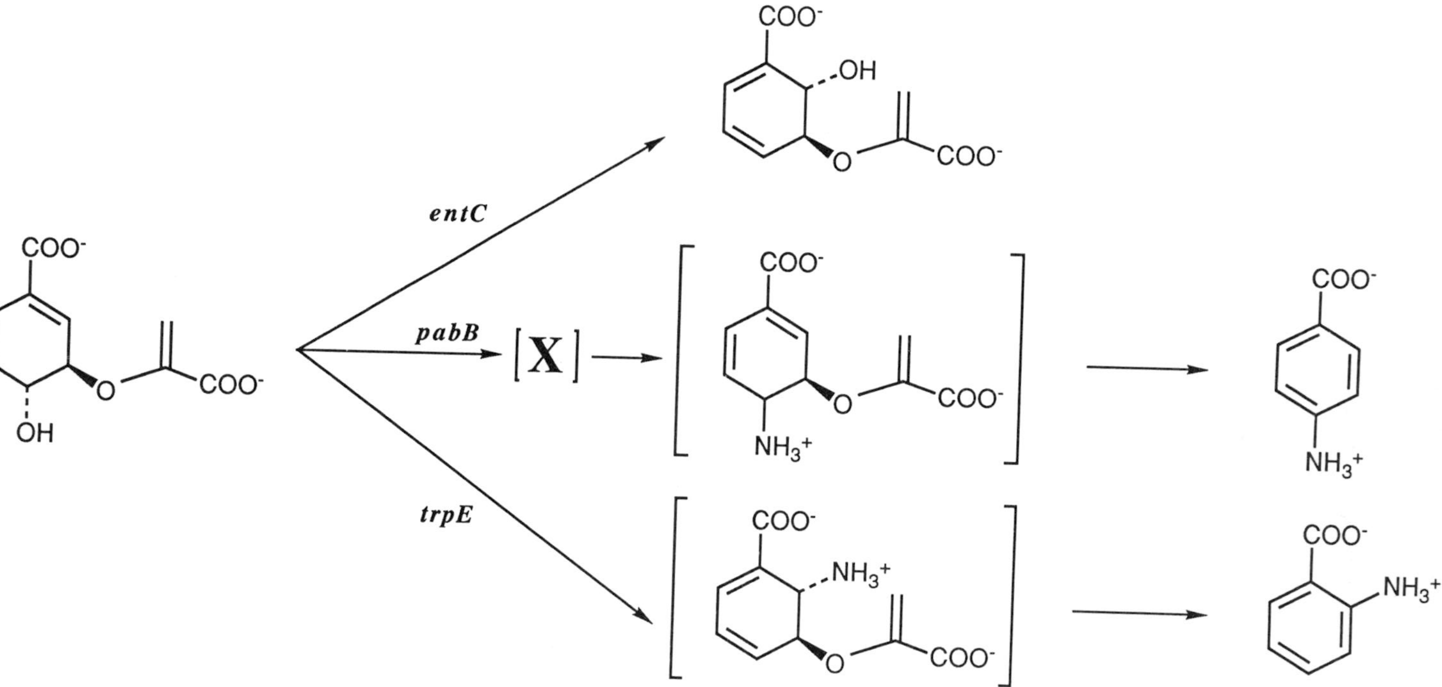

Figure 3: Proposed mechanistic similarity to EntC, PabB, and TrpE, enzyme action

Table 1: Properties of four enzymes involved in 2,3-dihydroxybenzoate biosynthesis and activation.

P/O entC entE entB entA entX

Ent	C	E	B	A
Gene Size (bp)	1173	1611	855	744
Protein Size (daltons)	43,000	59,000	33,000	26,000
Active Form	α	α_2	α_4	α_8
Catalytic Activity (min^{-1})	173	340	410	1520

the Massachusetts Institute of Technology[4,5] and were crucial in development of enzyme assays. Inspection of the *ent* gene order in Figure 1b reveals that the first four genes in the order *entCEBA* comprises one operon. We have subcloned this DNA, participated in its sequence determination, developed methods for efficient subcloning and expression of each gene, and have succeeded in purifying each of the four encoded enzymes to homogeneity for the first time.[6–8] Figure 4 documents the purity of each enzyme by SDS polyacrylamide analysis. Table 1 summarizes some relevant properties of *entC*, *E*, *B*, and *A* genes and enzymes. At this point one can begin to characterize the molecular aspects of catalysis in each step and salient features of the four enzymes are summarized below.

4 EntC (ISOCHORISMATE SYNTHASE)

With pure EntC enzyme available, we have been able to develop continuous coupled spectrophotometric assays of enzyme activity in both forward (chorismate to isochorismate) and backward (isochorismate to chorismate) directions and establish ready reversibility.[8] Steady state kinetic constants give a K_m of 14μM for chorismate, 5μM for isochorismate and k_{cat} values of 173 min^{-1} and 108 min^{-1} for k_{cat}/K_m values of 2 x 10^5 M^{-1} s^{-1} and 2.5 x 10^5 M^{-1} s^{-1}, respectively. These very similar catalytic efficiency ratios

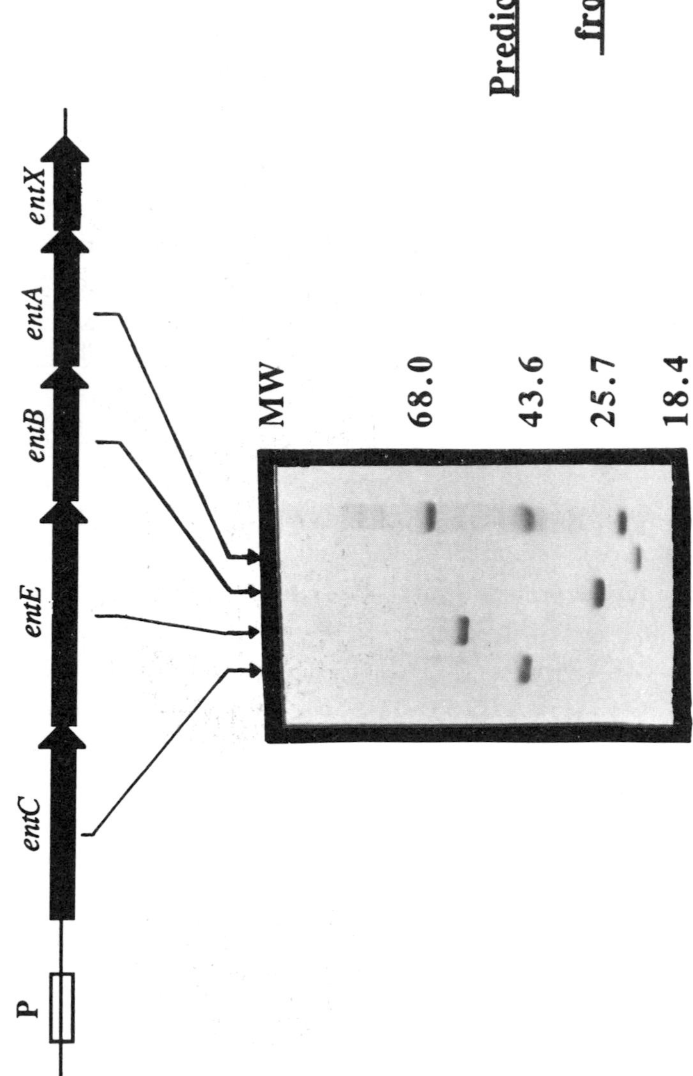

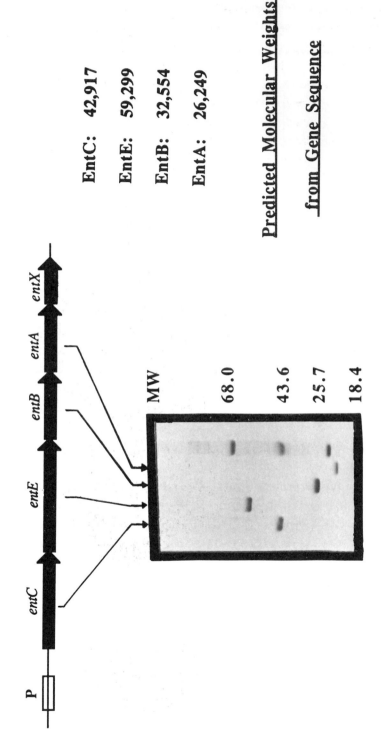

Figure 4: Sodium dodecyl sulphate–polyacrylamine gel electrophoresis of four purified enzymes in the *EntCEBA* operon

suggest equivalent pseudosymmetric recognition of each substrate isomer by the enzyme. In some agreement with this idea, a K_{eq} of 0.6, quite close to unity, was determined both by application of the Haldane equation and by 500 MHz ^1H NMR analysis of enzyme-catalysed approach to equilibrium. *Inter alia,* this suggests only a slight ΔG difference of 0.36 kcal mol^{-1} between chorismate and isochorismate structures. Enzymic incubation in $H_2{}^{18}O$ established that isomerization occurs with incorporation of a hydroxyl group from the solvent (eqn 1):

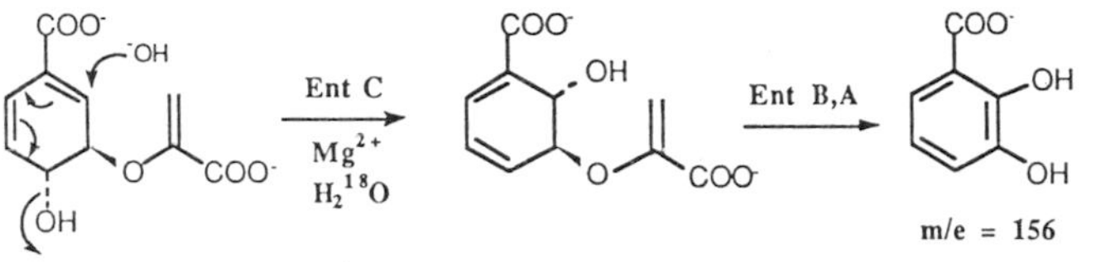

Ent C, Mg^{2+}, $H_2{}^{18}O$ → Ent B,A → m/e = 156 (eqn 1)

This result limits the mechanistic possibilities. Studies are underway to determine whether the mechanism of this and the analogous TrpE-mediated conversion is a concerted 1,5-addition/elimination or a stepwise process (addition of ^-OH followed by elimination of ^-OH or *vice versa*).

5 EntB (ISOCHORISMATASE)

The availability of synthetic[4] and now natural isochorismate (*via* preparative EntC incubations) has allowed us to characterize isochorismatase, the second enzyme in enterobactin construction.[7] This enzyme also catalyses a somewhat unusual biochemical reaction, the hydrolysis of the enolpyruvyl ether side chain, presumably *via* a tetrahedral intermediate as shown (eqn 2):

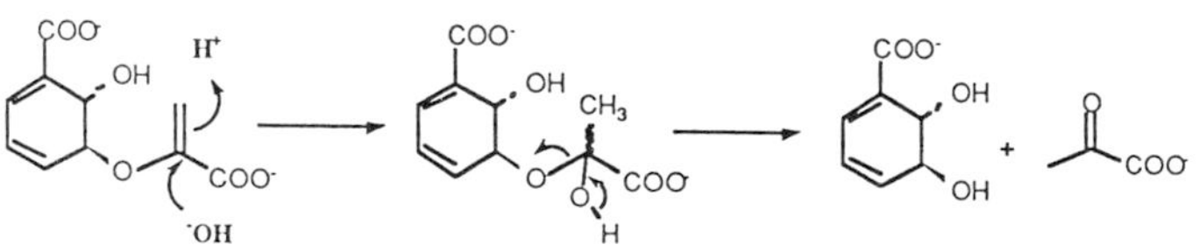

(eqn 2)

Two other enolpyruvyl transferases are known in bacterial metabolism. One is 5-enolpyruvylshikimate-3-phosphate (EPSP) synthase, which operates earlier in the chorismate biosynthetic pathway, and the other is the transferase catalyzing attack of the 3'-OH of UDP-*N*-acetylglucosamine on phosphoenolpyuvate in the first committed step in bacterial peptidoglycan assembly. The first enzyme is the target for the herbicide glyphosphate (*N*-phosphonomethylglycine) and the second for the antibiotic phosphonomycin. This class of enzymes therefore merits further mechanistic examination. While isochorismate is clearly the physiological substrate for EntB, we have determined[7] that chorismate is slowly converted into *trans*-3,4-dihydro-3,4-dihydroxybenzoate, a known natural product produced by bacteria under iron deficiency.

6 EntA (2,3-DIHYDRO-2,3-DIHYDROXYBENZOATE DEHYDROGENASE)

We have reported[6] the purification of this enzyme to homogeneity and observed a turnover number of 1500 min^{-1}. Oxidation of the dihydrodiol by this NAD-linked dehydrogenase likely involves the oxidation of either the C_2- or C_3-hydroxyl to give a hydroxycyclohexadienone which isomerizes non-enzymatically to the aromatic catechol tautomer. Recently we have obtained results which indicate that oxidation occurs at the C_3-hydroxyl, since EntA-catalyzed oxidation by NAD^+ occurs with two alcohols lacking the hydroxyl at C_2. These were generated *in situ* from compounds containing the enolpyruvylether side-chain at C_3 by the action of isochorismatase (EntB) (eqns 3 and 4):

COO⁻ ... O ... COO⁻ —Ent B→ COO⁻ ... OH —Ent A (NAD⁺ → NADH)→ COO⁻ ... O (eqn 3)

COO⁻ ... O ... COO⁻ —Ent B→ COO⁻ ... OH —Ent A (NAD⁺ → NADH)→ COO⁻ ... O (eqn 4)

Trans-dihydrodiol dehydrogenases are found in mammalian liver metabolism[9] while *cis*-glycol specific dehydrogenases are known in bacterial conversions of such substrates as benzene and toluene into the corresponding catechols.[10]

7 EntE (2,3-DIHYDROXYBENZOATE-AMP LIGASE)

After dihydroxybenzoate has been produced from chorismate by consecutive action of EntC, B, and A, the remaining tasks in assembly of enterobactin are construction of three amide bonds (dihydroxybenzoylserine) and then three head-to-tail serine ester linkages in macrocyclic trilactone formation, probably in that order. Amide formation requires activation of three dihydroxybenzoate molecules, while serine ester condensation requires activated L-serine species. ATP is employed to generate mixed acylphosphate anhydrides with EntE acting on dihydroxybenzoate and EntF on L-serine. Attack of substrate carboxylates on the α-phosphate of ATP yields the enzyme bound acyladenylate (acyl-AMP) species. We have validated this expectation with pure EntE and shown that the enzyme retains dihydroxybenzoyl-AMP, presumably waiting for transfer to a serine moiety at the active site of EntF.[11]

Purification of EntF is underway but earlier reports on partially purified EntF[12] suggest that the initial L-seryl-AMP is converted into a covalent L-seryl-X-EntF species, imposing a directionality to the enterobactin scaffolding. Thus, one expects that an EntE-DHB-AMP, EntF-X-ser complex will engage in directional transfer of the activated DHB moiety to the seryl-NH_2 group and lead to amide formation and DHB-serine covalent tethering in the active site of EntF. Presumably this is iterated twice and then the three DHB-seryl-EntF intermediates are converted into an enzyme bound linear trimeric ester and cyclized to enterobactin before release, under the aegis of the still uncharacterized EntD enzyme.

The molecular architecture of enzyme complexes that assemble macrocyclic lactones, depsipeptides, or peptides is not known in any detail, and the later steps of enterobactin assembly provide an opportunity for genetic and biochemical structure/function studies of a representative member of these macrocycles.

ACKNOWLEDGEMENTS

We would like to thank Dr I G Young of the Australian National University, Dr A McIntosh of the University of Missouri-Columbia, and Dr C F Earhart of the University of Texas-Austin for providing us with several of the plasmids used in this study. J Liu acknowledges a DuPont graduate fellowship from the Massachusetts Institute of Technology. This research was supported by NIH grant GM20011 (C T Walsh) and NIH Postdoctoral Fellowship GM12806-02 (F Rusnak).

REFERENCES

1 B A Ozenberger, T J Brickman, and M A McIntosh, *J Bacteriol,* 1989, **171**, 775.

2 M F Elkins and C F Earhart, *FEMS Microbiol Letts*, 1988, **56**, 35.

3 J B Neilands, T J Erickson, and W H Rastetter, *J Biol Chem*, 1981, **256**, 3831.

4 F R Busch and G A Berchtold, *J Am Chem Soc*, 1983, **105**, 3346.

5 R M DeMarinis, C N Filer, S M Waraszkiewicz, and G A Berchtold, *J Am Chem Soc,* 1974, **96**, 1193.

6 J Liu, K Duncan, and C T Walsh, *J Bacteriol*, 1989, **171**, 791.

7 R Rusnak, J Liu, N Quinn, G A Berchtold, and C T Walsh, *Biochemistry*, 1990, in press.

8 J Liu, N Quinn, G A Berchtold, and C T Walsh, *Biochemistry*, 1990, in press.

9 D M Jerina, J W Daley, and B J Witkop, *J Am Chem Soc*, 1967, **89**, 5488.

10 D T Gibson, *Crit Rev Microbiol*, 1971, **1**, 199.

11 F Rusnak, W S Faraci, and C T Walsh, *Biochemistry*, 1989, **28**, 6827.

12 G F Bryce and N Brot, *Biochemistry*, 1972, **11**, 1708.

ANTIBIOTIC BIOSYNTHESIS: ORIGINS AND POSSIBLE MECHANISMS OF FORMATION OF OBAFLUORIN[1]

Richard B Herbert and Andrew R Knaggs

School of Chemistry, The University, Leeds LS2 9JT, UK

1 INTRODUCTION

Obafluorin (1) is a novel β–lactone antibiotic produced by *Pseudomonas fluorescens* (ATCC 39502),[2] that was first isolated during the screening of bacteria for β–lactam natural products. It is highly susceptible to hydrolysis by β–lactamases and was found to possess weak antibacterial activity against a wide range of bacteria. The mode of antibiotic action of obafluorin is unknown but the reported loss of activity on hydrolysis of the β–lactone implicates this functionality in antibiotic action.[2] Alternatively, the presence of the dihydroxybenzoyl moiety and the latent β–hydroxyamino–acid fragment indicates a possible biochemical role as a siderophore.

The β–lactone functionality is rarely encountered among natural products, although when present it is usually associated with biological activity.[2,4] The stereochemistry of the substituents on the β–lactone ring of obafluorin (1) is interestingly the same as the corresponding stereochemistry in the penicillins and cephalosporins. The *X*–ray crystallographic analysis of (1) reveals a folded structure in which the two aromatic rings lie close to one another.[2]

Like the β–lactone functionality, the nitro group seen in obafluorin (1) is rarely found amongst natural products.[3] Chloramphenicol (4) is a clinically useful antibiotic which is elaborated by *Streptomyces venezuelae*; it contains such a nitro group and inspection of the structures (1), *cf* (3), and (4), *cf* (5), for obafluorin and chloramphenicol reveal further structural similarities. However, whereas (4), *cf* (5), has an aromatic residue with a C_3 side chain, the corresponding fragment in (1) is intriguingly C_4, *cf* (3).

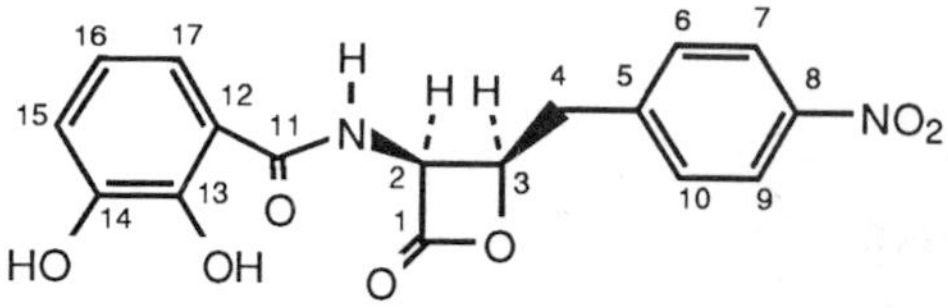

(1) Obafluorin

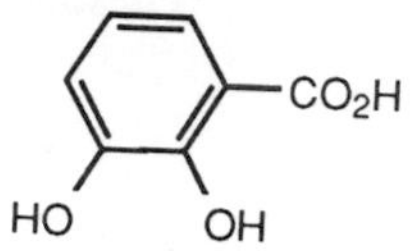

(2)

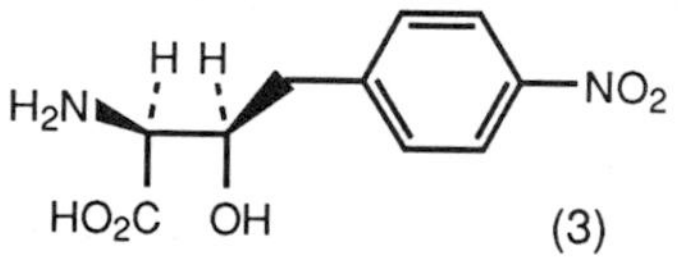

(3)

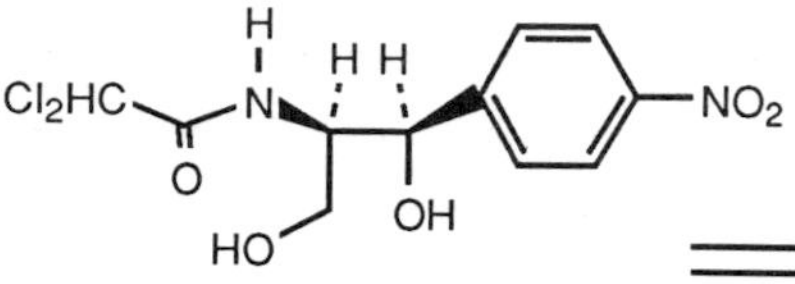

(4) Chloramphenicol

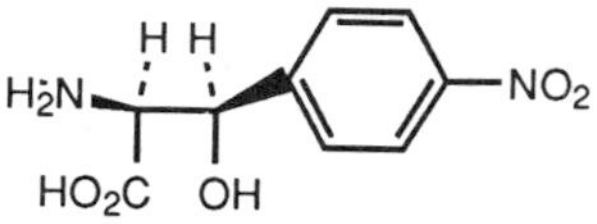

(5)

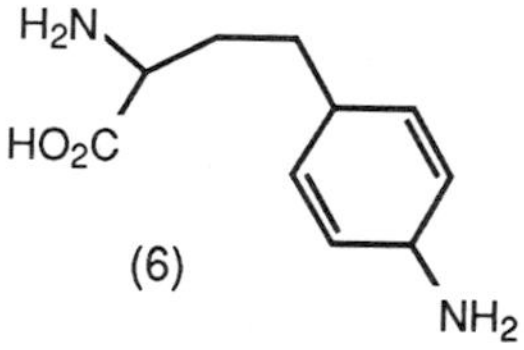

(6)

It is interesting to note further that a number of antibiotics is known which are based on amiclenomycin (**6**),[5] with a C_4 side chain, and some of these antibiotics are elaborated by *S venezuelae*. In summary, the unusual structural features found within obafluorin (1) make it a suitable metabolite for biosynthetic study.

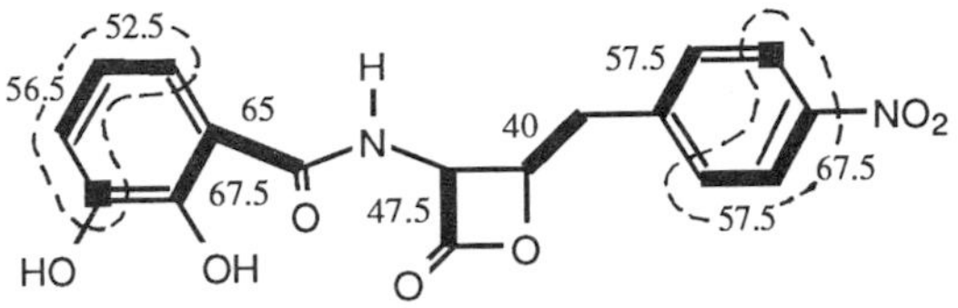

Figure: Labelling of obafluorin (1) by [U-^{13}C]glucose; the numbers are the observed $^{13}C-^{13}C$ coupling constants in Hz.

2 BIOSYNTHESIS OF OBAFLUORIN: 4-AMINOPHENYLALANINE AS A KEY PRECURSOR

Several biosynthetic mechanisms can be sketched to account for the unusual fragment (3) of obafluorin (1). D-[U-^{13}C]Glucose has been used ingeniously in several cases to define the separate units which are used to construct a complete metabolite *in vivo*.[6] We have used D-[U-^{13}C]glucose (each carbon atom : 98.2% ^{13}C) to narrow the number of possible ways in which obafluorin could be biosynthesised. Thus, an aqueous solution of D-[U-^{13}C]glucose was added to *Ps fluorescens* cultures after 8 and 10 hours of growth and the obafluorin (0.93% incorporation) was isolated after a further 3 hours. The antibiotic was purified by HPLC (Polymer Laboratories PLRP-S100 column using $CH_3CN:H_2O$, 1:1) and it was analysed by ^{13}C NMR spectroscopy at 100 MHz. The separate units which constitute the antibiotic could be deduced clearly (see Figure) from the observed couplings and the pattern in the two aromatic rings is consistent with the expected biosynthesis *via* the shikimate pathway (Scheme).[6-8] The occurrence of an intact C_3 unit (see Figure) for C-11 through C-13 of (1) confirmed that the dihydroxybenzoyl moiety (2) ≡ (11) is formed by the known pathway from shikimic acid (8) *via* isochorismic acid (10).[9]

Notably C-3 and C-4 of obafluorin (1) (see Figure) constitute a C_2 unit as (less clearly) do C-1 and C-2. The pattern for C-3 through C-10 is that expected of biosynthesis *via* an aromatic amino acid such as phenylalanine. Chloramphenicol (4) is known to arise from 4-aminophenylalanine (**14**) rather than phenylalanine, *ie* the aromatic ring is aminated at a stage prior to aromatisation [other evidence indicates that amination occurs on chorismic acid (**9**)].[10] A similar pathway *via* 4-aminophenylalanine (**14**) is thus suggested for obafluorin (**1**).

Nitration[11] of L-[2,6-$^3H_2$]phenylalanine gave L-[2,6-$^3H_2$]4-nitrophenylalanine [as (**13**)], which upon catalytic hydrogenation gave L-[2,6-$^3H_2$]4-aminophenylalanine (**14**). This was found to be a very efficient precursor (10.43%) for obafluorin (**1**), whilst L-[2,6-$^3H_2$]4-nitrophenylalanine and L-[2,6-$^3H_2$]phenylalanine were poor precursors (0.17% and 0.2%, respectively). We conclude therefore that L-4-aminophenylalanine (**14**) is a key precursor in the biosynthesis of obafluorin (**1**) and, as found for chloramphenicol (**4**), neither phenylalanine nor 4-nitrophenylalanine are involved in the antibiotic's biosynthesis. It is clear that, as for chloramphenicol:

(i) oxidation of the aromatic amino group only occurs after modification elsewhere in (**14**) and,

(ii) amination occurs at a pre-aromatic stage [isochorismate (**10**)/chorismate (**9**)?].

The congruence between the biosynthesis of chloramphenicol and obafluorin in *Streptomyces* and *Pseudomonas* species, respectively, is notable.

The simplest prediction of the labelling patterns in the two aromatic rings of obafluorin following the incorporation of [U-^{13}C]glucose is as shown in (**12**); the C_4 units are each provided by way of an intact molecule of erythrose 4-phosphate (**7**). The pattern observed was, however, largely C_3 + C_1 for both aromatic rings (see dotted lines in Figure). Although this could be attributed to cycling through the pentose phosphate pathway,[8] it is well established that *Pseudomonas* species catabolise glucose mainly *via* the Entner-Doudoroff pathway[12] (instead of by glycolysis) which involves an obligatory split of glucose into two C_3 units. Hence, ultimately the synthesis of erythrose 4-phosphate through the pentose phosphate pathway would be from C_3 units and would give rise to a net C_3 + C_1 pattern in erythrose 4-phosphate, *ie* as in (**15**), and thus in the two aromatic rings of obafluorin, *cf* Scheme and Figure. Our results then are consistent with the functioning of the Entner-Doudoroff pathway as a major route from glucose.

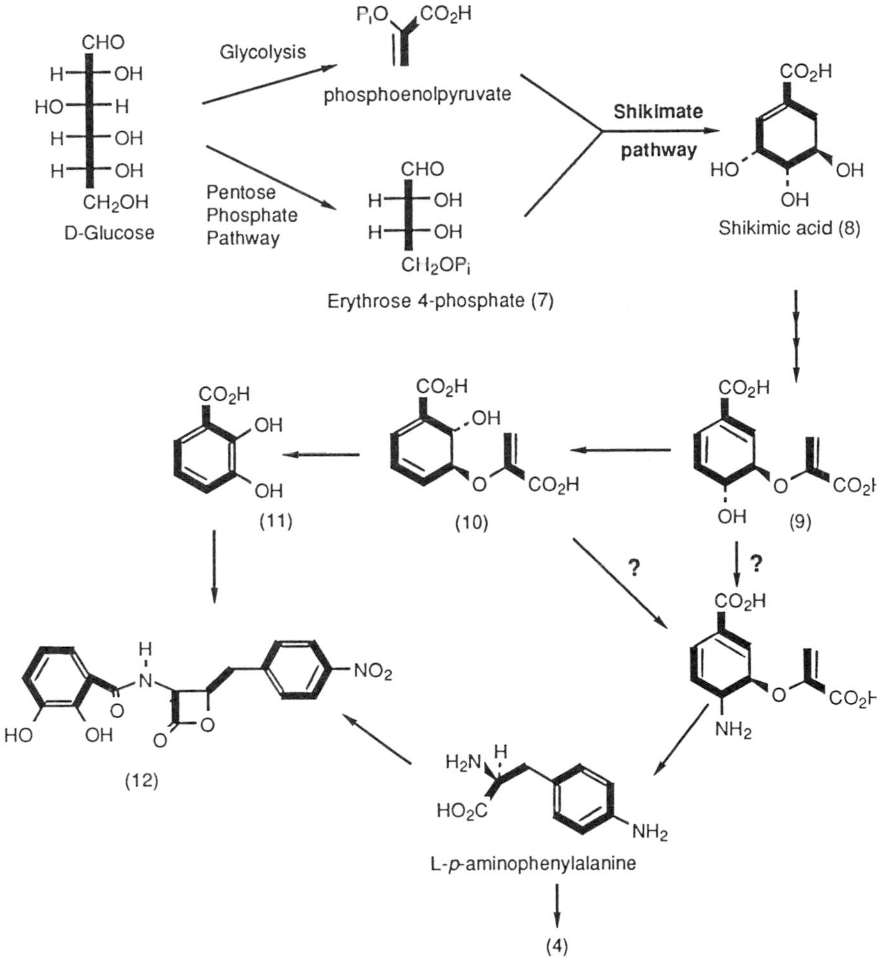

Scheme: Predicted labelling pattern of obafluorin (12) derived from D-[U-^{13}C]glucose.

3 FROM 4-AMINOPHENYLALANINE TO OBAFLUORIN

What of the steps of biosynthesis beyond 4-aminophenylalanine which lead to obafluorin (1)? We have found that 4-nitrophenylacetic acid (16) and the corresponding alcohol (20) are co-metabolites with obafluorin in *Ps fluorescens* and the structures of these metabolites seem to point to what these steps of biosynthesis are. We can hypothesise that the acid (21), very possibly as its CoA ester (22), or

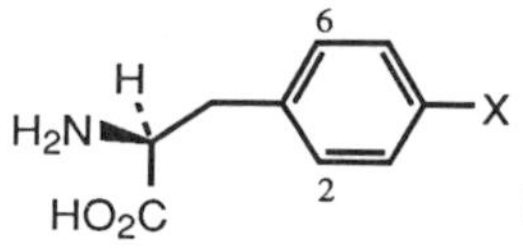

(13) X=NO_2
(14) X=NH_2

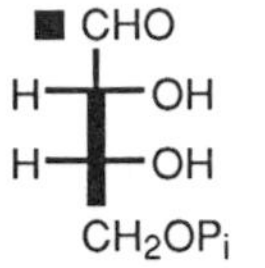

(15)

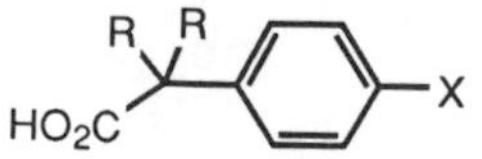

(16) X=NO_2 , R=H
(17) X=NH_2 , R=H
(18) X=NO_2, R=^2H
(19) X=NH_2, R=^2H

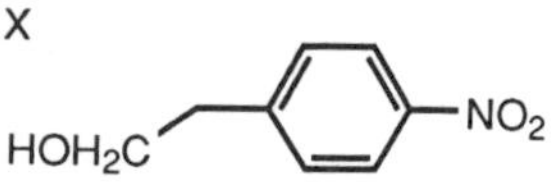

(20)

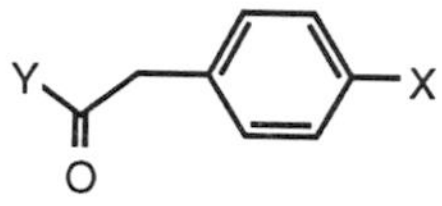

(21) Y=OH, X=NH_2 or NO_2
(22) Y=CoA-S-, X=NH_2 or NO_2
(23) Y=H, X=NH_2 or NO_2

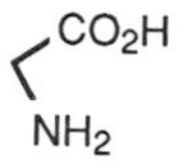

(24)

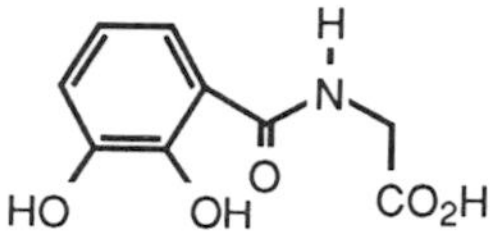

(25)

the aldehyde (**23**), condenses in a pyridoxal-phosphate mediated reaction with a C_2 unit, *eg* glycine (**24**).

So far with cultures of *Ps fluorescens*, we obtained negative results with ^{14}C-labelled glycine, serine, pyruvate, and acetate. These findings may be because at the time the feeds had to be done the cultures were actively growing. Thus, the use of cell-free preparations may give more useful biosynthetic data. It is possible that 2,3-dihydroxybenzoylglycine (**25**) rather than glycine is a precursor for obafluorin and we are currently examining this possibility.

Like obafluorin, 4-nitrophenylacetic acid (**16**) is biosynthesised from L-4-aminophenylalanine (7.44% incorporation, as against 0.29% for the corresponding nitro-compound). However, neither of the deuteriated compounds (**18**) and (**19**) afforded labelled obafluorin (**1**). Interestingly, the 4-nitrophenylacetic acid which was isolated in the experiment with (**19**) (98% $^2H_2$, 2% $^2H_1$) contained deuterium, but label had been lost to a large extent (13% $^2H_2$, 31% $^2H_1$), whereas acid isolated in the experiment with (**18**) retained most of the deuterium label. The benzylic hydrogen atoms in (**16**) are obviously much more acidic (exchangeable) than those in (**17**). We conclude that (**19**), but not (**18**), is transported into the cells where it is oxidised to (**18**), which suffers exchange within the cells to yield

4-nitrophenylacetic acid with a low level of label; extracellular acid does not undergo exchange. Drawing together the various strands of evidence we hypothesise that it is (**22**) or (**23**) formed directly from (**14**) which is involved in the biosynthesis of obafluorin (**1**) and which condenses with a C_2 unit. No doubt there will be many more interesting findings before the full detail of the biosynthesis of obafluorin is known.

REFERENCES

1 Part of this work has been published previously: R B Herbert and A R Knaggs, *Tetrahedron Lett*, 1988, **29**, 6353.

2 A A Tymiak, C A Culver, M F Malley, and J K Gougoutas, *J Org Chem,* 1985, **50**, 5491; J S Wells, W A Trejo, P A Principe, and R B Sykes, *J Antibiot*, 1984, **37**, 802.

3 'Dictionary of Antibiotics and Related Substances', ed B W Bycroft, Chapman and Hall, London, 1989.

4 H Tomoda, H Kumagai, Y Takahashi, Y Tanaka, Y Iwai, and S Omura, *J Antibiot*, 1988, **41**, 247.

5 A Kern, U Kabatek, G Jung, R G Werner, M Poetsch, and H Zähner, *Liebigs Ann Chem*, 1985, 877 and refs cited therein.

6 *eg* K L Rinehart Jr, M Potgieter, D L Delaware, and H Seto, *J Am Chem Soc*, 1981, **103**, 2099; S J Gould and D E Cane, *ibid*, 1982, **104**, 343 (see also W R Erickson and S J Gould, *ibid*, 1987, **109**, 620 and refs cited); K L Rinehart Jr, M Potgieter, and D A Wright, *ibid*, 1982, **104**, 2649.

7 R B Herbert, 'The Biosynthesis of Secondary Metabolites', Chapman and Hall, London, 1989, 2nd edition.

8 L Stryer, 'Biochemistry', W H Freeman, New York, 1988, 3rd edition.

9 U Weiss and J M Edwards, 'The Biosynthesis of Aromatic Compounds', Wiley, New York, 1980.

10 C-Y P Teng, B Ganem, S Z Doktor, B P Nichols, R K Bhatnagar, and L C Vining, *J Am Chem Soc*, 1985, **107**, 5008 and refs cited therein.

11 E Erlenmeyer and A Lipp, *Liebigs Ann Chem*, 1883, **219**, 213.

12 T G Lessie and P V Phibbs Jr, *Ann Rev Microbiol*, 1984, **38**, 359.

STEREOCHEMICAL COURSE OF THE BAKER'S YEAST HYDROLYSIS–REDUCTION OF 2,3–EPOXYKETONES

Gerda Fouche,[a] R Marthinus Horak,[b] and Otto Meth–Cohn[c]

a Chemistry Department, University of South Africa, Pretoria 0001, Republic of South Africa
b Division of Food Science and Technology, CSIR, Pretoria 0001
c Present address: Sterling Organics Ltd, Newcastle upon Tyne NE3 3TT, UK

1 INTRODUCTION

The increased application that optically active epoxides have recently found in the total synthesis of natural products can be ascribed to the facile preparation of these useful synthons by the Sharpless asymmetric epoxidation of allylic alcohols.[1] However, even this highly successful synthetic procedure has limitations arising from steric problems, particularly when substituents occupy the 1–position of the allylic alcohol. We have attempted to develop an alternative synthetic method for such substituted, allylic alcohols based on the microbial reduction of racemic *trans*–2,3–epoxyketones[2] such as (1) to give the corresponding optically active epoxy alcohols (2). Baker's yeast was chosen for this investigation because of its relaxed substrate specificity and the fact that the required cofactors (NADH or NADPH) are easily recycled.[3] In the course of this study it was found that baker's yeast catalyses the stereospecific hydrolysis–reduction of epoxyketones (3) into triols (4). We now report the results of our investigation of the unusual stereochemical and mechanistic details of this reaction.

2 REACTIONS OF EPOXYKETONES WITH BAKER'S YEAST

In a typical reaction the epoxyketone (1.0 g) was subjected to an actively fermenting suspension of baker's yeast (12.5 g) in water (200 ml) containing 20 g of sucrose and allowed to ferment at 30°C under anaerobic conditions. After 24 hours, additional sucrose (20 g) was added and at the end of the incubation period (48 hours) the

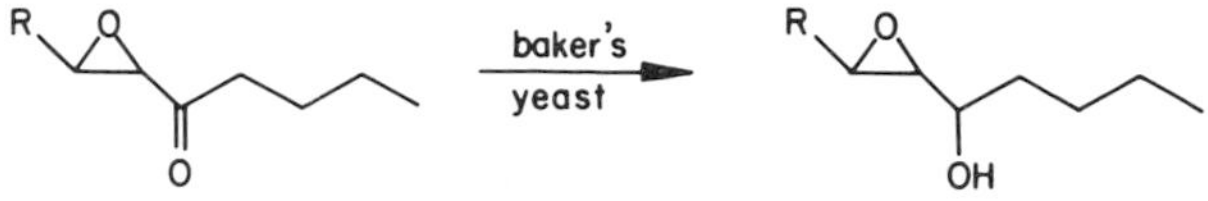

baker's yeast

R
(1) methyl
(13) 4-nitrophenyl

R
(2) methyl
(14) 4-nitrophenyl

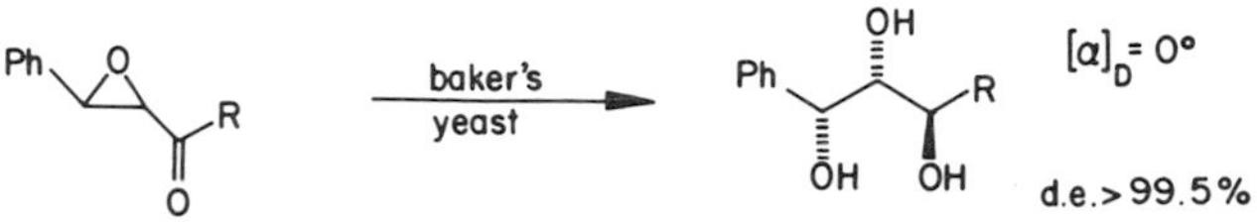

baker's yeast

$[\alpha]_D = 0°$

d.e.>99.5%

R
(3) n-butyl
(5) n-propyl
(6) ethyl
(9) methyl
(10) phenyl

R
(4) n-butyl
(7) n-propyl
(8) ethyl

Table: ^{13}C NMR Data of the Triol (**4**)

Carbon atom	δ_C/ppm[a]
	144.40
Phenyl	128.60
	127.54
	127.43
1	73.43
2	78.90
3	72.90
4	33.78[b]
5	28.66[b]
6	23.44[b]
7	14.37

a Recorded with a Bruker WM-500 spectrometer; solvent acetone-d_6

b May be interchanged

transformed product as well as any unreacted epoxyketone was recovered by continuous extraction of the cell mass with chloroform. According to this procedure epoxyketone (**3**) gave the triol (**4**), 74% yield, as white needles, melting point 156–157 °C and with zero optical rotation. However, despite the lack of optical resolution in this reaction the ^{13}C NMR data (see Table) of the triol indicated that it consists of a single diastereomer. This unusual stereochemical outcome of the baker's yeast catalysed reaction implies that the reduction of the carbonyl group of the two enantiomeric epoxyketones proceeds in the opposite stereochemical sense, but in each case with complete stereospecificity.

A number of epoxyketones were reacted with baker's yeast and it was found that substrates with simple alkyl chains, as opposed to a phenyl ring adjacent to the oxirane are not hydrolysed. *Trans*–2,3–epoxy–octan–4–one (**1**), for example was treated with baker's yeast to give *trans*–2,3–epoxy–4–hydroxyoctane (**2**), in 73% diastereomeric excess at 50% conversion. The hexan– (**5**) and pentan–3–one (**6**) analogues of epoxyketone (**3**) gave the expected triols (**7**) and (**8**), respectively, whereas the butan–3–one analogue (**9**) and the chalcone epoxide (**10**) did not undergo the baker's yeast catalysed hydrolysis–reduction reaction.

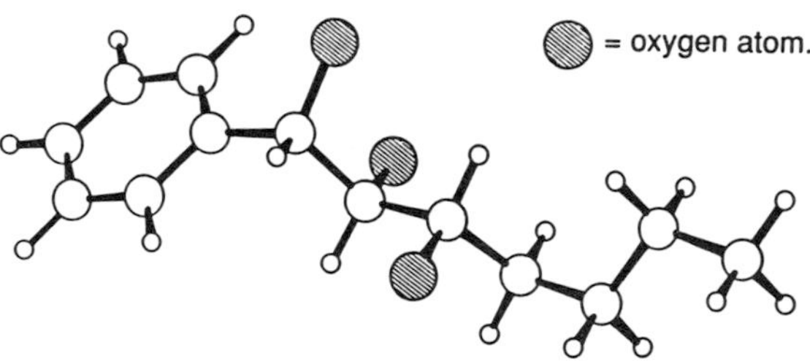

Figure 1: Perspective drawing of the triol (**4**) from crystal structure analysis

	R^1	R^2	R^3
(12)	phenyl	OH	=O
(15)	4-nitrophenyl	H	OH

3 MECHANISM OF THE HYDROLYSIS–REDUCTION OF EPOXYKETONES BY BAKER'S YEAST

The relative configuration of the triol (**4**), as determined by an X–ray crystallographic analysis (see Figure 1, details to be published elsewhere), indicated that the epoxide moiety of the epoxyketone (**3**) was hydrolysed with overall *syn* stereochemistry. This result is unusual and is in contrast with previous reports on epoxide hydrolases that transform oxiranes into diols with Walden inversion.[4] Two mechanisms (Schemes 1 and 2) can be proposed to explain the unexpected *syn*–opening of the epoxide moiety. A non–concerted hydrolysis of the epoxide and stabilization of the intermediate carbocation (**11**) by a negatively charged residue on the enzyme is depicted in Scheme 1. The subsequent attack of water on the carbocation is therefore under enzymatic control and retention at C–1 is achieved. Alternatively, an X–group mechanism which comprises two S_N2 reactions will lead to overall retention of stereochemistry at C–1.

In the second proposed mechanism (Scheme 2) net retention of stereochemistry during hydrolysis of the epoxide is the result of an enzyme–catalysed Payne rearrangement[5] which presupposes the initial reduction of the ketone. However, all attempts to detect an intermediate reaction product failed and even at low substrate conversion only racemic triol (**4**) was produced. The epoxyketone (**3**) was therefore reduced with sodium borohydride and the resultant diastereomeric mixture (diastereomeric excess 54.6%) of epoxy alcohols, upon treatment with baker's yeast, gave the triol. Conversely, the triol (**4**) was transformed into a five–membered ketal involving C–1 and C–2, oxidized with pyridinium dichromate and deprotected to give the racemic diolketone (**12**), which was not a substrate for baker's yeast. These results indicate that the reduction of the carbonyl group precedes hydrolysis of the epoxide moiety, a prerequisite for but not proof of the mechanism based on a Payne rearrangement.

The two possible mechanisms were distinguished by identifying the carbon–oxygen bond which is cleaved during hydrolysis using the epoxide (**3**) labelled with oxygen–18 in the oxirane ring. The labelled epoxide was synthesized by the Darzen's method[6] starting

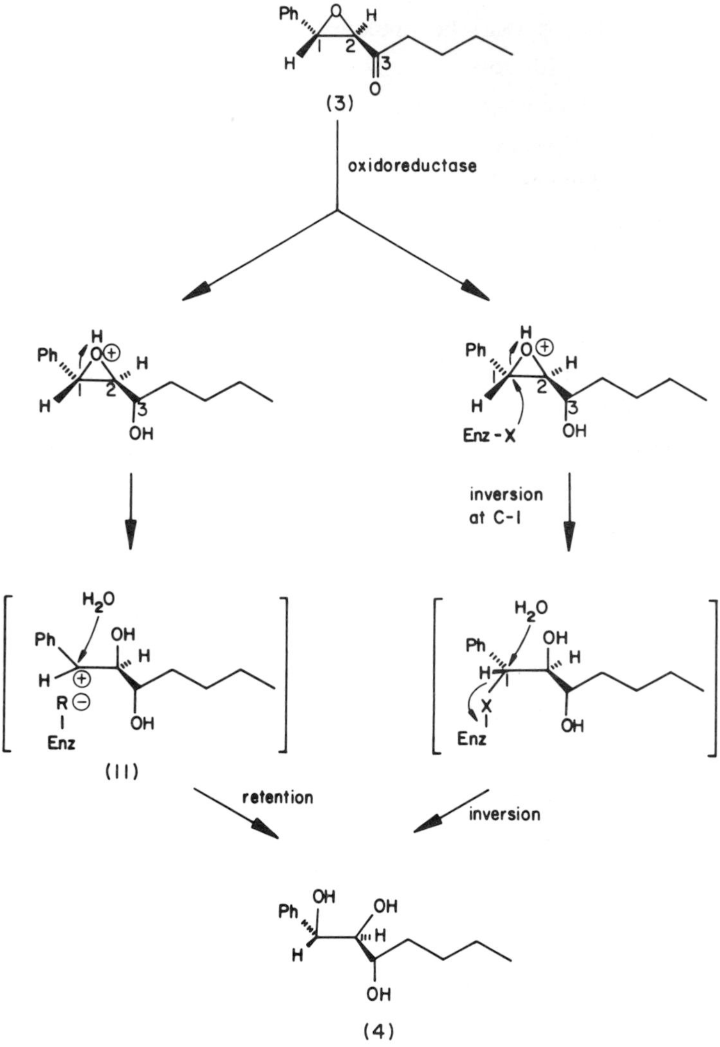

Scheme 1: Mechanism for the formation of the triol (**4**) based on a high energy intermediate

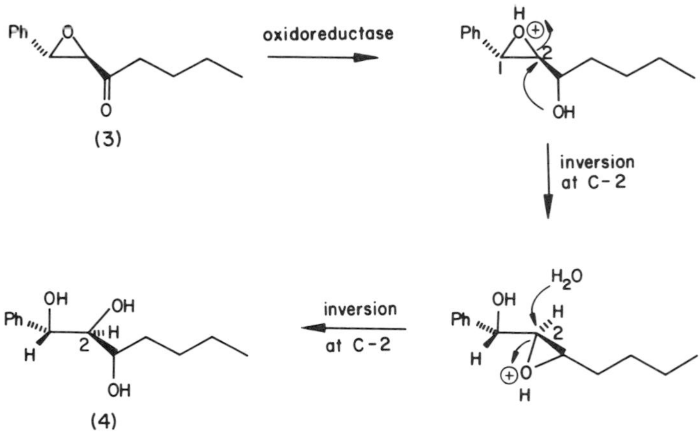

Scheme 2: Mechanism for the formation of the triol (**4**) based on a Payne rearrangement

from labelled benzaldehyde (prepared by partial exchange with oxygen–18–labelled water)[7] and *N*,*N*–diethyl α–chloroacetamide in the presence of potassium–*t*–butoxide. The *trans*– and *cis*–2,3–epoxyamides were formed in a 2:1 ratio (85% yield) and after separation the *trans* isomer was reacted with *n*–butyllithium to give the epoxyketone in 84% yield.[2] This material was subjected to fermenting baker's yeast and the resultant oxygen–18–labelled triol was analysed by ^{13}C NMR spectroscopy. The assignment of the C–1, C–2 and C–3 resonances of the triol (**4**) (see Table) is based on an analysis of the ^{1}H NMR spectrum and subsequent correlation of the ^{1}H and ^{13}C signals in a 2–dimensional (^{13}C, ^{1}H) chemical shift correlation experiment.[8] The characteristic upfield α–isotope shift of 0.02 ppm[7,9] observed for the C–2 signal in the proton–decoupled ^{13}C NMR spectrum of the labelled triol is due to the presence of oxygen–18 at this position (Figure 2) and proves that the opening of the epoxide is not the result of a Payne rearrangement.

It is now proposed that the overall *syn*–stereochemistry observed for the hydrolysis of the epoxide can be explained by the mechanism illustrated in Scheme 1, which proceeds either by a double inversion or by retention. Evidence for the latter, which implies the existence of a high–energy intermediate (**11**), is provided by the fact that the 4–nitrophenyl analogue (**13**) which is expected to lead to a relatively

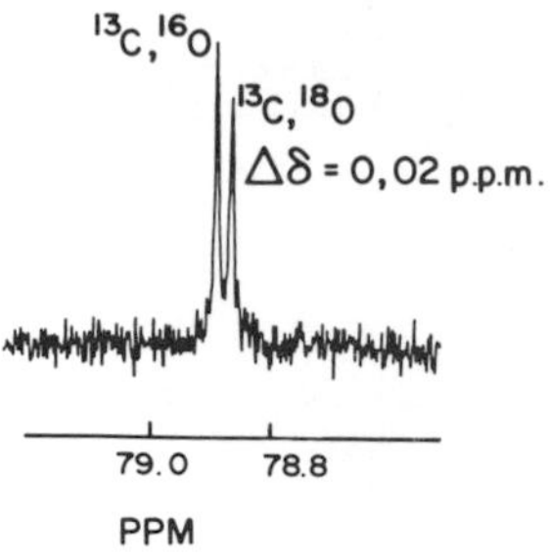

Figure 2: Section of the proton decoupled ^{13}C NMR spectrum of ^{18}O–labelled (4) showing the C–2 signal

destabilized intermediate carbocation, resists hydrolysis of the epoxide moiety and the intermediate epoxyalcohol (**14**) was now isolated for the first time. This result indicates that stabilization of the incipient carbocation (**11**) by the adjacent phenyl ring[10] is a critical feature of the enzyme–catalysed hydrolytic opening of the epoxide moiety.

The above mechanistic and stereochemical assignments of the hydrolysis–reduction of epoxyketone (**3**) imply hydride delivery to the *Re*–face of the carbonyl moiety of (1*S*,2*R*)–(**3**) and to the *Si*–face of the enantiomeric epoxyketone. The subsequent hydrolysis of the enantiomeric oxirane moieties is equally stereospecific leading in each case to a single diastereomer of the triol (**4**). The formation of the racemic triol could be the result of two enzyme systems with antipodal substrate and product stereospecificities[4] or a single enzyme which is capable of catalysing the enantiomeric hydrolysis–reduction of the antipodal epoxyketones (**3**). Precedent for this unusual enzymatic process is provided by a report that a cell–free preparation of bornyl pyrophosphate cyclase from sage, which catalyses the cyclization of (3*R*)–linalyl pyrophosphate to (+)–bornyl pyrophosphate, can also produce (–)–bornyl pyrophosphate if presented with the enantiomeric substrate (3*S*)–linalyl pyrophosphate.[11] In addition, the lack of optical resolution during the transformation of (**3**) into (**4**) leads to the rather unusual but inevitable conclusion that the catalytic efficiency parameters, V_{max}/K_M,[12] for the two antipodal, baker's yeast catalysed processes are identical.

The baker's yeast catalysed reaction of the 4–nitrophenyl analogue (**13**) (see above) yielded an unexpected diol (**15**) in 12% yield. This product represents the formal reduction of the oxirane, which could conceivably be due to a novel baker's yeast–catalysed hydride transfer from NADH or NADPH to the epoxide moiety. However, an alternative which should be considered is the regiospecific enzyme–catalysed opening of the epoxide to generate a benzylic carbocation which is quenched by an intramolecular 1,2–hydride shift. This would generate a new carbonyl moiety at C–2 which can be reduced by an oxidoreductase. The details of this novel, baker's yeast catalyzed reaction are currently under investigation.

4 SUMMARY

It was found that baker's yeast catalyses the stereospecific *syn*–hydrolysis–reduction of certain *trans*–2,3–epoxyketones to give single diastereomers of racemic triols. It has been determined that this unusual stereochemical result is not dictated by a reaction mechanism such as the Payne rearrangement, but is due to an intrinsic and previously unreported feature of this biocatalyst.

ACKNOWLEDGEMENTS

We thank Dr J L M Dillen for the determination of the *X*–ray crystallographic structure of the triol.

REFERENCES

1 T Katsuki and K B Sharpless, *J Am Chem Soc*, 1980, **102**, 5974.
2 G Fouche, Ph D Dissertation, University of South Africa, Pretoria, in preparation.
3 For example: B Zhou, A S Gopolan, F Van Middlesworth, W–R Shieh, and C J Sih, *J Am Chem Soc*, 1983, **105**, 5925; G Fronza, C Fuganti, P Grasselli, G Poli, and S Servi,

J Org Chem, 1988, **53**, 6153; C Fuganti, P Grasselli, H–E Hogberg, and S Servi, *J Chem Soc Perkin Trans 1*, 1988, 3061; and R Chênevert and S Thiboutot, *Can J Chem*, 1986, **64**, 1599.

4 D Wistuba and V Schurig, *Angew Chem Int Ed Engl*, 1986, **25**, 1032.

5 G B Payne, *J Org Chem*, 1962, **27**, 3819.

6 G Darzens, *Compt Rend*, 1904, **139**, 1214; J March, 'Advanced Organic Chemistry', Wiley, New York, 3rd edn, 1985, p843.

7 J Diakur, T T Nakashima, and J C Vederas, *Can J Chem*, 1980, **58**, 1311.

8 A Bax and G A Morris, *J Magn Res*, 1981, **42**, 501; G Bodenhausen, and R Freeman, *ibid*, 1977, **28**, 471.

9 J M Risley and R L van Etten, *J Am Chem Soc*, 1980, **102**, 4609.

10 J March, 'Advanced Organic Chemistry', Wiley, New York, 3rd edn, 1985, p145.

11 R Croteau, D M Satterwhite, D E Cane, and C C Chang, *J Biol Chem*, 1986, **261**, 13438.

12 I H Segel, 'Biochemical Calculations', Wiley, New York, 2nd edn, 1976, p218.

4-OXONORVALINE: ARTIFACT OR INTERMEDIATE OF 5-HYDROXY-4-OXONORVALINE BIOSYNTHESIS?

Robert L White,[a] Kevin C Smith,[a] Shi-Yu Shen,[a] and Leo C Vining[b]

a *Department of Chemistry, Acadia University, Wolfville, NS, Canada B0P 1X0*

b *Department of Biology, Dalhousie University, Halifax, NS, Canada B3H 4J1*

1 INTRODUCTION

The biosynthesis of the non-protein amino acid, 5-hydroxy-4-oxonorvaline (HON), has been investigated in *Streptomyces akiyoshiensis* using labelled substrates.[1] Incorporation of ^{13}C label from [4-^{13}C]aspartic acid, [1-^{13}C]acetic acid and [2-^{13}C]acetic acid and the intact incorporation of the ^{13}C-^{15}N unit from [2-^{13}C,^{15}N]aspartic acid have demonstrated that the nitrogen atom and four of the carbon atoms of HON are derived from aspartic acid and that the fifth carbon is supplied by the methyl group of acetic acid.

A hypothesis for the biosynthesis of HON that is consistent with these incorporation results is shown in Figure 1. The first step in the proposed pathway is a condensation between a derivative of acetic acid (*eg* acetyl CoA or malonyl CoA) and an activated form of aspartic acid (1), such as aspartyl phosphate or aspartic semialdehyde,

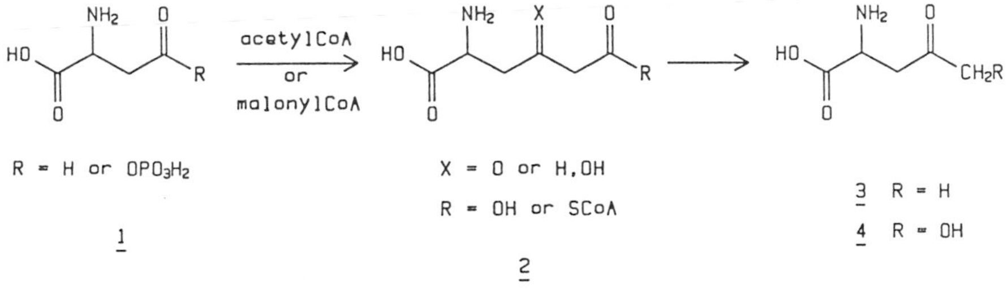
NH2
O
HO
R
O
acetylCoA
or
malonylCoA
NH2
X
O
HO
R
O
NH2
O
HO
CH2R
O
R = H or OPO3H2
1
X = O or H,OH
R = OH or SCoA
2
3 R = H
4 R = OH

Figure 1: Suggested pathway for the biosynthesis of HON.

to form a β-keto acid (or thioester) or a β-hydroxy acid (or thioester) as a six-carbon intermediate (2). A similar condensation between γ-glutamylphosphate[2] or glutamic semialdehyde[3] and the methyl group of acetyl CoA (or malonyl CoA) has been proposed as the first step in the biosynthetic pathway of the carbapenem antibiotics. Conversion of the six-carbon intermediate (2) into HON (4) requires decarboxylation and hydroxylation. 4-Oxonorvaline (3) would be formed as an intermediate if decarboxylation and hydroxylation occurred in sequential steps.

The possibility that 4-oxonorvaline is formed in cultures of *S akiyoshiensis* has been investigated as the initial phase in the identification of intermediates of HON biosynthesis.

2 RESULTS AND DISCUSSION

The identification of intermediates along the proposed biosynthetic pathway to HON (Figure 1) was approached from two directions: synthesis of the potential intermediate, 4-oxonorvaline, and examination of culture broths by HPLC for the accumulation of constituents other than HON.

HPLC Conditions and Chemical Synthesis of 4-Oxonorvaline

Amino acids in culture broths were analysed as their fluorescent *o*-phthalaldehyde derivatives using published HPLC conditions[4] for the separation of acidic and polar amino acids on a reversed-phase column by gradient elution with 0.1 M sodium acetate (pH 6.2)-methanol- tetrahydrofuran (900:95:5) and methanol.

The chemical synthesis of 4-oxonorvaline from benzamide, glyoxylic acid, and ethyl acetoacetate was carried out according to literature procedures.[5,6] An aqueous solution of the product was decolourized with charcoal and with C_{18} reversed-phase silica. After removal of the water *in vacuo*, 4-oxonorvaline hydrochloride was recrystallized from 99% ethanol-petroleum ether:
mp 156-158°C; ^{1}H NMR (360 MHz, D_2O) δ: 4.15 (t, 1H, J 5 Hz), 3.19 (d, 2H, J 5 Hz), 2.12 (s, 3H); ^{13}C NMR (90 MHz, D_2O) δ: 207.83, 169.70, 46.84, 40.31, 27.21.

As it lacks the terminal hydroxyl group of HON, 4-oxonorvaline exhibits a longer retention time than HON (1.5 *versus* 1.2 min) and peak area measurements showed a linear fluorescent response from 15 to 1500 nmol/ml.

4-Oxonorvaline as a Culture Constituent

HPLC analysis of broths of *S akiyoshiensis* grown in a medium containing starch and Pharmamedia[1] showed the accumulation of HON (retention time 1.2 min) and an unidentified metabolite with a longer (1.5 min) retention time (Figure 2a). The retention time of the metabolite was similar to that of 4-oxonorvaline, which, lacking the terminal hydroxyl group of HON, was retained more strongly than HON on the reversed-phase column. When the metabolite and synthetic 4-oxonorvaline were co-chromatographed, they were inseparable under the conditions used (Figure 2b).

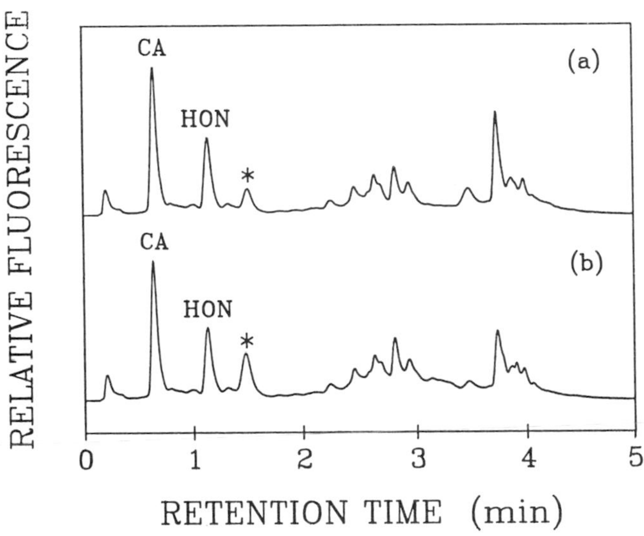

Figure 2: Analysis by HPLC of a 141 h culture broth showing:
(a) accumulation of HON and a peak at 1.5 min (*) and
(b) a symmetrical peak with increased size at 1.5 min (*) when the broth sample was supplemented with 4-oxonorvaline. CA represents the internal standard, cysteic acid.

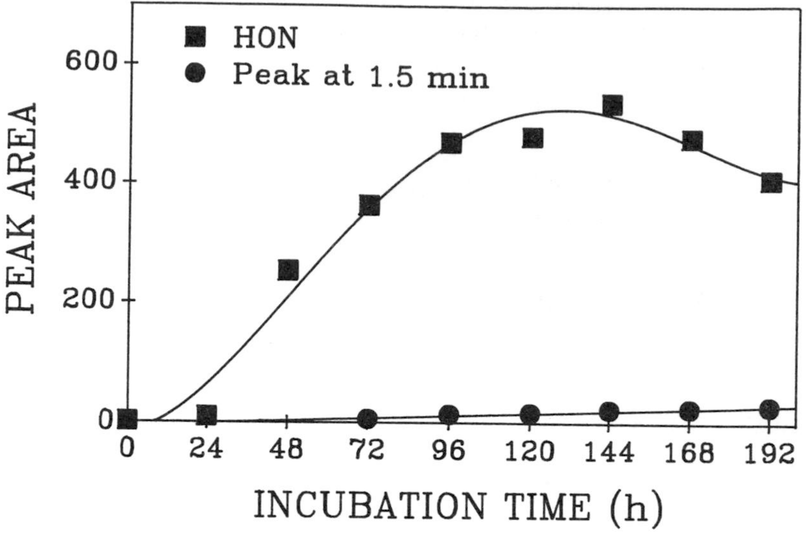

Figure 3: Accumulation of HON and the unidentified metabolite (retention time 1.5 min) in a culture broth.

Although the chromatographic evidence suggested that the metabolite was 4-oxonorvaline, the two substances exhibited different chemical behaviour. When treated with $NaBH_4$, 4-oxonorvaline was reduced quantitatively to a product, presumably 4-hydroxynorvaline, which chromatographed with a retention time of 2.8 min. However, the size and the retention time of the metabolite peak were not affected when culture broth was treated with $NaBH_4$, although the HON present in the culture broth was reduced quantitatively.

Examination of the production of the metabolite and HON in a culture of *S akiyoshiensis* (100 ml medium/2l Erlenmeyer flask) by HPLC also was inconsistent with the metabolite being an intermediate in HON biosynthesis. As shown in Figure 3, HON was formed rapidly between 24 and 96 h, with little or no subsequent production. The first appearance of the metabolite in the broth lagged that of HON by about 24 h; the metabolite was then produced at a fairly constant low rate from 48 h to the last sampling at 192 h. Accumulation of a biosynthetic intermediate of HON would be expected to precede HON production. Furthermore, HON production in a culture with limited oxygenation (500 ml medium/2l Erlenmeyer flask) was markedly reduced, but that of the metabolite was not significantly affected. 4-Oxonorvaline would be expected to accumulate under these conditions if oxygen was required for its conversion to HON. Large amounts of intracellular glutamic acid (retention time 1.4 min) masked the presence of the metabolite (retention time 1.5 min) in chromatograms of ethanol extracts of cells.

No evidence has been found for the presence of 4-oxonorvaline in cultures of *S akiyoshiensis*. Furthermore, no increase in the rate of HON production or the total produced was detected by HPLC of cultures after addition of unlabelled 4-oxonorvaline at 78 h after inoculation. However, HPLC analysis of the broth and ethanolic cell extracts showed that very little or no 4-oxonorvaline was taken into the cells in these experiments.

Isolation of the Metabolite

To obtain additional information about the metabolite and to confirm that it differed from 4-oxonorvaline, the broth from a 1.2l culture was fractionated by ion-exchange chromatography. Amino acids and the metabolite in the broth were retained on cation-exchange résin (Amberlite IR-120, hydrogen form) and were eluted with 0.3 M ammonia. The metabolite and the neutral and basic amino acids in the cation-exchange eluate were separated on anion-exchange resin (Dowex 1-X8, acetate form, equilibrated in water) using 0.5 M acetic acid as the eluent. Further anion-exchange chromatography employing slightly different conditions (Dowex 1-X8, acetate form, equilibrated and eluted with 0.5 M acetic acid) separated the mixture of metabolite and acidic amino acids obtained in the previous anion-exchange eluate.

HPLC analysis of the purified metabolite (37 mg) showed one major peak (1.5 min retention time) but an indication that the sample was not pure was obtained from ^{1}H NMR which showed signals due to aromatic protons as well as a complex series of peaks between 0.7 and 4.2 ppm. The metabolite was stable to hydrolysis in 6 M HCl at 110°C, but a number of amino acids, including aspartic acid and glutamic acid, appeared in the hydrolysate. It was concluded that peptidic material that did not yield a peak in HPLC analyses was present in the sample. This would account for the complex ^{1}H NMR obtained. The peptide contaminant was not separated from the metabolite by gel filtration chromatography on Sephadex G-10.

Although the structure of the metabolite has not been elucidated a number of conclusions regarding the functional groups present can be made on the basis of the above information. Since the metabolite formed a derivative with *o*-phthalaldehyde it must possess a primary amine; this substituent would also account for its retention on the cation exchange resin, Amberlite IR-120. The behaviour of the metabolite on Dowex 1-X8 was characteristic of an acidic amino acid and, since it was stable to hydrolysis, it is unlikely to be a peptide. Functional groups reduced by borohydride, *eg* ketone, and functional groups cleaved by periodate, *eg* diol or hydroxyketone, are not part of the metabolite's structure since it is not affected by these reagents.

ACKNOWLEDGEMENTS

We thank the Natural Sciences and Engineering Research Council (NSERC) of Canada for financial support and Dr D L Hooper and the Atlantic Region Magnetic Resonance Centre for providing NMR spectra. We are indebted to the Agricultural Research Station, Kentville, NS, for providing temporary laboratory space.

REFERENCES

1 R L White, A C DeMarco, and K C Smith, *J Am Chem Soc*, 1988, **110**, 8228.
2 J M Williamson, E Inamine, K E Wilson, A W Douglas, J M Liesch, and G Albers-Schonberg, *J Biol Chem*, 1985, **260**, 4637.
3 B W Bycroft, C Maslen, S J Box, A G Brown, and J W Tyler, *J Chem Soc Chem Commun*, 1987, 1623.
4 R L White, A C DeMarco, and K C Smith, *J Chromatogr*, 1989, in press.
5 U Zoller and D Ben-Ishai, *Tetrahedron*, 1975, **31**, 863.
6 D Ben-Ishai, J Altman, Z Bernstein, and N Peled, *Tetrahedron*, 1978, **34**, 467.

INDEX